DRINKING WATER SUPPLY AND AGRICULTURAL POLLUTION

ENVIRONMENT & POLICY

VOLUME 11

The titles published in this series are listed at the end of this volume.

Drinking Water Supply and Agricultural Pollution

Preventive Action by the Water Supply Sector in the European Union and the United States

Edited by

Geerten J.I. Schrama

University of Twente,
The Netherlands

This research project has been funded, under contract number EV5V-CT94-0368, by:

EUROPEAN COMMISSION
DIRECTORATE-GENERAL XII
SCIENCE, RESEARCH AND DEVELOPMENT
Directorate D - RTD Actions: Environment
Research into economic and social aspects of the environment

SPRINGER SCIENCE+BUSINESS MEDIA, B.V.

A C.I.P. Catalogue record for this book is available from the Library of Congress.

ISBN 978-94-010-6145-2 ISBN 978-94-011-5106-1 (eBook)
DOI 10.1007/978-94-011-5106-1

Printed on acid-free paper

TABLE OF CONTENTS

PART ONE

1. INTRODUCTION

GEERTEN J.I. SCHRAMA
Center for Clean Technology and Environmental Policy, University of Twente, Enschede, The Netherlands.

1.1 Water management[1]

The importance of water for life on earth can hardly be underestimated. Too much water at the same place is a problem, often a disaster. The same goes for too little water. Water management is one of the oldest tasks to be taken on by government. The management of water is not only a matter of quantity: water pollution is a great threat to ecological systems as well as a threat to many applications, such as its use by humans as drinking water.

1.1.1 SCARCITY AND POLLUTION OF DRINKING WATER RESOURCES

The mean annual water abstraction per capita in the European Union of 12 member states (EUR 12) has increased from 590 m^3 in 1970 to 790 m^3 in 1985, an overall increase of 35% in 15 years. There are great regional variations in water abstraction, ranging from below 200m^3 per capita per year in Luxembourg to about 1000 m^3 per capita per year in Spain. On average, 70% of the water abstraction is drawn from surface sources, groundwater being by far the next most important source. For Europe as a whole, 53% of the abstracted water is used for industrial purposes, 26% for agriculture and only 19% for domestic purposes. Water use in agriculture is mainly for irrigation and livestock, while domestic uses of water include drinking and sanitation purposes. So, just a small part of the water resources is actually used as drinking water (Kallis, 1995: 9).

Groundwater. In the European Union, 65% of the water delivered by public supply and suited for human consumption stems from groundwater sources (Kallis, 1995: 9).

[1] This section contains several contributions from Christa Oude Nijhuis (Center for Clean Technology and Environmental Policy) and Kenneth Hanf (Erasmus University Rotterdam).

Geerten J.I Schrama (ed.), Drinking Water Supply and Agricultural Pollution, 3-18.

Groundwater runs in permeable rock formations or unconsolidated deposits, known as aquifers. The main characteristics of groundwater systems are:

- relative isolation;
- low flow rates, sometimes only a few meters per year, and lengthy residence times;
- and magnitude of the reservoirs.

Groundwater resources are of great importance for a number of vital functions as mentioned above. At the same time groundwater is necessary for the diversity, sustainability and survival of many nature reserves and ecosystems, and as a main source of river discharge. Groundwater resources could be of sound quality due to the purifying capacity of the soil and the protection afforded by geological conditions. If necessary, drinking water can be extracted from layers at great depth that have been unaffected for a very long time, sometimes over a thousand years. Nowadays, clean water is increasingly becoming a scarce resource. Each form of human application has to compete with many alternative options. The present scale of groundwater abstraction is causing serious problems of dehydration in many areas, as it affects the diversity of the local flora and fauna, and increases the need for irrigation in agricultural areas.

Moreover, human activities tend to have serious effects on the quality of groundwater. The quality of groundwater, and therefore the quantity available for use, is particularly threatened by anthropogenic pollution. There are two different sources of pollution: diffuse and point. Point sources are primarily related to municipal and industrial activities. Given the present extent of environmental awareness, groundwater contamination from point sources must be caused either by incidents or by negligence. The latter is related to leaching from a large variety of sources, such as industrial installations and storage depots, waste water treatment installations, and landfills. Most groundwater contamination, however, stems from diffuse sources: (1) atmospheric deposition of polluting, mainly acidifying substances in the atmosphere, carried by precipitation and subsequently percolating through the soil into the groundwater; (2) drainage of contamination from adjacent surface water; and (3) application of manure, fertilizers and pesticides to the land (cf. Kallis, 1995: 12). This last type of contamination is predominantly agricultural, but not exclusively so.

Groundwater contamination can be alleviated through natural processes of degradation (although pesticides in particular may be very persistent), by circulation (although flow rates may be very low; residence times in water catchment areas can be up to 25 years), and by remedial activities (which are often drastic and costly affairs, usually requiring the abstraction of large amounts of groundwater).

Some of the most important substances causing surface water pollution are (cf: Kallis, 1995: 21-22):

1. Organic matter: excessive discharges from human activities to water bodies, mainly from domestic and industrial waste water, can lead to oxygen depletion and undesirable changes in the composition of aquatic biota. There is an evident relationship between the concentration of organic matter in a river and the population density of its catchment. The practicing of waste water treatment has improved the overall situation with regard to organic matter concentrations.

2. Nutrients: excessive input of nutrients (nitrogen and phosphorous) into water bodies can result in a series of adverse effects known as cultural eutrophication. Eutrophication, which is the nutrient enrichment leading to excessive photo-plankton growth, can significantly affect the use, ecological functioning, and aesthetic quality of a lake. It can also lead to taste and odor problems in potable surface water resources, substantially increasing treatment costs. Although phosphorous is the prime nutrient responsible for freshwater eutrophication, nitrogen loading of surface waters, derived primarily from agricultural activity, can also be a problem. This nitrogen comes especially from the use of nitrogen fertilizers and manure. There is a clear relationship between nitrogen levels and the percentage of the catchment that is agricultural land. The increase in the use of nitrogen fertilizers in the last 10 to 15 years has lead to a significant increase of nitrate concentrations. The nutrient levels in many areas of Europe are still too high (Kallis, 1995: 14). Unless drastic efforts are made to reduce inputs of nutrients, eutrophication is likely to continue to be an important environmental issue.
3. Acidifying substances, such as SO_2 and NO_2 from atmospheric deposition, which affect aquatic ecosystems in several different ways.
4. Heavy metals and their compounds, derived from industrial activities.
5. Micro organisms originating from industry, urban waste and agriculture.

Surface Water. In general, the production of drinking water from fresh surface water requires much more treatment than its production from groundwater. General procedures are sedimentation and purification in reservoirs by natural means, such as vegetation and microbes, and filtration, which can be done either in water purification plants or by infiltration in dunes or other types of soil.

1.1.2 AGRICULTURAL WATER POLLUTION

The use of land for agriculture in Europe is in the order of 57% of the total land mass of 226 million hectares; 52% of the cultivated area is arable and 37% is used for grass and green fodder production. Approximately 10% of the area is used for the production of fruit, wine grapes, and vegetables. Although the contribution of agricultural production to the GNP of several countries has fallen in relative terms, the production in absolute value has increased enormously. Technological developments, i.e. increased mechanization and use of fertilizers and pesticides, have led to an increasing production. At the same time, most forms of agricultural land use provide an important source for diffuse contamination of soil and groundwater. The use of manure and fertilizers may lead to leaching of nitrates (NO_3), potassium (KNO_3, also called saltpeter or nitric acid), and phosphate (PO_4) into groundwater. Emission to the air - notably ammonia (NH_3), ammonium (NH_4), and NO_x - may contaminate surface and groundwater through atmospheric deposition. Phosphates in surface water is a major cause of eutrophication. However, this constitutes no major problem for drinking water production, in contrast to nitrates. The use of pesticides for all kinds of crops results in an accumulation in the top soil and leaching of persistent substances towards groundwater (c.f. RIVM/RIZA, 1991).

Nitrogen Compounds. Nitrogen compounds are an important type of pollutants and are due to three types of non-point sources:

- application of manure to arable and grassland;
- use of fertilizers in agriculture; and
- deposition of NO_x and NH_y from the atmosphere.

Within the European Union the production of dairy products, meat and eggs is characterized by the high livestock densities on farms. In a number of member states, such as Belgium, Denmark, France, the Netherlands, and the UK, the growth and intensification of animal husbandry has led to the complete separation of crop and pasture production. In these areas excessive manuring leads to serious environmental problems on a regional scale.

TABLE 1.1: Estimated manure production and nitrogen load per country per hectare cultivated soil in 1989.

Country	N production (in mln. kg)	Agricultural area (in mln. ha)	Manure load (N in kg/ha/y)
Belgium	203	1.4	143
Denmark	254	2.8	89
France	1,680	28.5	59
Germany	1,240	11.9	104
Greece	280	4.1	68
Ireland	470	5.0	94
Italy	850	15.6	54
Luxembourg	10	0.1	100
Netherlands	500	2.0	250
Portugal	180	3.3	54
Spain	940	23.5	40
UK	1,250	16.8	74

Source: Eurostat (1991).

These figures indicate that the highest manure production rates and nitrogen loads occur in parts of the Netherlands, Belgium, and Germany.

Since 1950, the use of fertilizers, and notably nitrogen, has shown a sharp increase in most countries as a result of the expansion of intensive crop production. Great differences exist regarding the amounts of nitrogen in fertilizers per hectare cultivated area.

The data indicate that present agricultural practice may lead to nitrate pollution of groundwater in at least nine of the twelve EU member states. In about 43% of the area nitrate concentrations between 25 and 50 mg/l have been computed. In 25% of the area nitrate levels, exceeding 50 mg/l may be expected. The greatest problems occur in Northwestern Europe (France, England, Belgium, Netherlands, Northern Germany and Denmark). This is either a result of large surplus in the regional nitrogen balance (input greater than crop uptake), a high sensitivity of the soil or a combination of these two.

From the data it can be concluded that in Denmark, Germany, the Netherlands and Belgium a combination of high nitrogen input and great sensitivity of the soil leads to nitrate concentrations in soil above the drinking water standard of 50 mg/l for the greatest part of the country. In France, Italy, and to a lesser degree in England, the areas with high nitrate concentration show a more localized distribution. Model computations of nitrate leaching from agricultural soils indicate that 87% of the agricultural area in Europe has nitrate levels above the EU target value (25 mg NO_3/l) and 22% above the drinking water standard (50mg NO_3/l) (Kallis, 1995: 12).

Nitrogen compounds appear in water mainly as nitrate or ammonium. Both are readily soluble in water, and are absorbed by vegetation. The human body transforms nitrates into nitrites and nitrosamines, mainly by the saliva. In human organs, nitrites have an effect on the oxygen carrying capacity of the haemoglobin - which may, for instance, cause the 'blue-baby syndrome' - and the functioning of the thyroid. Nitrosamines are assumed to be carcinogens. As a result, legal standards have been laid down for the nitrate contents of food, beverages and drinking water (cf. Van den Nieuwenhof, 1995: 9-10; Copius Peereboom and Reijnders, 1986: 222-225).

On the other hand, nitrates and nitrites are part of the natural nitrogen cycle and have an important function in the nitrogen supply to a large variety of organisms. Nitrogen is among the principle elements of living organisms and nitrates are basic components of natural and artificial fertilizers. Rising concentrations of nitrates and nitrites in soil and groundwater, caused by manure surpluses and the over generous application of artificial pesticides, have influenced biodiversity, leading to a loss of plants and micro organisms that depend on a lean environment or specific minerals and salts.

The rates of nitrate movements and the amounts which are ultimately leached into the groundwater are determined by the concerted effect of many different conditions, such as soil type, hydraulic conductivity of the soil, temperature, soil moisture, vegetation or crop type, precipitation, and fertiliser application. With the exception of ammonium ions, which may be absorbed by clay minerals, most nitrogen compounds appear to move freely through the saturated zone of an aquifer without much change in their total concentrations.

Pesticides. Pesticides is the common term for agro-chemicals used to protect crops and soil from weeds (herbicides), insects (insecticides), or infections (fungicides). There are a large number of types and brands and a large number of active substances. Of these active substances, which are often toxic to humans, only a certain proportion is absorbed by the organisms that are meant to be destroyed. The remainder is partly degraded in the soil and transformed into harmless substances or into metabolites with uncertain effects, and part is washed away by rain or irrigation into the groundwater and possibly into adjacent surface waters, or blown away by the wind to unknown destinations. Our knowledge about the behavior and effects of the numerous components of pesticides is limited, and there are very few data about their concentration in groundwater and soil.

Groundwater is threatened by pesticides, however, in all EU member states. This is obvious from the available monitoring data and calculations on pesticide load, on soil sensitivity and leaching. Very high pesticide loads are often used on highly sensitive soils. It has been calculated that in 65% of all agricultural land the EU standard for the sum of pesticides (0,5μg/l) will be exceeded. In approximately 25 % of the area this standard will be exceeded by more than 10 times (Kallis, 1995: 12).

The presence of pesticides in environmental areas may constitute a risk to human health and to natural ecosystems. In general the environmental impact of pesticides depends on:

- chemical structure and properties of the chemical, like rate of biodegradation;
- mobility in the environment and toxicity to non-target organisms;
- application factors, like dosage per hectare, and method of application;
- environmental conditions, like soil properties and climate (Reus et al., 1994: 3).

Chemicals (including their degradation products) having a low degradation rate and high mobility can easily leach into groundwater. Other products are immobile and persistent and may accumulate in the soil. It may take several years before pesticides reach the groundwater, especially in the case of lower groundwater levels. As yet little is known of the degradation rate of pesticides in the water saturated zone (Reus et al., 1994: 7).

The most important factors to influence the diffuse pesticide contamination of surface water are the physical and chemical properties of the pesticide, the local geological, hydrological and meteorological conditions, and the method of application. Aerial spraying, for example, results in large emissions to surface waters in the neighborhood of agricultural plots.

When pesticides have landed on surface water they become subject to degradation at a rate that depends on the properties of the chemicals and the environmental conditions. Pesticides may also settle down to the sediment layer, resulting in a long-term source of contamination (Reus et al., 1994: 11).

The concentrations in surface water and the toxicity of the pesticides to aquatic organisms determine the impact on the aquatic ecosystems and the sediments in the water. High concentrations in surface waters adjacent to agricultural plots may cause acute toxic effects. A prolonged period of lower concentrations in water systems may result in chronic effects. Some pesticides accumulate in organisms and have a toxic effect through the food chain. Chronic effects may occur in and outside agricultural areas as pesticides can be transported over long distances. Aquatic organisms are usually exposed to more than one pesticide at a time, but the effects of simultaneous exposures have hardly been examined (Reus et al., 1994: 11).

The European Community is the largest user of agro-chemicals in the world and five of the individual countries are within the top 10. There are approximately 600 different pesticides applied as part of European agriculture which pass through the subsurface environment and are transformed into an unknown number of metabolites. Knowledge on the fate and effects of these components is limited and there is extremely limited data on their concentration in groundwater or arable land. Also data about pesticide use per country is scarce and unreliable.

TABLE 1.2: Annual sales of pesticides in agriculture by product group (kg of active ingredients per hectare of arable land cq. land under permanent cropping)

	herbicides	fungicides	insecticides	nematicides	other	total
Belgium	3.5	4.5	0.5	1.2	1.0	10.7
Denmark	1.3	0.7	0.1	<0.1	0.1	2.2
France	1.4	2.3	0.3	0.1	0.2	4.4
Germany	2.3	1.5	0.2	-	0.5	4.4
Greece	0.9	2.7	0.8	0.1	1.6	6.0
Ireland	1.1	0.7	0.1	0.1	0.3	2.2
Italy	0.9	4.8	0.9	0.8	0.2	7.6
Luxembourg	1.5	1.4	0.1	-	0.1	3.1
Netherlands	3.3	4.6	0.6	7.4	1.6	17.5
Portugal	0.4	1.2	0.2	-	0.1	1.9
Spain	0.1	1.6	0.1	0.5	0.2	2.6
UK	2.0	1.0	0.2	-	0.5	3.6
EUR 12	**1.1**	**2.3**	**0.4**	**0.4**	**0.4**	**4.5**

Source: Brouwer, Terluin, and Godeschalk, 1994: 15.

The Dutch LEI-DLO report 'Pesticides in the EC' (Brouwer, et al., 1994) contains data on sales and uses of pesticide in the member states. Total sales of pesticides in the EUR 12 amount to some 5.8 billion ECU per year. The major markets of sales of pesticides are in France, Germany, and Italy. These three countries already cover about two thirds of the total sales of pesticides.

As can be seen from table 1.2, the use of pesticides per hectare of arable land and land under permanent crops ranges from less than 3 kg per hectare (Denmark, Spain, Ireland, and Portugal) to over 10 kg per hectare (Belgium, and the Netherlands). The use of pesticides is highly correlated to the output from crop production. Countries with a relatively high output level per hectare tend to have a high usage of pesticides.

Drinking Water Supply. The most worrying findings are the high concentrations of nitrates and pesticides in the soil above groundwater, although their patterns of percolation to the groundwater and their effects on humans and ecosystems are not yet well understood. Available data indicates that present agricultural practice may lead to nitrate pollution of groundwater in at least nine of the then twelve EU member states. In about 43% of the area nitrate concentrations between 25 and 50 mg/l have been computed. In 25% of the area nitrate levels, exceeding 50mg/l may be expected (Kallis, 1995: 12).

Contamination of drinking water resources by nitrates and pesticides compel water supply companies to apply expensive treatment techniques or else to take other measures such as interruption of surface water inflow, or the abandonment of boreholes for groundwater abstraction. The various options are assessed based on their costs.

1.2 Steering problem

This book is about the agricultural pollution of drinking water resources, conceived as a steering problem. The goal is to get farmers to change their behavior, especially those whose farms are within or near drinking water catchment areas. This, in principle, is the responsibility of national governments, who have to develop a water quality management policy. As a consequence of the contribution of the agricultural sector to the causes of the problem, this is partly, and perhaps predominantly, a matter of agricultural policy.

Regulating agricultural pollution is a complex matter, as most EU member states have for a very long time adopted a protective attitude to the agricultural sector, encouraging development in the direction of enlargement of scale and intensification. Agricultural policy used to be formulated in consensus between politicians, civil servants, and organizations representing the agricultural interest, together constituting tight and closed networks. These networks have come under great pressure and are starting to disintegrate, largely through the opening up of the markets for agricultural products, the termination of the EU subsidy schemes, and the emergence of environmental issues that the sector could no longer ignore.

At the national levels, governments encounter great difficulties in reformulating their agricultural policies, in particular the implementation of environmental standards. Agricultural interest organizations are not used to having legal regulation imposed on them, certainly not when the measures really hurt farmers. Moreover, in accommodating the pace of the implementation of legal regulation national governments have less discretion than they used to, since they are themselves under pressure from the strict implementation schemes of the European Union. Further, the relationship between the agricultural interest organizations and their clientele is becoming looser. The distance between policy makers at the national level and individual farmers has been widened to an unprecedented degree as these developments have accumulated.

Meanwhile, some water supply companies have tried to cope with the pollution of their natural resources by means of preventive actions, often in co-operation with the regional water authorities. For this purpose they utilize their contacts with individual farmers and regional and local farming associations. These contacts are often the offspring of special legal regulations concerning groundwater protection zones (in the Netherlands and Germany), which impose extra restrictions on farmers in these areas, with an associated right to compensation payments, which ultimately have to be paid by the beneficiaries, in most cases the water supply companies themselves.

The national levels show a trend towards more impositional policy styles based on legal regulation - the outcome of changing circumstances, rather than of deliberate choices - while the regional and local levels show a variety of examples of consensual approaches based on economic and communicative steering by water supply companies and regional water authorities. In terms of a policy network approach, the latter concern regional and local networks with the water supply sector, guided by their own mission and orientation, actively trying to steer farmers' behavior. Considered from a broader perspective, the very same water supply sector is acting more or less as an intermediary in the implementation of national policy, supplementing to a considerable extent the attempts being made to steer farmers directly by legal regulation within water catchment areas.

1.3 Content of the research project

1.3.1 RESEARCH QUESTIONS

The research project is aimed at the water supply companies and the regional water authorities - who are considered to be the constitutive elements of the water supply sector at regional and local levels - and their ability to exercise some degree of control over farmers within or near drinking water catchment areas. The regional water authorities are (semi-) governmental bodies charged with legal authority with respect to water quality management.[2] This ability is referred to as 'control capacity', a concept that is explained and elaborated in the next chapter on the theoretical framework. The central research questions concern the actual situation and the possibilities for the national and European governments to reinforce this control capacity:

1. To what extent can the water supply sector act as environmental regulator in order to prevent agricultural pollution of drinking water resources?
2. By what means can the European and the national environmental policy makers strengthen the control capacity of the water supply sector in this regard?
3. What lessons can be learned for environmental policy makers who want to affect the behavior of target groups by means of indirect regulation and 'network management'?

1.3.2 RESEARCH DESIGN

The empirical part of the research project consists of five case studies: three EU member states, the European Union, and - for comparative purposes - the United States. These case studies are embedded in a theoretical framework that serves as a common

---

2 The term 'regional water authorities' as used in this study is a general one. It is assumed that in the countries under study such an agency is designated for each relevant area. It should not be confused with the agencies in the United Kingdom acting between 1974 and 1989, which are the predecessors of the present private water service companies as well as the NRA (cf. Gray, 1994:8).

perspective. Conclusions and policy recommendations are based on a comparative analysis of the case studies.

The case studies are based as much as possible on a common design. For the national case studies the situation at the national levels is taken as point of departure: the national policies on agricultural pollution in general and on water quality management, as well as the relevant policy networks and policy styles. Next, an assessment is made of the extent to which the agricultural pollution of drinking water resources is a distinctive (national) policy issue, and of the policy instruments applied to this issue, either directly aimed at farmers within or near drinking water catchment areas, or indirectly through the water supply sector. The actual control capacity of the water supply sector at the regional and local levels is then examined, and finally the reinforcement options are presented.

The core of the project is constituted by the three comparative national case studies of EU member states: Germany, the Netherlands, and the United Kingdom. The level of analysis is the national level, although each case study is focused on specific parts of the countries, according to the relevance to the research question. In Germany, water quality management is the competence of the *Länder* or states. Therefore, substantial differences in regulatory systems do occur within the country. The German case study presents an overall review of the situation in former Western Germany with accents on three extreme cases among the 16 states. The drinking water supply in these states originates mainly from groundwater sources. In the Netherlands, the study is concentrated on the five provinces - out of twelve - where the problem of agricultural pollution of drinking water resources is most urgent. In these provinces drinking water is predominantly taken from groundwater sources, which are rather vulnerable due to the permeability of the soil. In the United Kingdom, the case study deals with the ten large-scale, private water supply companies in England and Wales which share a common regulatory system, different from those in Scotland and Northern Ireland. In this case equal attention is paid to drinking water extraction from groundwater and surface water.

The national case studies concern three EU member states in the northwestern part of the Union. These countries are characterized by a tradition of intensive agriculture, as evidenced, for instance, by the average amounts of pesticides used per unit of arable land. Another distinctive feature shared by these three is the relative absence of quantitative problems with the drinking water supply, especially when compared with southern member states, such as Spain and Portugal. More important than differences between the regions of the EU in the urgency of various problems with drinking water supply, is the fact that this study concerns the implementation of important EU legislation concerning the quality of drinking water as well as groundwater and surface water. In this respect the study is relevant to the entire European Union, while the selected cases represent three member states in which this issue ranks high on the agenda.

The European Union case study first of all offers a framework and point of reference for the three national case studies, by providing a review of the present EU policies and their origins. The second part of the case study focuses on the actual contribution of the European Union to the capacity to control of the water supply sector and the possibilities of its reinforcement.

The American case study is meant as parallel study to give an impression of the way similar problems are treated within a different context. As far as the scale and level of analysis is concerned, the comparison is between the United States and the European Union. For this purpose the first part of the study addresses the urgency of the problem, the regulatory framework, and the policy networks from a national perspective. The differences within the United States are larger than those within the European Union, not only because of the varied natural and geological conditions, but also due to the wider discretion that American states enjoy in the formulation of policies. This leads to the description in the second part of four - unrepresentative - cases at the state and regional levels of successful initiatives taken by the water supply sector to deal with farming communities. Elaborating from the four cases, the present control capacity of the water supply sector is analyzed in terms of available organizational resources. The concluding section gives a number of options for reinforcing this control capacity.

1.3.3 RESULTS OF THE PROJECT

The comparative analysis of the case study outcomes yields some generalizations about the way the problem is dealt with in the three EU member states under study, at the national levels (policy networks, issues, styles and instruments) as well as at the regional and local levels (actual control capacity of the water supply sector). Two themes deserve special attention.

The first theme is the crucial function of legal regulation in complex network constellations. The case studies highlight a number of initiatives of the water supply sector, where there has been a successful appeal to farmers in respect of the effects of their agricultural practices. These initiatives are characterized by a consensual approach, where the water supply sector and the farmers are seeking to optimize their mutual interests. At the same time these initiatives are a special form of self-regulation, involving steering of social actors (farmers) by quasi-governmental or private organizations (water supply sector). Yet government is not completely absent from the scene, as the entire context is permeated by legal regulation, and it is completely clear that nothing would have happened without strict legal regulation and credible enforcement. This seems to be the decisive driving force for farmers as well as for the water supply sector, which calls for a reconsideration of the concept of 'retreating government'.

The second theme is constituted by the opposite trends at the national and the regional and local levels. It has been mentioned above that the traditional national policy networks that used to shape agricultural policy are eroding. Consensus building is increasingly giving way to top-down regulation, which is affecting the legitimacy and acceptance of the policy formulated for of the agricultural sector. At the same time successful initiatives are being developed at the regional and local levels by the water supply sector in co-operation with farmers and their associations. These initiatives are highlighted in all of the four national case studies.

1.3.4 POLICY RELEVANCE

The research outcomes apply in the first place to the prevailing issue of agricultural pollution of drinking water resources. They contain examples of ways in which water supply companies and regional water authorities can be more active in a preventative capacity. National governments have several options to reinforce the control capacity of the water supply sector, either through extra support for organizations in this sector, or through direct steering of the agricultural sector.

The examples of successful co-operation between the water supply and agricultural sectors at the regional and local level can be proliferated horizontally to other places, and perhaps also vertically to the national and European levels. Direct contacts with farmers and their organizations at the regional and local level seem to be a precondition for concrete results, but reinforcement of the contacts between national associations of all sectors involved - or perhaps between their European counterparts in Brussels - can be beneficial, too. For instance, the available knowledge and expertise accumulated within the sectors may be used more effectively, and a certain degree of mutual commitment may be created between water quality managers, the water supply industry and the agricultural sector at all levels, from the European down to the local level. In this way, the success stories described in the case studies will no longer just be occasional events, as intersectoral co-operation will become more common practice. A particular merit of this approach is that the negative effects of the stalemates encountered in the national agricultural policy formulation processes - as reported in several case studies - may to some extent be neutralized.

Moreover, the project outcomes may have a wider application than the prevailing policy issue. Agricultural pollution of drinking water resources is just one of many policy issues within the field of agricultural policy. The case studies show that policy networks have developed around this particular issue, in this case with the involvement of the water supply sector. Similar network structures are likely to have been developed around other policy issues, with other sectors trying to influence farmers' behavior in the protection of their specific interests, such as the resistance of supermarket chains and consumer organizations to pesticide use in horticulture. Where similar network configurations exist, this study has relevance too.

1.3.5 THEORETICAL FRAMEWORK

Although the emphasis of the project is on empirical research, its design is based on an explicit theoretical framework, which is introduced and explained in Chapter 3. This belongs in the realm of the policy network approach. Initially it elaborates on the European tradition and the debates on 'policy communities' and 'issue networks' with authors such as Richardson and Jordan (1979) and Rhodes and March (1992). Apart from network characteristics, relevant subjects are policy styles and the choice of policy instruments. In addition, a link is established with some American literature, notably the

'Advocacy Coalition Framework' developed by Sabatier (1988) and Sabatier and Jenkins-Smith (1993).

A policy community is a challenging concept for the analysis of the current question of the agricultural sector, which contains an archetype of this concept. The application of available theoretical knowledge to this policy problem is not an obvious procedure. The first complication is how to deal with disintegrating policy communities: are they still acting as such? Is there a regression to the status of issue networks? Or is this a new type of network configuration? Further, there is the issue of the delimitation of policy networks, as the prevailing policy issue concerns at least two major policy fields - agricultural policy and water quality management - while the participation of the water supply industry has to taken into consideration, too. The fact that no existing policy network structure is able to claim this policy instrument as belonging exclusively to their domain calls into question the meaning and nature of complex, impermanent network structures consisting of actors who often know each other from other situations.

Moreover, the complexity of the policy issue is extended over at least three levels: the European Union, the national, and the regional or local levels. This is not simply a matter of policy networks engaged in policy formulation or implementation. On the one hand there is the issue of the implementation of EU directives, which is the responsibility of the national governments, which is a particularly sensitive subject in regard to water quality management. On the other hand there is the tension between policy formulation at the national level and implementation at lower levels, requiring skillful network management and the mobilization of intermediaries: the water supply sector in this case.

The theoretical relevance of the project concerns the field of conceptions of steering within complex networks. Network steering is not conceived to be an alternative to direct regulation, at the discretion of the policy maker. Complex network structures are considered to be given contingencies for many cases of public policy, although network characteristics may differ, with policy communities and issue networks being two extreme types. In any case, policy makers have the option of direct steering of the target group, in this case the agricultural sector. Policy makers often have the alternative of indirect steering, activating intermediaries to address the target group, in this case the water supply sector. In the theoretical framework applied in this project, direct as well as indirect steering of target groups is conceived in terms of appealing to specific motivational driving forces, with a basic classification of economical, legal and communicative steering.

1.4 Organization of the research project

The research reported in the book is the outcome of a project of international co-operation of a qualified and approved research team of policy scientists. This group - a 'researchers' network' - has existed in more or less the same composition since 1990, and has previously conducted research into water management networks. The results are published in two volumes: Bressers and O'Toole (1992) and Bressers, O'Toole and Richardson (1995).

The research for the present project was conducted between April 1994 and July 1997. The project is financed by the European Commission, Directorate General XII, Science, Research, and Development.[3] The co-ordinating institute is the Center for Clean Technology and Environmental Policy (CSTM) at the University of Twente in the Netherlands. The project supervisor is Professor Dr. Hans Th.A. Bressers, scientific director of the CSTM. The project's managing co-ordinator is Dr. Geerten J.I. Schrama, senior research fellow at the CSTM. The Dutch case study has been conducted by Drs. Stefan M.M. Kuks. The partners in this co-operation are:

- Professor Dr. Jeremy J. Richardson, Department of Government, University of Essex, UK.
- Dpl.-Ing. R. A. Kraemer, Ecologic gGmbH in Berlin, Germany (formerly at the Institute for European Environmental Policy in Bonn).
- Professor Dr. Kenneth I. Hanf, Nijenrode Business University at Breukelen, the Netherlands and Erasmus University Rotterdam, the Netherlands.
- Professor Dr. Laurence J. O'Toole, Jr., Department of Political Science, The University of Georgia, USA.[4]

1.5 Layout of the book

The book consists of four parts. Part 1 contains the introductory chapter and Chapter 2 on the theoretical framework, both by Geerten J.I. Schrama. The latter has sections on policy networks, policy styles, policy instruments, and finally on the control capacity of the water supply sector.

Part 2 reports four case studies, each split up in two chapters. Although the EU case study has a distinct nature compared to the case studies of individual EU member states, its outline and the sequence of subjects is uniform, as much as possible. The EU case study is presented first (Chapters 3 and 4, by Ellis L.M. Brand, Kenneth I. Hanf, and Jos Lipman), followed by the German case study (Chapters 5 and 6, by R. Andreas Kraemer and Ralph Piotrowski), the Dutch case study (Chapters 7 and 8, by Stefan M.M. Kuks) and the case study on England and Wales (Chapters 9 and 10, by Jeremy J. Richardson). In all cases the first chapter is entitled 'Policy Framework' and deals with the regulatory or legal framework, the policy networks and the relevant (national) policies. The second chapter is entitled 'Control Capacity of the Water Supply Sector' and deals with the

---

[3] The project is formally called 'Water supply authorities in Europe preventing agricultural water pollution. Lessons from integrating environmental policy instruments affecting behavior into a policy sector with a dominant technical approach'. It is part of the R&D Programme in the Field of Environment, research area III, Research on Economic and Social Aspects of Environmental Issues (in short: SEER II).

[4] The US case study conducted by Professor O'Toole has been funded by the Center for Clean Technology and Environmental Policy.

actual control capacity of this sector at the regional and local level and with the prospects of increasing this control capacity.

Part 3 presents the comparative analysis of the case studies, beginning with Chapter 11 'Comparative Analysis', by Geerten J.I. Schrama. Here, the outcomes of the three national case studies as well as those of the EU case study are taken together. In three distinctive steps the analysis focuses on the subject of the control capacity of the water supply sector, the dependent variable of this research project. The first step is to present an overview of the relevant policy issues - those concerning the curbing of agricultural pollution in general , and water pollution by nitrates and pesticides in particular - plus the policy networks and policy styles at the national level. The next step concentrates on the national policies that are intended to influence farmers' behavior, in particular those in or near water catchment areas whose activities may affect drinking water resources. The final step is the assessment of the actual control capacity of the water supply sector.

Part 4 introduces the United States. The US case study is reported in a similar fashion as the others, in Chapters 12 and 13 by Laurence J. O'Toole Jr.

The final piece (Chapter 14 'To Reinforce Control Capacity?' by Geerten J. I. Schrama) resumes the comparative analysis. In this chapter the additional research data from the United States are incorporated in the analysis and the conclusions and recommendations from the project are presented. Two main themes are elaborated: (1) the importance of credible legal regulation and enforcement as driving force for actors, and (2) issue networks to be constituted by the water supply and agricultural sectors on all relevant levels as a countervailing power to the disintegration of the agricultural policy communities. Recommendations concern the water supply sector itself, and the national and European governments.

References

Bressers, J.Th.A., and L.J. O'Toole, Jr. (eds.) (1992) *International comparative policy research.* Enschede: University of Twente, Center for Clean Technology and Environmental Policy.

Bressers, J.Th.A., L.J. O'Toole, Jr., and J.J. Richardson (eds.) (1995) *Networks for water policy. A comparative perspective.* London: Frank Cass.

Brouwer, F.M., I.J. Terluin, and F.E. Godeschalk (1994) *Pesticides in the EC.* The Hague: Agricultural Economics Research Institute.

Copius Peereboom J.W. and L. Reijnders (1989) *Hoe gevaarlijk zijn milieugevaarlijke stoffen?* Meppel/Amsterdam: Boom.

Eurostat (1991) *Environment Statistics.* Luxembourg: Eurostat Office for Official Publications of the European Communities.

Gray, N.F. (1994) *Drinking water quality. Problems and solutions.* Chichester: John Wiley & Sons.

Kallis, G. (1995) *Assessment of EU water quality policy,* Luxembourg: European Parliament, the STOA Programme (EP 165.512).

Reus, J.A.W.A., H.J. Weckseler, and G.A. Pak (1994) *Towards a future EC pesticide policy.* Utrecht: Centre for Agriculture & Environment.

March, D., and R.A.W. Rhodes (eds.) 1992 Policy networks in British government. Oxford: Oxford University Press.

Richardson, J.J., and A.G. Jordan (1979) Governing under pressure. The policy process in a post-parliamentary democracy. Oxford: Martin Robertson.

RIVM/RIZA (1991) *Sustainable Use of Ground water. Problems and Threats in the European Communities.* Bilthoven (The Netherlands): RIVM.

Sabatier, P.A. (1988) An advocacy coalition framework of policy change and the role of policy-oriented learning therein. *Policy Sciences*, 21: 129-168.

Sabatier, P.A., and H.C. Jenkins-Smith (eds.) (1993) *Policy change and learning. An advocacy coalition approach.* Boulder (CO): Westview Press.

2. THEORETICAL FRAMEWORK

GEERTEN J.I. SCHRAMA
Center for Clean Technology and Environmental Policy, University of Twente, Enschede, The Netherlands.

2.1 Introduction

The theoretical relevance of this study concerns the operation of intermediary organizations at the regional or local level in the implementation of national and European policies. In this chapter the theoretical framework is presented, which draws in general on the policy network approach. 'Control capacity' is the central concept of this research project. It concerns the water supply sector and its in relation to the protection of drinking water resources from agricultural pollution. Control capacity is the capability of influencing social processes, in this case the behavior of a particular group of farmers. The concept of control capacity is constituted by three aspects: (1) mission and orientation of the organizations involved, in this case the constituents of the water supply sector, with respect to the policy issue; (2) the available organizational resources and their willingness to apply them for the cause; and (3) the steering strategies chosen in the application of these resources. The mobilization of intermediaries and their control capacity runs parallel to direct attempts to influence the relevant social processes, which is represented in figure 2.1.

The plan of this chapter takes a stepwise zoom into the concept of control capacity. First, the policy networks at the national level are analyzed, being the policy context within which the control capacity of the water supply sector is established and applied. This analysis involves the relevant actors (Ministries of Agriculture and Environment), the representatives of the agricultural sector, and the national water supply sector. The policy networks at the national level are supposed to be the scene for the formulation of policy on agricultural water pollution in general, including specific elements aimed at farmers within or near drinking water catchment areas. Attention will be paid to policy network characteristics ('policy communities' versus 'issue networks') and policy styles.

Geerten J.I. Schrama (ed.), Drinking Water Supply and Agricultural Pollution, 19-42.

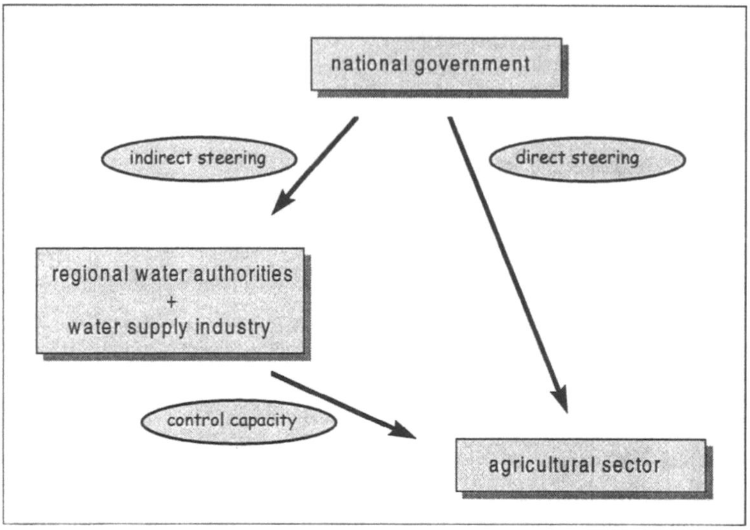

Figure 2.1: Graphic representation of the research model

The second step is still at the national level. It examines the public policy on agricultural pollution of drinking water resources in greater detail. A distinction is made between policies and instruments directly aimed at farmers within or near drinking water catchment areas, and indirect policies with the water supply sector as intermediary. Policy styles are analyzed in terms of 'regulation' versus 'self-regulation'. Individual policy instrument are reviewed with respect to their specific characteristics. In case of indirect steering, the policy instruments are also relevant in that they provide the water supply sector with organizational resources.

The third step concerns the control capacity of the water supply sector, which is constituted by the regional water authorities and the water supply companies, mainly organizations at the regional and local level, which often have direct contacts with farmers within or near drinking water catchment areas. The former are involved in the implementation and execution of (national) policies on qualitative water management, the latter have direct interests in the protection of drinking water resources, their basic raw material. As mentioned before, control capacity is assumed to be the result of three aspects: mission and orientation, organizational resources, and steering strategies.

The fourth step concerns the possible reinforcement of the control capacity of the water supply sector. This is not a major issue for this chapter, as it is foremost a matter of the conclusions to be drawn from the research project and the formulation of policy recommendations. Nevertheless, the structure of their presentation will be derived from the theoretical framework. The purpose of this chapter is to clarify this framework and to link it to the relevant literature.

2.2 Policy networks

The relevancy of the theoretical concept of policy networks first of all concerns the interaction between the actors at the national level, the Ministries of Agriculture and Environment, the organizations and representatives of the agricultural sector, and the water supply sector. In particular, the focus is on the formulation of policy on issues of agricultural water pollution by manure or nitrates and by pesticides, and notably the policy aimed at farmers within or near drinking water catchment areas.

2.2.1 DELINEATION OF THE CONCEPT

The present research project elaborates on a previous one, not just with respect to research questions and staffing; it also elaborates on the theoretical framework with policy networks as a central concept, which is presented by Bressers, O'Toole and Richardson (1994a)[1]. The flourishing of this concept in the field of policy studies can be explained by the growing awareness of multiple interdependencies in almost every aspect of governance (recently: Kickert, Klijn and Koppenjan, 1997). Bressers, O'Toole and Richardson (1994b: 4) refer to the multiple origins of the concept, such as issues of inter-organizational relations from organizational theory, and the heuristic concept of 'iron triangles', taken from the political sciences, as applied to the interrelatedness of administration, legislation and organized interests, as well as intergovernmental relationships. They employ the term policy network for 'the large class of multi-actor arrangements of interdependence in [the] varied phase of the policy process.' (p.5). To account for the use of a fashionable concept, Bressers et al. (1994b: 8-9) mention four arguments that go beyond the purely descriptive application of the term policy network: (1) it indicates that the interaction between the actors involved is not confined to what can be observed, while it may be affected by past and likely future interaction beyond the reach of the observer; (2) in addition to the previous point, interaction in a specific field or arena may be affected by interaction in other fields or arenas with an overlapping set of actors; (3) actors within the network may not act independently but under the influence of unobserved external stakeholders; and (4) finally, the network perspective directs our attention to lager structures of interdependency.

2.2.2 POLICY COMMUNITIES VERSUS ISSUE NETWORKS

An obvious way to apply the concept of policy network is to explain the policy outcomes (partly) in terms of basic characteristics of the network. For this purpose the typology of policy communities versus issue networks is often used. The term 'policy

[1] Also published as: Bressers, O'Toole and Richardson (1995)

community' was introduced first[2] to indicate stable and closed configurations of a limited number of actors, which managed to control specific policy sectors. They are characterized by a high frequency of interaction, a certain balance of power, as well as shared values and perceptions of situations, giving rise to consensus on major issues. Policy processes within a policy community are often characterized by what is called a 'consensual policy style' (see next section), little transparency of the decision making process, and little opportunity for outsiders to voice their interests. The archetype of a policy community is perhaps the kind of policy network that used to dominate the agricultural sectors in many EU member states (e.g. Bressers, 1993). Somewhat later Heclo (1978) used the term 'issue network' for constellations which are in every aspect the opposite of policy communities; networks with large numbers of heterogeneous actors emerging around specific policy issues.

The three European countries under study have known genuine agricultural policy communities, comprised by the Ministry of Agriculture, farmers' interest organizations and industries related to the agricultural sector, and certain members of representative bodies like the parliament. It is far from certain that these policy communities are currently still functioning in the old way. Several developments have put traditional ties under heavy pressure, such as the environmental issue related to unsustainable and polluting agricultural practices. Nowadays, national agricultural policies are inconceivable without due attention to the curbing of agricultural pollution. As a result the position of the Ministry of Environment has become very relevant. Did it succeed in getting access to the agricultural policy network? And what was the reaction of the Ministry of Agriculture? Does the latter stand firm as protector of the interests of the agricultural sector? Has it become a coalition partner with the Ministry of the Environment? Or is it trying to balance the conflicting interests, agricultural, economic and ecological? (For a more dynamic interpretation of the openness of policy networks, see Richardson, Maloney, and Rüdig, 1992.)

2.2.3 POLICY NETWORK CHARACTERISTICS

In addition to the original formulations, many other policy network characteristics have been distinguished, such as the number and type of network participants; degree of stability or institutionalization; level of formalization; degree of complexity; type of interdependence; degree of overlap; extent of centralization or concentration; degrees of dominance, accessibility or openness, and co-ordination; and type or scope of issue or coverage. Several reviews have been published, such as Jordan and Schubert (1992); Van Waarden (1992); O'Toole (1997).

Policy formulation and policy implementation. A more fundamental issue, which goes beyond the specific network characteristics, is the distinction between networks for

2 Richardson (1994:166) mentions Jack Wacker as the one who first used the term 'policy community'. Among the early adapters are Heclo & Wildavsky (1974) and Richardson and Jordan (1979).

policy formulation and for policy implementation. The literature often concentrates - and not always explicitly - on policy formulation. Consultation with target groups, for instance, is particularly a matter of policy formulation. Networks for policy implementation can often be conceived as implementation structures. Such structures are complex inter-organizational networks cutting across several levels, with indefinite lines of command, ambiguous goals and multiple decision making loci. The target groups for the policy are frequently involved in these networks. Policy implementation through these networks requires skillful network management. On the one hand, the distinctive individual units must be led on a voluntary base by incentives that match their own interests and objectives, or must be forced by the exercise of power, including governmental interference. On the other hand, the collective objectives, constituted by the policy targets, must be preserved and realized (Hanf and O'Toole, 1992; O'Toole, Hanf and Hupe, 1997). The present project deals with policy formulation (notably at the European and national levels) as well as policy implementation (notably at the regional and local levels). The water supply sector is clearly part of an implementation structure through its role as intermediary. The position of the water supply industry, which is not part of the formal implementation structure of the policy on water quality management, is an especially interesting one in this perspective.

Interrelatedness and mutual commitment. The concentration on the juxtaposition of policy communities and issue networks gave rise to the assumption that, although many relevant characteristics could be mentioned, they were nearly all indicators of one basic dimension, and that most policy networks can be arranged on a single continuum (e.g. Rhodes and March, 1992). Some authors think that more mutually independent dimensions of policy networks should be distinguished. A possible extension of the conceptual framework is suggested by Bressers (Bressers and Kuks, 1992; Bressers, 1993; 1994), in distinguishing two basic dimensions: 'interrelatedness' (the structural aspect, referring to the extent of intensity of interaction between individuals, groups and organizations within the network) and 'mutual commitment' (the cognitive aspect, referring to the distribution of objectives between actors, or the extent to which the objectives involved are conflicting). Policy communities usually rate high on both dimensions, and issue networks low. Meanwhile, one should be aware of the possibility of networks characterized by high interrelatedness and low commitment, or the other way around, as transitional or even permanent configurations in our case studies. (More about Bressers' theory can be found in the section on policy instruments.)

Advocacy coalitions. For our present purpose, a useful addition to the policy network approach is the 'advocacy coalition framework' (Sabatier, 1988; Sabatier and Jenkins-Smith, 1993; 1997). This is particularly focused on the conditions that facilitate policy change and policy oriented learning. The unit of analysis is the 'policy subsystem', which is more or less equivalent to the policy network (this also reflects the preference of British authors for the concept of policy networks and of American authors for the concept of policy subsystems - c.f. Klijn, 1997: 27). It is not only interests groups, administrative agencies, and legislative committees that are considered to be relevant actors, but also other levels of government, the media as well as researchers and analysts. Sabatier cs. picture policy subsystems explicitly as multi-level structures, by using such concepts as 'advocacy coalitions' and 'policy brokers', the latter being the intermediates between the advocacy coalitions. Advocacy coalitions are described as consisting of:

> *... people from a variety of positions (elected and agency officials, interest group leaders, researchers) who share a particular belief system - i.e. a set of basic values, causal assumptions, and problem perceptions - and who show a non-trivial degree of co-ordinated activity over time.* (Sabatier, 1988: 139)

It is interesting to figure out how these concepts relate to those of the policy network approach. A policy subsystem seems to resemble a policy community in the sense that both refer to a institutionalized and relatively stable context within which policy issues are dealt with. The difference is that a policy community is a tight and closed entity, while a policy subsystem may involve several rival advocacy coalitions. Advocacy coalitions, on the other hand, also resemble policy communities in several respects, but they do not constitute the entire subsystem, and they are not able to exclude other parties from the policy process. Another way to look at is, is to conceive advocacy coalitions as policy networks 'of the second order', situated at a lower level of analysis. The 'first order' is constituted by the policy subsystem, which can be described as an institutionalized network in which overt conflicts of interests and diverging values and perceptions may occur - typical characteristics of issue networks.

2.2.4 POLICY NETWORKS AND POLICY ISSUES

The present concern is not the whole field of agricultural policy, for we are dealing with some particular policy issues in the subfield of agricultural water pollution. So the question is, how does one delineate the prevailing policy network (or subsystem), and how can one decide whether this network actually exists? [3]

A first nuance to the theory is a rather obvious one. A large policy area, such as agriculture, may be dominated by a policy community, but this does not imply that all members are actually involved in all policy issues. Certain policy issues will concern only particular sectors of the policy area, such as dairy farming. In these cases the acting policy community is a subset or a derivative of the original one, where individual participants may be different since sectoral interests are represented rather than general ones. This point has already been dealt with by early authors, such as Richardson and Jordan (1979). However, this is not a sufficient answer to the problem raised, as we are dealing with policy issues that are not confined to the field of agriculture, but which

[3] The linking of policy issues with policy subsystems or policy networks is often taking for granted, although it is obvious that competing subsystems or networks are contesting each others competence to deal with specific issues. The problem definition, as accepted by the larger system, may be crucial, as this links policy issues to the domains of the contending subsystems or networks. Further elaboration, however, is beyond the scope of the present chapter.

transgress the boundary into the fields of environment and water management. This is not a question of a particular section of the agricultural policy field and it cannot be solved by the concept of sectoral policy network. Similar problems occur in the absence of a policy community, when an issue network has to be established for every new policy issue.

The ontological issue of policy networks - or in this case policy subsystems - is addressed by Sabatier and Jenkins-Smith (1997) in a paper in which the 'advocacy coalition framework' is further elaborated in reply to critics. As likely criteria for the outlining of policy subsystems they mention:

1. participants regard themselves as a semi-autonomous community;
2. they share a domain of expertise;
3. they seek to influence public policy within the domain over a fairly long period of time;
4. there exist specialized subunits within agencies at all relevant levels of government to deal with the topic; and
5. there exist interest groups, or specialized subunits within interest groups, which regard this as a major policy topic.

All of these conditions tend towards the familiar policy community concept, except the presence of concurring advocacy coalitions, whereas policy communities are characterized by strong mutual commitment between all participants. However, not all policy networks or subsystems are true communities, therefor the lasting pattern of interaction and the awareness of its existence in the participants' perceptions seem to be the critical criteria to accept the existence of policy networks.

The concepts introduced by Sabatier cs. may be useful in our project to deal with the new situation that has occurred after the iron triangle had lost its control over the policy area, while the contacts between the constituents may be continued in a more or less similar way, as an advocacy coalition operating within a larger policy network, to which other parties and other interests have gained access. The role of the Ministry of Agriculture is likely to be a crucial one. In terms of the advocacy coalition framework, it may have established a new advocacy coalition with the Ministry of Environment, in case the urgency of transformation to sustainable agriculture is too overwhelming to shield its 'belief system' from it, or it may have chosen to act more as a policy broker.

The central - and strong - argument put forward by Bressers (1993) is that social structures, such as policy networks, tend to reproduce themselves and that this fact is a major driving force underlying the selection of policy instruments. So policy communities can assign newly emerging policy issues to their domain, mold the policy process in a way that fits their particular constitution, and reconfirm their own composition, procedures, and domain by doing so in a successful, or at least in an unchallenged way. This image is based on the assumption that policy networks are indeed social structures, lasting phenomena although subject to reproduction, while policy issues are transient phenomena.

This idea of transience points to the question of the ontological status of an issue network. Is this a social structure that encompasses the interaction between the parties involved in just one particular policy issue? How much structure is involved? What was there before the policy issue arose, and what will remain when it has passed away? In most cases there is nothing that makes the particular set of actors involved a discernible entity, apart from the policy issue concerned. One way to deal with it is to accept the *contradictio in terminis* of a transient network, that emerges and vanishes with a specific policy issue, and is based on the notion that the participants are not selected completely at random. Some participants may have close connections with each other, constituting an advocacy coalition that exists apart from and long before the emergence of the policy issue. Other participants may be aware of each other and some of them may adjust their behavior towards others, because they have met before and expect to meet each other again. In this view the structural aspects of issue networks are attributes of a larger and context, while the specific configuration of actors is as transient as the policy issue at stake.

A final consideration on the nature of issue networks is that it can be linked either to the advocacy coalition framework, in the sense that the policy outcomes are contested by concurring coalitions (provided that its momentary nature is no impediment to this), or to the approach suggested by March and Rhodes (1992). On the occasion of a comparative case study research, the latter distinguish an inner and outer circle of policy networks (or core and periphery). Members of the inner circle are involved in the policy processes of a specific policy field on a day-to-day basis, while the involvement of the members of the outer circle depends on the prevailing policy issue and their particular interests that may be affected. The difference between both approaches is whether one particular group of actors (a policy community or an advocacy coalition) holds a central position or whether several groups are competing on a more or less equal basis. It is suggested that such a central position can be attained when a coalition succeeds in assigning an emergent policy issue exclusively to its own domain. This situation comes close to the traditional policy community, except that outsiders with a stake in the policy issue have access to the policy process.

The above argument gives rise to several questions concerning the subject of the research project. First, is agricultural pollution of drinking water resources considered to be a distinct policy issue? Second, which actors are involved in the policy process? In particular, what is the role of the water supply industry? Third, is the policy issue considered to belong to the field of agricultural policy and the domain of the (former) policy community? If yes, how strong is the position of the core coalition and which other actors have found access to the policy process? If no, which advocacy coalition, probably from the fields of agriculture, environment, and water management, is contesting the policy outcomes?

2.3 Policy styles

'Policy style' is a current term to characterize the way policies are formulated and implemented. An early and often quoted application of the concept can be found in Richardson (1982). It contains no explicit definition, but a description with several phrases referring to the behavior of policy makers, such as 'systems of decision making' and 'standard operating procedures for handling issues' (Richardson et al., 1982: 2-3).

The concept is generally used for comparative purposes, often between different countries in the sense of typical *national policy styles.* Differences in national policy styles are related to the national context of political institutions, political relationships, legal systems, traditions, and - possibly - fashions (e.g. Howlett, 1991). According to this perspective, policy makers make many actual decisions without having substantial discretion. An alternative approach focuses on differences in policy styles within national settings, such as between sectoral policy styles. In this perspective policy styles are adapted to characteristics of the specific policy sector, in order to match the policy network characteristics and - perhaps - optimize the policy's effectiveness, or the preferences of the actors involved. This view assumes somewhat more discretion for the policy makers.

Consensual versus impositional policy styles. As to the features of the policy styles in which they differ from each other, Huitema and Van Snellenberg (1997) stated in a recent review, that the dominant dimension in both main perspectives is the extent to which consensus with interest groups in society is sought during the policy formulation or implementation phase, which may involve negotiations. A 'consensual' or 'consensus oriented' policy style is the opposite of an 'impositional', 'regulatory', or 'command and control' policy style (e.g. Richardson et al., 1982). The other dimension distinguished by Richardson et al. (1982) is 'anticipatory and active problem solving' versus 'reactive problem solving'. They observe a trend in Western Europe (in the early eighties!) towards more consensual and reactive policy styles.

In the present research project the distinction between consensual and impositional policy styles is relevant to the subject of policy networks. The policy formulation concerning agricultural water pollution - direct steering of the agricultural sector as well as indirect steering through the water supply sector - can be analyzed in these terms. A relationship is assumed between the disintegration of former agricultural policy communities and a propensity away from strict consensual policy styles, with possible drives such as pressure to implement EU law, opposition to agricultural pollution, and the representation of more and diverging interests in the political arena.

Regulation and self-regulation. The second dimension of policy styles to be applied in the present research project concerns the steering paradigm on which policy is based, expressed as the opposition between 'regulation' and 'self-regulation'. This dimension is of particular relevance here, because of the duality of direct and indirect steering of the agricultural sector. Where the water supply sector is used as intermediary, legal regulation is less appropriate for addressing farmers than approaches based on self-regulation.

The whole affair is about curbing agricultural pollution of drinking water resources by influencing the behavior of individual farmers. The function of policy networks, styles, instruments is to communicate the right signals and incentives to this target group. Two concepts apply here: the forces driving the behavior of individuals or groups, and steering paradigms.

Three basic types of driving forces are distinguished: economic rationality, legal compliance, and social responsibility.[4] Economic rationality is the driving force underlying the behavior of individuals or organizations, which is based on the pursuit of well understood self interest. Although this may be the dominant driving force, it is not assumed to be the sole one. Legal compliance refers to individuals or organization who accept the rule of law, regardless of their own position or interests, as law abiding citizens who acknowledge the legitimacy of the legislators. Social responsibility is the driving force that motivates individuals or organizations to act according to social norms, either because these are imposed on them by the community, or because they are internalized through socialization. The latter includes care for the environment in the interest of present and future generations.

The link between driving forces and steering paradigms is established through incentives and signals transmitted by policy makers that are assumed to appeal to the forces driving the target groups. Again, the classification includes three basic types - juridical, economic, and communicative (e.g. Van der Doelen, 1989; Pröpper and Herweijer, 1992; Glasbergen, 1994) - but correspondence between both classifications is not unambiguous.

Traditionally, public policy is predominantly based on legal regulation, according to the so-called 'legal' or 'regulatory steering paradigm' (e.g. Drupsteen, 1994). In practice, most people comply with most legal demands in an obvious way, almost without reflection. This highlights two basic functions of legal regulation. First, the codification of social norms, which concerns a mixture of legal compliance and social responsibility. Second, it makes a moral appeal to civil loyalty, which is related to a more pure form of legal compliance. Apart from these functions, the legal system is provided with an extensive apparatus of control and sanctioning. When moral appeals are not sufficient, sanctions will be invoked that address the driving force of economic rationality. The proportionality of these sanctions (in terms of magnitude and chance) should make unlawful behavior less attractive to target group than lawful behavior. Moreover, to the extent that legal regulation reflects social norms, violations may also be penalized by social sanctions enacted by the community.

The opposite of legal regulation is self-regulation. This does not imply that government is declining its responsibility, or is taking a passive attitude and refraining from all attempts at social steering. Here, self-regulation means that government intends to bring about the desired behavior of the target group by other means than those predominantly based on legal regulation. In other words, it is applied in cases where appealing to economic rationality or social responsibility is assumed to be more effective.

The 'economic steering paradigm' (e.g. Vermeulen, 1994) is based on policies involving incentives that appeal to economic rationality. The application of certain

4 A similar classification is used by Bressers and Klok (1988) and Klok (1991). They use the term 'individual rationality' for economic rationality to stress the distinction from both other types that refer to group processes.

policy instruments implies that preferred behavior is rewarded, or extra costs are attached to unwanted behavior. In its pure form, no moral appeal (to legal compliance or social responsibility) must be present, leaving actors actually free to chose between alternative modes of behavior. As members of the target group are assumed to act rationally, they should prefer desired behavior above unwanted behavior and act accordingly (or at least a satisfying proportion of them should do so). However, pure economic or market conformity policies are scarce. Quite often negative incentives or actual sanctions are provided in case the initial policy formulation is not successful, usually in the form of looming legislation.

The 'communicative steering paradigm' (e.g. Van de Peppel and Herweijer, 1994) is more difficult to asses in terms of appeals to forces that drive target group behavior. Communicative steering involves the application of communicative policy instruments, the dissemination of information that should affect behavior. Referring to pure forms, Pröpper (1992) stresses the aspect of 'self commitment' based on information provided, with no decisive role being played by economic incentives or legal enforcement. The idea is that people are motivated by their sense of social responsibility, and that the policy aims to encourage these motivational processes. Pure communicative strategies, however, are rare and it is doubtful whether any communicative strategy that appeals solely to social responsibility, and not to legal compliance or economic rationality, can be effective at all (cf. De Bruijn and Lulofs, 1993). In practice, communicative policy instruments are not exclusively employed for communicative steering, but also to support legal or economic steering, by drawing attention to existing legal regulation or economic policy incentives. In fact, communication is a prerequisite for any successful policy.

Regulation versus self-regulation is presented here as the second dimension of the concept of policy style. The present research project deals with agricultural pollution in general - assessed in terms of policy network characteristics and consensual versus impositional policy styles - but focuses on a relatively small part of it: the policy on protection of drinking water resources. In this subfield direct regulation of the farmers' behavior is possible, while there is also the alternative of indirect steering through the water supply sector as intermediate. At this level the three steering paradigms will be applied as analytical tools to assess the various steering efforts covered by the national case studies.

2.4 Policy instruments

The issue of protection of drinking water resources from agricultural pollution is not only analyzed in terms of policy styles and steering paradigms. When they go into greater detail, the case studies also examine the policy instruments applied as part of the national policies, while the means and resources applied by the organizations of the water supply sector are discussed in later sections.

Many classifications of policy instruments have been proposed. An appealing one, one that fits in well with the theoretical framework presented here, is the triad of economic, juridical, and communicative instruments, formulated by Van der Doelen (1989), analogous to the three steering paradigms discussed in the previous section.

2.4.1 POLICY INSTRUMENTS AND ORGANIZATIONAL RESOURCES

In their theory of policy instruments Bressers and Klok define policy instruments in terms of effectuating changes in behavior in line with the policy objectives (Bressers and Klok, 1987; 1988; Klok, 1991; Bressers, 1994). From this perspective the choice of behavior by policy subjects - conceived as subjective rational actors - can be influenced in five distinct ways:

1. exclude behavioral options (prohibition, physical obstruction);
2. create behavioral options (facilitation);
3. affect the pros and cons of behavioral options (positive and negative incentives);
4. affect the relative importance of the pros and cons of behavioral options (persuasion);
5. provide information about the pros and cons of behavioral options (to affect the decisions of actors).

The application of most policy instruments can be conceived in terms of transactions of resources between legal authority and policy subjects (Klok, 1993). In organization theory the concept of organizational resource is usually connected with the resource dependency perspective (Pfeffer and Salancik, 1978). The idea is that all organizations are struggling for survival (by maintaining or increasing their size, domain, influence or power), which requires a specific supply of resources. In order to acquire these resources an organization has to manage its relationships with the environment. Benson (1975) introduced the basic classification of money and authority. This does not imply that relevant resources are always money and power in their most vulgar form; this should be understood in terms of material versus immaterial resources. A somewhat more elaborate list of organizational resources may be the following:

1. financial capacity (money or capital),
2. organizational capacity (manpower),
3. legitimacy and authority:
   - legitimacy based on legal or formal authority,
   - acknowledgment of organizational domain,
   - acknowledgment of expertise,
4. information:
   - professional knowledge, expertise, and skills,
   - information and Intelligence (about facts).

The perspective of organizational resources in the study of public policy is based on the assumption of members of the target groups, or policy subjects as subjective rational actors, which involves a focus on economic rationality as principle driving force underlying their behavior. The classification of policy instruments according to the type of transfer of resources starts with the distinction between positive and negative incentives; positive when actors are supplied with resources, negative when resources are abstracted from them. The resources involved are assessed in terms of their economic value to the policy subjects, who are assumed to make a rational choice, pondering all pros and cons of the available behavioral options. The second distinction is between conditional and unconditional application (Klok, 1993). In the case of positive incentives, resources may be unconditionally granted to members of a specific target group, as a stimulus for behavioral adaptation on a voluntary basis (reduction of income taxes causes increase of consumptive expenditures). A conditional positive incentive implies that the transfer of resources will take place only when the policy subject acts in a certain way. Subsidies are the most familiar instrument of this kind. Negative incentives are usually conditional. Levies and fines are attached to unwanted behavior. The choice from among behavioral alternatives can be influenced by conditional negative incentives: a fine in case the unwanted behavior is prohibited, a levy when it is not. Proportional negative incentives usually have the form of levies or taxes, varying with the extent of the behavior (more discharges into surface water means higher levies). Note that control over behavior can result from the application of policy instruments as unintended or non-deliberate effects. These effects can be considered positive or negative. The former was the case with the introduction of proportional levies on pollution of waste water by companies in the Netherlands (studied by Bressers, 1983).

Obviously, organizational resources are not solely of a material nature, and their value is not only economic. Organizational resources can be of an immaterial nature too, shaped in terms of legitimacy or information. The perspective of organizational resources is also compatible with the driving force of legal compliance. The crucial issue is the lawfulness of certain activities. Legitimacy as an organizational resource can be derived in a passive manner from prevailing legal regulation, but it can also be actively obtained from a legal authority, for instance through the granting of a permit. Information is also a relevant resource with respect to legal compliance. Any actor who tends to assess his behavior against prevailing legal regulation will attach a positive value to information about *ist* and *soll*. As the achievement of legal compliance often requires the investment of time and money, these resources are relevant, too.

With respect to the third driving force of social responsibility, the exchange of organizational resources through policy instruments is not so obvious. One option is to provide subsidies for socially responsible behavior, where it is impeded by financial problems. If one looks at pure communicative policy instruments, assessment in terms of transfers of resources may be misleading, as subsidies, for instance, do not belong to this category. Nevertheless government resources or budgets are applied in attempts to influence the views and behavior of target groups, for instance through campaigns. However, even if the communicative instruments are effective, which is a difficult achievement, it is unlikely that the target groups will acknowledge that they have provided valuable informational resources.

It should be mentioned that an organizational resources perspective on policy instruments is not simply a matter of the transfer of resources between policy maker and target group. A subsidy scheme, for instance, not only requires a budget to draw the money from; its execution also imposes other costs, in terms of financial and organizational resources; 'transaction costs', so to speak. With respect to legitimacy, one could argue that the application of particular policy instruments involves the application of organizational or financial capacity to 'produce' legitimacy for the target group. In this view, resources of one type are transformed into another type, which is not a very illuminating metaphor. It must be clear that target groups do not always gain or lose something in terms of resources, and they certainly do not always perceive it as such.

2.4.2 POLICY NETWORK CHARACTERISTICS AND POLICY INSTRUMENTS CHOICE

Bressers (1993) has explored the relationship between network characteristics (the dimensions of interrelatedness and mutual commitment, as discussed in a previous section) and the typical choice of policy instruments.[5] In connection with the five ways of influencing target group behavior presented above, Bressers has distinguished six major policy instrument characteristics (Bressers and Kuks, 1992; Bressers, 1993; Bressers, 1994; Bressers, Huitema and Kuks, 1994):

Moral appeal to the target group. This may involve an appeal to legal compliance (juridical instrument) as well as an appeal to social responsibility (communicative instrument). An economic incentive (appeal to economic rationality) implies the absence of any moral judgment about the behavioral choice of the actors involved, let alone the possibility of sanctions in case of unwanted behavior. Policy makers are inclined to apply instruments with this property in case of (relative) absence of mutual commitment.

Fine tuning to target group behavior. Fine tuning may be applied to achieve a balance between target group behavior and administrative (re-)action. To the extent that policy instruments are less generic and more specific, fine tuning to individual target group members becomes more feasible. Strong commitment is an important motive for fine tuning, while interrelatedness is a precondition.

Enlarging versus restrictive instruments. This indication refers to the instrument's effect on the behavioral options of the target group. Enlarging instruments are also referred to as positive stimuli or the (conditional) provision of organizational resources, while restrictive instruments are related to negative stimuli or the (conditional) abstraction of organizational resources. Economic and juridical instruments can be enlarging as well as restrictive, while communicative instruments are always enlarging (unless propaganda and other forms of manipulative information are considered as communicative instruments, and this can be restrictive). A positive relation is assumed between both interrelatedness and commitment and favoring incentives over sanctions.

Discretion for the target group. This feature is also referred to as facultative versus obligatory instruments. Some policy instruments involve freedom of choice to the target group to accept the 'proposal' or not (e.g. subsidies), while application of other

---

5 It should be mentioned that the network characteristics distinguished by Bressers are mainly aimed at the bilateral relationships between policy makers and target groups or other policy subjects, and less at other relationships between the network participants.

instruments does not depend on the acceptance by the policy subjects (e.g. regulation). High interrelatedness is assumed to be an inclination towards the use of facultative instruments, while they might be successful in cases when commitment is lacking.

One-sided versus negotiated instruments. The so-called 'two-sided instruments' require negotiations between policy makers and target group before application (e.g. voluntary agreements). High interrelatedness is assumed to be an inclination towards the use of negotiated instruments, while high commitment may make negotiations superfluous.[6]

Integration versus separation of policy formulation and implementation. To the extent that policy formulation and implementation are more integrated, the policy makers have a bigger role in the policy execution. A positive relation is assumed between both interrelatedness and commitment and integration of policy formulation and execution.

2.4.3 APPLICATION TO THE PRESENT POLICY ISSUE

In the present research project, not all of these six policy instrument characteristics will receive similar amounts of attention. The focus will be on three characteristics. The first one is the distinction between direct and indirect steering. This is more than the role of the policy makers in the implementation stage, as it also involves the option of policy instruments directed at intermediaries to encourage them to attempt to influence the behavior of the real target group (in this case instruments directed at the water supply sector). Facilitating of intermediary organizations is the ultimate form of separation of policy formulation and implementation, as the original policy makers transfer much of the policy formulation, too. A classification of policy instrument, which stresses the roles of the policy makers and the water supply sector, is as follows:

- instruments directly aimed at farmers within or near drinking water catchment areas with no role for the water supply sector in their implementation;
- instruments directly aimed at farmers within or near drinking water catchment areas to be implemented by the water supply sector;
- instruments not directly aimed at farmers within or near drinking water catchment areas, but aimed at the reinforcement of the control capacity of the water supply sector.

The second characteristic is the presence of a moral appeal to the target group, although this will rather be conceived as a characteristic of the policy style (and the steering strategy of the water supply sector, as discussed in a later section). The third instrument characteristic is a combined one, under the assumption that the numbers three and four in Bressers' classification usually coincide: enlarging and facultative versus restrictive and obligatory instruments.

---

6 One should be aware of the difference between a consensual policy style and the instrument characteristic mentioned here, which refers to the necessity of consultation with the target group for its application.

2.5 Control capacity of the water supply sector

In the third step of the theoretical framework the focus is on the actual control capacity of the water supply sector. The first one concerned the national policies on agricultural water pollution in general, the prevailing policy networks and policy styles. The second step concerned the issue of agricultural pollution of drinking water resources with a special target group of farmers within or near drinking water catchment areas. Policy styles were further examined in terms of regulation versus self-regulation, and attention was paid to individual policy instruments and the likely role of the water supply sector as intermediary.

Here, the control capacity of the water supply sector is examined as the dependent variable in the research model. The term 'control capacity' has in fact been coined for the present research project to indicate the extent to which the water supply sector is able to exercise control over the causes of agricultural pollution of drinking water resources. The same term, however, has been used before, with a similar meaning. Kickert (1993: 275) used it to indicate the extent to which government is able to control social processes in general. Here, the application of the term control capacity is based on the assumption, that the extent to which the water supply sector is able to keep control is the combined result of (1) the mission and orientation of the organizations that constitute the water supply sector, (2) the available organizational resources, and (3) the choice of steering strategies. These factors will be discussed below in separate subsections.

2.5.1 MISSION AND ORIENTATION

The water supply sector as referred to in this project is constituted by the water supply companies and the regional water authorities. They are involved in the issue of agricultural pollution of drinking water resources first of all because it affects their self-interest. To the former it is a hindrance to their production of good quality drinking water, for the latter it concerns their primary function as water quality managers. In order to act as intermediaries for public policy they should have an orientation on this issue that is in accordance with this policy. Therefore, mission and orientation constitute the first aspect of control capacity. In this particular case orientation refers to the organization's vision on the problem, in particular the importance attached to prevention over remediation. Prevention means controlling social processes - influencing or steering farmers' behavior - while remediation focuses rather on technical processes. According to their mission, some water supply companies may act completely as private enterprises, and treat the issue from an economic perspective in terms of the costs and revenues of preventive actions. Others may see themselves to some extent as public organizations with a function in the chain of water quality management. Finally, social responsibility may be a relevant driving force for water supply companies which are publicly owned. For regional water authorities similar driving forces may be involved in varying proportions, depending on the extent to which the task of water quality management is imposed on them from above or the issue is internalized and felt to be part of the organizational mission. Presumably, the discretion left to water supply companies for operating in this issue is generally larger than for regional water authorities. And finally, the control capacity in a particular region depends on the extent to which the water supply sector operates as an advocacy coalition in the sense of Sabatier, with shared 'belief systems' and frequent interaction.

2.5.2 ORGANIZATIONAL RESOURCES

The concept of organizational resources has been discussed in a previous section. Here, attention is drawn to the aspects of resources available to the water supply sector that should enable them to control the polluting behavior of farmers. The basic classification of organizational resources applies, while some accents will be added in the case studies.

Financial capacity refers to the means available to water supply companies to cover drinking water production costs (which are reflected in consumer prices) or to finance investments (which must be earned back over a certain period of time). Whether any water supply company deems it useful to apply its financial resources for the purpose of protecting drinking water resources depends on its mission and orientation, as well as on its financial situation, which should allow investments with rather uncertain returns. Regional water authorities may have budgets for their tasks of monitoring water quality and enforcement of related legal regulation.

Organizational capacity concerns the manpower required for the exercise of control capacity. The direct relation between organizational and financial capacity is obvious, but organizations - and certainly (semi-) governmental agencies, such as the regional water authorities - do not always have the discretion to employ financial capacity for creating organizational capacity. Water supply companies are run more in a business like way, and the application of organizational capacity is rather based on economic considerations. Economies of scale may play a significant role, as the creation and employment of expertise on the topic of agricultural pollution may be more efficient for large companies than for smaller ones.

Legitimacy and authority are a generic term for a class of immaterial organizational resources. Usually, formal or legal authority is the prerequisite of governmental agencies. Where it concerns steering of farmers' behavior this is often relevant, either with respect to the option of direct interference in farming practices, or with the respect to the option of imposing legal sanctions. Informal kinds of legitimacy and authority concern the acknowledgment of organizational domains or stakes (such as the stakes of water supply companies in the protection of their natural resources), the acknowledgment of specific expertise ('they know what they are talking about'), or recognition based on confidence or trust gained from other parties. Legitimacy and authority are required to establish contacts with farmers or their representatives. The informal forms are not effective when it comes to enforcement, but they may be very useful for making a good start when (voluntary) co-operation is sought.

Information concerns availability of, or access to specific knowledge or facts that are meaningful in specific situations. A distinction can be drawn between expert knowledge, which has a wider applicability (which also has an aspect of legitimacy), and factual knowledge, which is related to the uncertainties in specific situations. The latter applies to the present issue, especially where threats to underground aquifers are concerned: the actual water quality, the amounts of pollution on its way to the wells, and the agricultural practices of farmers in the neighborhood. Apart from the chances of actually changing farmers' behavior, the water supply sector may benefit from more accurate knowledge about the condition of their natural resources and the amounts of pollution that can be expected to show up at their boreholes in the coming years.

Time can finally be treated as an organizational resource in its own right (cf. Klok, 1993), in the sense of absence of time pressure on organizational decision-making and activities. Time can also be conceived as an asset that may enable organizations to make optimal use of their resources rather than a resource in the strict sense.

2.5.3 STEERING STRATEGIES

Social steering (or 'governance') is not the prerogative of government. Other organizations or institutions, such as the water supply sector, are also involved in social steering, undertaking attempts to influence social processes. In this respect there is a counterpart of policy instruments, in terms of the organizational resources available for such purposes.[7] These instruments can be characterized in terms of 'facultative and enlarging' in the case of positive incentives and 'obligatory and restrictive' in the case of negative incentives. The application of facultative and enlarging instruments require predominantly financial and organizational capacity, to furnish the means, to manage the process and to monitor or enforce contractual arrangements when they apply. The application of obligatory and restrictive instruments requires predominantly some kind of legal or formal authority, which exists in the hierarchical context of a formal organization, but which is the privilege of governmental agencies in a societal context. In addition, some organizational capacity is required for monitoring and enforcement. To the extent that the attempts to influence social processes are deliberate and consistent, the concept of steering strategies applies. There are some similarities with the concept of policy style, such as the distinction between direct and indirect steering and the way the basic driving forces of the target groups are addressed. Below, three basic steering strategies are discussed, classified according to the way the target groups are addressed.

Economic strategies address economic rationality as a force driving the target group's behavior. Positive economic incentives can be generated by everyone possessing the required financial and organizational capacity. In fact, this kind of steering comes down to paying the target group for changing its behavior. Negative economic incentives, communicated through obligatory and restrictive instruments, require some kind of legal

---

7 De Bruijn and Ten Heuvelhof (1997) refer to these as 'governance instruments'.

or formal authority to impose economic sanctions on actors.[8] In the case of the water supply sector this option is only open to the regional water authorities.

Legal strategies first of all address legal compliance as a force driving the target group's behavior. The idea behind this type of steering strategy is that the target group is pressed or encouraged to comply with the prevailing legal regulation. It is not very likely that non-governmental organizations would follow such a strategy. When the water supply sector is applying legal strategies, they will be framed around the legal authority of the regional water authorities. Water supply companies can be involved by supplying additional resources, such as organizational capacity, expertise, and - perhaps - confidence. Legal strategies are predominantly obligatory and restrictive, but they may be facultative and enlarging, a permit system, for instance, is devised for the application of manure and pesticides within groundwater protection zones.

TABLE 2.1: Steering Strategies, instrument characteristics, and required organizational resources

Steering strategies	Instrument characteristics	Organizational resources
Economic	Obligatory/Restrictive	Legal authority Organizational capacity
	Facultative/Enlarging	Financial capacity Organizational capacity
Legal	Obligatory/Restrictive	Legal authority Organizational capacity
	Facultative/Enlarging	Legal authority Organizational capacity
Communicative	Obligatory/Restrictive	Financial capacity Organizational capacity
	Facultative/Enlarging	Financial capacity Organizational capacity

Communicative strategies are usually facultative and enlarging, and in their pure form they address the driving force of social responsibility. It is not unlikely that the water supply sector will apply such a strategy in order to increase farmer's awareness of the issue. But while communicating, offering information, and - perhaps - training, farmers will not only consider their social responsibility; they will also consider the economic consequences of more sustainable practices and methods, as well as the issue of legal

[8] The distinction between obligatory and restrictive economic and legal instruments may be somewhat unclear. A principle distinctive feature is that in the case of legal instruments, non-conformance is a legal offense, while in the case of economic instruments the target group is formally free to disregard the negative incentive of the measure.

compliance. Restrictive forms of communicative strategies, such as propaganda, are not supposed to be chosen by the water supply sector and are not mentioned in the case studies.

2.6 Summary

The purpose of this section is to present a survey of the theoretical framework presented in this chapter, including relevant links to the comparative analysis that follows after the chapters reporting on the case studies.

Policy issues, policy networks and policy styles on the national level. The first step of this comparative analysis is to present a review of the policy networks and policy styles at the national level. The relevant policy issues concern the curbing of agricultural pollution in general, and water pollution by nitrates and pesticides in particular. The arenas in which these policies are brought about, are mainly the traditional networks constituted by the respective Ministries of Agriculture and agricultural sectors. The following issues will be addressed:

- the relevant policy issues;
- the characterization of the policy networks in terms of policy community versus issue network, and the predominant policy style, notably the extent of the consensus base;
- the position of the Ministries of Environment and the way they have succeeded in getting the environmental issue on the agricultural policy agenda;
- special attention will be paid to the specific policy efforts directed at farmers within (ground-)water protection zones and to the roles of the organizations from the water supply sector or their representatives at the national level.

Policy instruments. The second step concentrates on the national policies intended to influence farmers' behavior, in particular farmers in or near water catchment areas, who may affect the drinking water resources. Three policy instrument characteristics are selected:

1. The first distinction concerns policy instruments with farmers as direct and principle addressees versus those aimed at the role of the water supply sector as intermediary. In fact, three classes can be distinguished:
   - policy instruments aimed directly at farmers, with the water supply sector not involved in the implementation;
   - policy instruments aimed directly at farmers, with a role for the water supply sector in the implementation; and
   - policy instruments not aimed directly at farmers, which may contribute to the control capacity of the water supply sector.
2. The second instrument characteristic - whether the instrument involves a moral appeal to the target group - is related to the extent of self-regulation as characteristic of the prevailing policy style. Again, three classes can be distinguished:
   - at one extreme is a moral appeal to legal compliance (including the option of sanctions in case of non-compliance);

- the other extreme is a hint (rather than an appeal) at economic rationality (self-regulation without any form of legal regulation or sanctions);
- the third form is self-regulation based on social responsibility, where government can either choose to apply truly neutral communicative steering without any moral appeal, or to make a moral appeal to the target group to adjust its behavior for the sake of some general interest, bur without legal regulation or sanctions.

3. The third policy instrument characteristic is a combination of provision or extraction of organizational resources (enlarging versus restrictive instruments), and the discretion left to the target group (obligatory versus non-obligatory instruments). All types of policy instruments, juridical, economic, and communicative, can be either restrictive and obligatory, or enlarging and non-obligatory.

Actual control capacity of the water supply sector. The third analytical step - after national policy issues, networks, and styles, and after individual policy instruments - is the assessment of the actual control capacity of the water supply sector. The latter is the designation used for the water supply industry and the regional water authorities. Control capacity is the sum total of:

1. Mission and orientation of the organizations involved: the relation of the issue of protection of drinking water resources to the mission of these organizations, and their driving forces to contribute to the solution of the problem.
2. Available organizational resources:
   - financial capacity (money or capital),
   - organizational capacity (manpower),
   - legitimacy and authority, and
   - information.
3. Selected steering strategies:
   - economic,
   - legal, and
   - communicative.

To reinforce the control capacity? The fourth step concerns the possible reinforcement of the control capacity of the water supply sector, which is foremost a matter of the conclusions to be drawn from the research project and the formulation of policy recommendations. Some directives for their formulation can be derived from the framework presented here. First, there is the distinction between the European and national levels on the one hand, and the regional and local levels on the other. The former is engaged in policy formulation, and although policies aimed directly at the target group of farmers are possible, there is no direct contact with farmers, apart from the national and European interest organizations. On the other hand, direct contact with farming communities and individual farmers makes the water supply sector a high-profile intermediary. Apart from the options of direct and indirect attempt to influence farmers' behavior, the selection of policy styles and policy instruments - or the steering strategies applied by the water supply sector - is related to policy network characteristics (in terms of interrelatedness and mutual commitment) and to the available organizational resources (legal authority, financial and organizational capacity). Policy recommendations can be based on: (a) promising courses of action not chosen although allowed by the circumstances, and (b) prevailing circumstances that may be changed to create better conditions for certain courses of action. In both cases this may lead to the reinforcement of the control capacity of the water supply sector.

References

Benson, J.K. (1975) The interorganizational network as a political economy. *Administrative Science Quarterly*, 20: 229-249.

Bressers, J.Th.A. (1993) Beleidsnetwerken en instrumentenkeuze. *Beleidswetenschap*, 7: 309-330.

Bressers, J.Th.A. (1994, 4th ed.) Beleidsinstrumenten in het beleidsproces. In: P. Glasbergen (ed.), *Milieubeleid. Een beleidswetenschappelijke inleiding*, pp: 125-142. 's Gravenhage: VUGA.

Bressers, J.Th.A., D. Huitema, and S.M.M. Kuks (1994) Policy networks in Dutch water policy. *Environmental Politics*, 3 (4): 24-51.

Bressers, J.Th.A., and P.J. Klok (1987) Grondslagen voor een instrumententheorie. *Beleidswetenschap*, 1: 77-97.

Bressers, J.Th.A., and P.J. Klok (1988) Fundamentals for a theory of policy instruments. *International Journal of Social Economics*, 15 (3/4): 22-41

Bressers, J.Th.A., and S.M.M. Kuks (1992) The agricultural policy-network and environmental policy. Paper presented at the *EGPA Conference in Pisa (Italy) from 2 to 5 September 1992*. Enschede: University of Twente, Centre for Clean Technology and Environmental Policy.

Bressers, J.Th.A., L.J. O'Toole, Jr., and J.J. Richardson (eds.) (1994a) Networks for water policy. A comparative perspective. Special issue of *Environmental Politics*, 3 (4).

Bressers, J.Th.A., L.J. O'Toole, Jr., and J.J. Richardson (1994b) Networks as models of analysis. Water policy in comparative perspective. *Environmental Politics*, 3 (4): 1-23.

Bressers, J.Th.A., L.J. O'Toole, Jr., and J.J. Richardson (eds.) (1995) *Networks for water policy. A comparative perspective*. London: Frank Cass.

Bruijn, J.A. de, and E.F. ten Heuvelhof, (1997) Instruments for network management. In: W.J.M. Kickert, E.H. Klijn, and J.F.M. Koppenjan (eds.), *Managing complex networks. Strategies for the public sector*, pp: 119-136. London: SAGE Publications.

Bruijn, T.J.N.M. de, and K.R.D. Lulofs (1993) De vergunningverlening, vanuit de bedrijfsvoering beschouwd. *Beleidswetenschap*, 7: 158-176.

Doelen, F.C.J. van der (1989) Beleidsinstrumenten en energiebesparing. De toepassing en effectiviteit van voorlichting en subsidies gericht op energiebesparing in de industrie van 1977 tot 1987. Enschede: Ph.D. Thesis University of Twente.

Drupsteen, T.G. (1994, 4th ed.) Het juridische sturingsmodels. In: P. Glasbergen (ed.), *Milieubeleid. Een beleidswetenschappelijke inleiding*, pp: 143-158. 's Gravenhage: VUGA.

Glasbergen, P. (red.) (1994, 4th ed.) *Milieubeleid. Een beleidswetenschappelijke inleiding*. 's Gravenhage: VUGA.

Hanf, K.I., and L.J. O'Toole, Jr. (1992) Revisiting old friends. Networks, implementation structures and the management of inter-organisational relations. *European Journal of Political Research*, 21: 163-180.

Heclo, H. (1978) Issue networks and the executive establishment. In: A. King (ed.), *The new American political system*, pp: 87-124. Washington D.C.: American Enterprise Institute.

Heclo, H., and A. Wildavsky (1974) *The private government of public money. Community and policy inside British politics.* London: MacMillan.

Howlett, M. (1991) Policy instruments, policy style, and policy implementation: National approaches to theories of instrument choice. *Policy Studies Journal*, 19 (2): 1-21.

Huitema, D. and A.H.L.M. van Snellenberg (1997) Beleid in stijl. *Beleidswetenschap*, 11: 55-72.

Jordan, A.G. and K. Schubert (1992) A preliminary ordering of policy network labels. *European Journal of Political Research*, 21: 7-27.

Kickert, W.J.M. (1993) Autopoiesis and the science of (public) administration: Essence, sense and nonsense. *Organization Studies*, 14: 261-278.

Kickert, W.J.M., E.H. Klijn, and J.F.M. Koppenjan (eds.) (1997) *Managing complex networks. Strategies for the public sector.* London: SAGE Publications.

Klijn, E.H. (1997) Policy networks. An overview. In: W.J.M. Kickert, E.H. Klijn, and J.F.M. Koppenjan (eds.), *Managing complex networks. Strategies for the public sector*, pp: 14-34. London: SAGE Publications.

Klok, P.J. (1991) *Een instrumententheorie voor milieubeleid. De toepassing en effectiviteit van beleidsinstrumenten.* Enschede: Ph.D. Thesis University of Twente.

Klok, P.J. (1993) Beleidsinstrumenten als stromen hulpbronnen. In: J.Th.A. Bressers, P. de Jong, P.J. Klok, and A.F.A. Korsten (red.), *Beleidsinstrumenten bestuurskundig onderbouwd*, pp: 147-157. Assen: Van Gorcum.

O'Toole, Jr., L.J (1997) Treating networks seriously. Practical and research-based agendas in public administration. *Public Administration Review, 57(1)*: 45-52.

O'Toole, L.J., K.I. Hanf, and P.L. Hupe (1997) Managing implementation processes in networks. In: W.J.M. Kickert, E.H. Klijn, and J.F.M. Koppenjan (eds.), *Managing complex networks. Strategies for the public sector*, pp: 137-151. London: SAGE Publications.

Peppel, R.A. van de, and M. Herweijer (1994, 4th ed.) Het communicatieve sturingsmodel. In: P. Glasbergen (ed.), *Milieubeleid. Een beleidswetenschappelijke inleiding*, pp: 189-207. 's Gravenhage: VUGA.

Pfeffer, J., and G.R. Salancik (1978) The external control of organisations. A resource dependence perspective. New York: Harper & Row.

Pröpper, I.M.A.M. (1992) Communicatieve sturing. In: Pröpper, I.M.A.M., Herweijer, M. (red.), *Effecten van plannen en convenanten.* Deventer: Kluwer.

Pröpper, I.M.A.M., and M. Herweijer (red.) (1992) *Effecten van plannen en convenanten.* Deventer: Kluwer.

Rhodes, R.A.W., and D. March (1992) New directions in the study of policy networks. *European Journal of Political Research*, 21: 181-205.

Richardson, J.J. (ed.) (1982) *Policy styles in Western Europe.* London: Allen & Unwin.

Richardson, J.J. (1994) EU water policy. Uncertain agendas, shifting networks and complex coalitions. *Environmental Politics*, 3 (4): 139-167.

Richardson, J.J., G. Gustafsson, and A.G. Jordan (1982) The concept of policy style. In: J.J. Richardson (ed.), *Policy styles in Western Europe*, pp: 1-16. London: Allen & Unwin.

Richardson, J.J., and A.G. Jordan (1979) Governing under pressure. The policy process in a post-parliamentary democracy. Oxford: Martin Robertson.

Richardson, J.J., W.A. Maloney, and W. Rüdig (1992) The dynamics of policy change: Lobbying and water privatization. *Public Administration*, 70(2): 157-175.

Sabatier, P.A. (1988) An advocacy coalition framework of policy change and the role of policy-oriented learning therein. *Policy Sciences*, 21: 129-168.

Sabatier, P.A., and H.C. Jenkins-Smith (eds.) (1993) *Policy change and learning. An advocacy coalition approach.* Boulder (CO): Westview Press.

Sabatier, P.A., and H.C. Jenkins-Smith (1997) The advocacy coalition framework. An assessment. *Paper presented at a symposium at the Department of Political Science, University of Amsterdam, February 4, 1997.* Davis (CA) and Albuquerque (NM): University of California, Davis and University of New Mexico.

Vermeulen, W.J.V. (1994, 4th ed.) Het economische sturingsmodel. In: P. Glasbergen (ed.), *Milieubeleid. Een beleidswetenschappelijke inleiding*, pp: 159-187. 's Gravenhage: VUGA.

Waarden, F. van (1992) Dimensions and types of policy networks. *European Journal of Political Research*, 21: 29-52.

PART TWO

3. POLICY FRAMEWORK - EUROPEAN UNION

ELLIS L.M. BRAND
Center for Clean Technology and Environmental Policy, University of Twente, Enschede, The Netherlands.

KENNETH I. HANF
Erasmus University Rotterdam, Department of Public Administration, Rotterdam, the Netherlands.

JOS LIPMAN
Bureau for Contract Research, Rotterdam, the Netherlands.

3.1 Introduction

The protection of water from pollution and the maintenance of good quality of water resources have been key issues of the EU environmental policy. There exist now about 20 relevant legislative measures aimed at the control of discharges into the aquatic environment and the setting of standards concerning the quality of inland and sea waters. Today, the water policy of the European Union is at a critical point. In the coming months decisions are due to be taken which will set EU water legislation for years to come. EU water policy is undergoing dramatic reorganization, trying to adapt to the increased importance given by the Union to the protection of the environment and to the 'subsidiarity principle'. Existing directives are being up-dated and refined while new key framework directives are being introduced or prepared. It is intended that these directives form part of a larger integrated structure for the management of water resources.

The Commission has presented proposals for revision of the directives on bathing and drinking water quality and a proposal for a new framework Directive on the Ecological Quality of Water. A draft proposal for the revision of the Directive concerning discharges to the aquatic environment is available, while a Groundwater Action Plan is soon to be published. Discussions continue on the proposal for a Directive on Integrated Pollution Prevention and Control. Recently, a proposal for a new framework Directive on the Ecological Quality of Water has been presented by the Commission, together with proposals for revisions of the Directives for the Quality of Bathing Water and Water

Geerten J.I. Schrama (ed.), Drinking Water Supply and Agricultural Pollution, 45-72.

Intended for Human Consumption. A draft concept for revision of the Framework Directive of the Discharge of Dangerous Substances to the Aquatic Environment is also available. A debate has also been launched for the preparation of a Groundwater Action Plan, which is intended eventually to lead to a new Groundwater Directive.

In this chapter we will consider briefly what the EU has been doing to combat pollution of drinking water sources through agricultural activities, and at the kinds of actions that could be promoted at and through the European level to mobilize the joint resources of actors drawn from the traditionally separate water quality and agricultural networks to deal more effectively with this problem.

In addition to reports by various parties to pending decisions, information for this report has been gathered from interviews with people involved with the water quality policy field in Brussels. Interviews were held with individuals from: [1]

- the European Commission, DG VI (agriculture) and DG XI (environment);
- the European Parliament, the Environment committee;
- the Economic and Social Committee (ESC); and
- several interest organizations - Union of National Associations of Water Suppliers (EUREAU), European Chemical Industry Council (CEFIC), European Crop Protection Association (ECPA), European Fertilizer Manufacturers Association (EFMA), European Environmental Bureau (EEB) and the Committee of professional Agricultural Organizations in the EC (COPA).

During these interviews several issues were discussed:

- the EU policy on the protection of the resources for drinking water production against agricultural pollution (nitrate, pesticides);
- the involvement of the organization with the policy;
- the network(s) of actors involved and the position of the organization in the network(s); and
- the ideas of the respondent on the possible role of the EU in supporting water supply authorities in preventing agricultural water pollution.

The chapter is organized as follows. First we take a look at the regulatory framework (section 3.2) presently in place, including a brief examination of some of the changes in these directives presently under consideration. Thereafter we describe the various actors involved with this policy issue (section 3.3), paying particular attention to the interactions among those concerned with the two sub-areas of drinking water quality and agriculture. We conclude (section 3.4) with a discussion of the nature of the contribution that the EU could make to dealing with the threats to water quality from agricultural activities.

3.2 Regulatory framework

In this section the policy framework will be described. The policy framework consists of policy programs and statements of the EU, for example the Environmental Action Programs, and EU legislation. Besides the environmental policy also the agricultural policy

[1] A list of respondents for the EU case study is included as annex to Chapter 4.

of the EU is considered. In describing the policies/regulations recent developments and discussions are taken into account.

3.2.1 WATER QUALITY POLICY OF THE EU

Water was one of the first sectors of environmental concern to which the European Community directed its attention. Some of the earliest measures in this area date from the First Action Program on the Environment (1973-1976). During the subsequent action programs, the protection and management of water remained a priority and is now probably the environmental sector with the most extended coverage under community law (Kallis, 1995: 23).

In the broadest sense, water quality policy of the EU includes more than the specific pieces of water legislation. Like other policy actions dealing with sectors of the environment (or with specific environmental issues and sources of environmental pollution), water quality management policy is based upon the more fundamental principles and objectives underlying the EU's environmental policy. These are contained in the relevant articles of the Single Europe Act and the Maastricht Treaty. In addition, the more specific policy principles and objectives, as well as the instrumental strategies to be applied, have been formulated for the medium term in the individual Environmental Action Programs. For the present period, the Fifth Environmental Action Program is the authoritative statement outlining the course of action and general measures to be taken in pursuit of the EU's environmental (water quality) objectives. In the specific case of pollution of drinking water resources by agricultural activities, EU policy with regard to agriculture is important. Of particular interest in this regard are the measures be contemplated to integrate environmental concerns into the reform of agricultural practices in the Union. And, of course, the binding statements of EU policy in the form of directives and other legislative measures constitute the core of official EU policy with regard to water quality and environmentally-sound agricultural practices. In the following paragraphs we consider briefly each of these elements of EU water quality policy.

General principles of EU environmental policy. Of course, it is to be expected that any action undertaken by the EU regarding the quality of water, should be consistent with the guiding principles for EU environmental policy as laid down in Art. 130R[2] These principles are:

- the polluter pays;
- the precautionary principle especially important when taking action in face of uncertainty regarding health effects/ecological impacts of certain substances or combinations thereof;

---

2 It could be argued that it would not really make much difference, in this regard, whether the action taken is based on 100A or 130R, since a Commission proposal based on the 'highest level of environmental protection' would have to be consistent with the general principles of EU environmental policy. Still, some observers argue that using 130R as the legal basis for water directives would strengthen their environmental functions.

- the principle of pollution reduction at the source: with regard to drinking water this would imply that pollution of the water resources should be prevented (already implied in relation between drinking water quality and quality of raw water used for human consumption), which will have consequences for the practices in those sectors of society responsible for pollution, especially industry and agriculture; and
- the principle of integration - environmental objectives should be integrated into Drinking Water Directive, or at least, as the European Environmental Bureau puts it, 'the effects of (new) parameters in a Drinking Water Directive on the protection of the environment should be taken into account.' (EEB, 1993)

Fifth Environmental Action Program. In 1993 the Commission launched its Fifth Environmental Action Program (FEAP), entitled 'Towards Sustainability' (CEC, 1993). In the FEAP, the EU has formulated a new strategy for its environmental policy: the focus is on the active involvement of all parties, a bottom-up approach instead of a top-down approach. In practice this means that the European Commission seeks to involve interested parties/groups in the drafting of policies/regulation or - as one respondent formulates it - the European Commission now tries to create a 'social acceptance' more than it used to do. An important aspect of the FEAP concerns the subsidiarity principle. This principle means that the primary responsibility and decision making competence should rest with the lowest possible level of authority of the political hierarchy, whether this is the community, the national or the local level.

The Fifth Environmental Action Program sets the framework for integrated water resources management. In the FEAP, among other things, the European policies regarding the 'management of the water resources' (one of the seven themes laid out in the program) and 'agriculture' (one of the five target groups of the FEAP) are set out. The program states that the EU should 'secure sufficient water of adequate quality without upsetting the natural equilibrium of the environment.' A primary objective in this connection is the 'prevention of pollution of fresh and marine surface waters and groundwater, with particular emphasis on prevention at the source.' The principle that pollution should be reduced at the source is on the one hand an operationalization of the prevention principle. It reflects the fact that pollution reductions generally are more cost-effective and technically more efficient close to the source. This principle also provides a link between environmental policy and other policy fields, the source reduction principle has lead to the development of the principle of integration. Furthermore, natural ground and surface waters are to be restored to 'an ecologically sound condition' so as to ensure, among other things, 'a suitable source for the extraction of drinking water.' Any proposals for the revision of existing directives or for the development of new measures would have to be judged with regard to their consistency with these objectives.

For the 'management of the water resources', ground- and surface water, nitrates and pesticides are considered as threats to the quality of the water. In order to prevent groundwater pollution, a strict implementation of the Nitrate Directive and a gradual replacement of harmful pesticides are announced. For nitrate the following actions have been announced:

1. strict implementation by the member states of the Nitrate Directive (from 1994);

2. the development of regional emission standards for new cattle breeding (on-going); and
3. phosphate reduction programs (from 1995).

Regarding pesticides, the EU aims at a reduction of pesticide use and the conversion to agricultural methods of integrated pest control. This goal is translated into the three action points:

1. registration of sales and use of pesticides (on-going);
2. promotion of 'integrated control' and organic farming (from 1992); and
3. control of sales and use of pesticides (from 1995).

Nitrate and pesticides form important issues in the environmental policy directed at the agricultural sector. With regard to the 'source' agriculture, the action program points out that both the environmental damage caused by present farming practices as well as the high costs to the community budget due to the surpluses generated by the present CAP system makes it not only environmentally desirable but also sound agricultural and economic sensible 'to seek to strike a more sustainable balance between agricultural activity and the natural resources of the environment' (CEC, 1993: 37). In order to achieve this objective, the Environmental Action Program sets targets for the year 2000 to include, among other things, a stand-still or reduction of nitrate levels in groundwater together with a reduced incidence of surface waters with a nitrate content exceeding 50 mg/l (or with amounts giving rise to eutrophication of lakes and seas). Furthermore, significant reductions of pesticide use per unit of land under production are to be realized. This goal is to be furthered by getting farmers to convert to methods of integrated crop control, at least in environmentally sensitive areas. The action program lists a number of actions that are to be taken, by both the EU, member states and affected parties, to achieve these objectives.

In outlining its environmental policy objectives with regard to agricultural activities, the Commission notes that an appreciation of the need to promote sustainable agricultural practices is also 'reflected in recent deliberations on the reform of the Common Agricultural Policy' (CEC, 1993: 37), which challenge the previous commitment to producing more through intensification of agriculture. The Commission's 1991 Reflection Paper on the development and future of the CAP recognized the:

> *need to encourage extensification with the object of not only reducing surplus production but also of contributing to an environmentally sustainable form of agricultural production and food quality....*

In this way, the dual role of farmers as 'food producers and guardians of the countryside' would be formalized.

The Hague Declaration. In the course of any single planning period, the Environmental Ministers of the member states will have different occasions to go on record with some kind of official statement or declaration in which they spell out the general direction, and guiding principles, of actions to be taken with regard to a particular environmental problem. The basis for the Community's water policy in the 90's was set at the Environment Ministers' Seminar on 'Community Water Policy for the Nineties' held in Frankfurt during the summer of 1988. The six main areas of work for the Commission identified by the Ministers were:

1. legislation aimed at improvement of the ecological quality of surface waters;
2. development of waste water treatment;
3. speeding up of measures dealing with the control of discharges of dangerous substances;
4. control of pollution from diffuse sources, particularly from agriculture;
5. development of an overall policy for water, integrating water resources and water quality management; and
6. integration of water policies into other environmental policies.

At a another ministerial seminar, held in The Hague on 26 and 27 November 1991, the environment ministers adopted a declaration which 'provides a firm link between guaranteeing the quality of drinking water and the protection of water resources' (Kraemer, 1994: 7). In The Hague Declaration the environment ministers of the EC called for establishing 'a program of action' to be implemented by 2000 at national and community level aiming at sustainable management and protection of freshwater resources. The Hague Declaration also laid down the principles for the development of groundwater protection measures and the revision of the Groundwater Directive. In particular, the ministers noted, among other things, that 'pollution by fertilizers, including livestock manure, and pesticides used for agricultural and non-agricultural purposes' is to be considered as a threat to groundwater resources thereby endangering the drinking water supply.

In this declaration, the ministers outlined the intended future development of community water policy. In this prospective policy, a very high priority was given to protecting water resources. One way of doing this was to make the integration of environmental protection into other community policies a key principle of community water policy. In this connection the ministers stressed:

> *that water management policies should be integrated within the wider environmental framework as well as with other policies dealing with human activities such as agriculture, industry, energy, transport and tourism.* (Kraemer, 1994)

Indeed, the requirement to integrate environmental protection into other sectoral community policies (already enunciated in the SEA, confirmed in the Maastricht Treaty and occupying a central place in the Fifth Environmental Action Program), means that there is a 'general obligation to further environmental protection and sustainable development through changes in these other policies...' (Kraemer, 1994: 2)

The precautionary principle was agreed on as a guiding principle for the sustainable management and protection of freshwater resources in the Hague Declaration.

3.2.2 WATER QUALITY LEGISLATION

In order to sketch the legislative framework we first look at the water legislation of the EU. The Community has adopted a wide range of provisions for the protection and improvement of the aquatic environment. In fact the water pollution policy is the oldest and the most completely developed sector of community environmental policy. More than 25 directives or decisions have been adopted covering both fresh and sea water pollution

(Hanf, 1992: 89). Of course, in the scope of this research is not necessary to study them all. The focus will be on those directives that are relevant for the issue: protecting the resources for drinking water production (means surface and groundwater) against agricultural pollution (pesticides, nitrates, phosphates). Besides the water directives several directives concerning nitrates and pesticides belong to the framework.

Legislation for the protection of water from pollution started with two key directives, issued in 1975 and 1976. The first deals with the principles and the standards necessary to improve and sustain the environmental quality intended for abstraction as drinking water. The second regulates the discharge of dangerous substances into the aquatic environment. A number of other directives have been derived from the two distinctive philosophies on which the initial directives were based: water use directives which are concerned with the quality of water intended for particular uses; and water pollutant directives which are concerned with the emissions of particular pollutants to water.[3]

Drinking Water Directive. There are two directives that regulate the quality of the 'tap'. The Drinking Water Directive (80/778/EEC)[4] sets standards for the quality of water intended for drinking or for use in food and drink manufacture in order to protect human health.[5] The directive covers all water for human consumption including public and private water supplies, bottled water not recognized as mineral water or medicinal water by individual member states, and water used in the food industry. The original directive lays down a set of some sixty-two mandatory quality standards which have to be implemented by the member states in order to obtain a high quality of drinking water in the entire Community. Regular water quality monitoring is to be carried out by individual member states, according to community-wide guidelines, and samples must be taken at the point at which the water is made available to the user.

Three types of standards are used: the Guide Level (GL), the Maximum Admissible concentration (MAC), and the Minimum Required Concentration (MRC). Annex II deals with water quality monitoring and the manner in which it should be conducted so as to ensure that the standards of Annex I are complied with. Annex III at least contains reference methods of analysis. For all of the parameters reference methods of analysis are not available yet. This annex must be adhered to as far as practicable. Derogations are allowable in certain circumstances as long as there is no danger for public health. The member states should inform the European Commission of these situations (Haigh, 1987: 42-43). Parameter 55 in Annex I is entitled 'Pesticides and related products'. The maximum admissible concentration is 0.1 μg/l for each pesticide. The standard for the total of pesticides in water is 0.5 μg/l. These standards are based on the prevention principle, the

---

3 There is a third type of legislation which consists of EC decisions on the protection of regional water bodies, often in the form of ratification of international conventions (STOA, 1995: 23).

4 OJ L229 30.8.80.

5 Although the Directive flows from the First Action Program on the Environment, it originated in the Health Protection Directorate; the standards contained in the Directive are based on the WHO 1970 Drinking Water Standards.

idea that pesticides do not belong in the water. Parameter 20 of the Annex concerns nitrates: the maximum admissible concentration is 50 mg/l and the Guide Level is 25 mg/l. It had long been recognized that excessive quantities of nitrate in drinking water can be harmful to health. For this reason, the directive set a standard of 50 mg/l whereas the WHO recommended a level of less than that. According to Haigh (1987), the directive, with its sharper limits, has 'had the effect of focusing attention of supplies that had not previously been a cause of contention.' In some countries, for example the UK, there was concern that a continuation of existing trends in the application of fertilizers would lead to an increase of nitrate concentration in surface waters and that this would, consequently, lead to a considerable increase in the proportion of abstractions exceeding the Directive limit. As far as the preparation of drinking water is concerned, the remedial actions available to water suppliers when dealing with high nitrate supplies include: blending different supplies to reduce the nitrate level of the water put into the distribution system; treating water to reduce nitrate levels; and exploiting new sources with lower nitrate levels. The courses of remedial action open to water authorities when dealing with high nitrate supplies are nearly all costly. It is for this reason that control of nitrate has become a sensitive political issues. Still, as Haigh (1987) points out, it is surprising that 'the nitrate parameter was not widely regarded as a problem during the negotiations....', probably because the imprecise language used, suggested a great amount of flexibility in applying these parameters in the specific situations of the different member states.

Directive Surface Water for Drinking. In addition to the Drinking Water Directive, there are also directives for the protecting the quality of both ground and surface waters. The Directive Surface Water for Drinking (75/440/EEC)[6], the first proposal for a directive concerning water following adoption of the first environmental action program in 1973, has two purposes: to ensure that surface water abstracted for use as drinking water meets certain standards[7] and, if needed, is given adequate treatment before being put into public supply; and thereby to improve rivers or other surface waters used as sources of drinking water. The member states are required to classify the sources of surface water for the abstraction of drinking water into three categories: A1, A2 and A3 corresponding to the three standards methods of treatment required to transform the surface water into drinking water (Haigh, 1987: 34). According to one of the respondents this directive is rather old and primitive and it is likely that with the 'new' Directive on Ecological Water Quality this directive will be withdrawn.

Groundwater Directive. Then there is the Groundwater Directive (80/68/EEC).[8] Seventy per cent of the Community's drinking water is extracted from groundwater. In order to protect exploitable groundwater sources, which are very difficult to restore once polluted, both direct and indirect discharges of dangerous substances are to be prohibited or

6 OJ L194 25.7.75.

7 The standards set for pesticides and nitrates are equal to the standards of the Drinking Water Directive.

8 OJ L20 26.1.80.

regulated (Haigh, 1987: 79). A List I and II of families and groups of dangerous substances are given in the Annex of the directive. member states are to 'prevent' the introduction into groundwater of List I substances and to 'limit' the introduction of those on List II so as to avoid pollution. In principle, all direct discharges of List I substances are to be prohibited, while direct discharges of List II substances are to be subjected to investigation before being authorized. Any disposal on land of either List I or List II substances which might lead to indirect discharges is to be subject to investigation before being authorized. Any other activity likely to lead to indirect discharges of List I substances is also to be controlled. Concerned then with the activities, such as farming, that led to indirect discharges of substances such as nitrates and pesticides into groundwater.

Pesticides and nitrates are also included in this directive as well, although it is hard to tell what consequences the directive has for agricultural activities. According to Krämer many member states have had substantial problems in incorporating the directive into national legislation. It would appear that groundwater pollution by agricultural activities - nitrates and pesticides - is a growing problem in most countries (Krämer, 1990: 5). The implementation of the groundwater directive is not going well because of a lack of concrete measures. This is why in November 1991 the environmental ministers agreed on the development of a EU groundwater action program. We will discuss this program in more detail later.

Directive on Dangerous Substances in Water. The Directive on Dangerous Substances in Water (76/464/EEC)[9] sets a framework for the elimination or reduction of pollution of inland, coastal and territorial waters (surface water) by particularly dangerous substances. Subsequent daughter directives are to set standards for particular substances. The directive is also to ensure consistency in implementing various international conventions and to reduce distortion to conditions of competition. An annex has a list I and a list II of families and groups of dangerous substances selected on basis of their toxicity, persistence, etc. List II contains possibly less dangerous substances. Member states are to take appropriate steps to eliminate pollution by list I substances and to reduce pollution by list II substances. Discharges of both list I and II substances are to be subject to prior authorization by a competent authority, but these authorizations are arrived at in different ways (Johnson, 1983).

Nitrate Directive. From this general framework directive we turn to a specific source of pollution, nitrates. In 1991 the Council adopted the Nitrate Directive (91/676/EEC). The purpose of the directive is to decrease water pollution caused by 'the application and storage of inorganic fertilizers and manure on farmland.' It is aimed, according to Haigh (1992), at both safeguarding drinking water supplies and preventing wider ecological damage in the form of the eutrophication of freshwater and marine waters in general. It is up to the member states to identify waters that are actually or potentially affected by pollution from nitrates. These are to include surface freshwaters, in particular those used

9 OJ L20 26.1.80.

for the abstraction of drinking water; groundwaters actually or potentially exceeding the 50mg/l nitrates; and other bodies of water endanger of eutrophication.

By December 1993 all known areas of land which drain into water identified in this way and contribute to pollution (were) to be designated by the member states as 'vulnerable zones.' For these areas, action programs are to be developed (by December 1995). There is a number of mandatory measures that these action programs are to contain which, among other things: specified periods during which certain fertilizers are prohibited; limits on quantities of fertilizers that may be applied; a limit on the application of livestock manure per hectare; conditions relating to available storage capacity on farms for livestock manure; and a code of good agricultural practice. In areas other than these vulnerable zones, the code of practice is to be implemented voluntarily by farmers; training and information programs are to be made available to them.

It is interesting to note, as Haigh (1992) points out, that this directive was 'foreshadowed in the Fourth Environmental Action Program, which promised action to limit pollution from livestock effluent and the excessive use of fertilizers.' More in connection with the concern with eutrophication in parts of the North and Baltic Seas, the Commission was charged with coming up with proposals for controlling diffuse pollution from agriculture, as well as for improving the treatment of municipal waste water. In its draft the Commission went much further in limits to application of livestock manure and stringent Action programs for vulnerable zones, but scrapped due to resistance from various governments. As this strong preventive approach makes clear, the problem of the impact of agriculture on water quality had been recognized and the Commission attempted to create the basis for taking action to alter or restrict farming practices. It became equally apparent that an attempt to deal with the pollution of water by agricultural activities in this way could also have serious social and economic impacts on agricultural sector and certain areas in some countries.

In order to establish a common protection level for all waters (not only the vulnerable zones) the member states were suppose to develop 'Codes of Good Agricultural Practices' (before the end of 1993) to be implemented by the farmers voluntarily. If necessary the member states are to develop a educational program in order to promote the GAPs. This directive also required member states to declare the areas to be designated as Vulnerable Zones (i.e. with water with nitrate excess of 50 mg NO_3/litre) before the end of 1993. The control of land use within them, including nitrate applications, will be determined by the action plans which have to be implemented within the zones. These actions plans have to be delivered to the European Commission before the end of 1995 and be implemented by the member states within a four year period. The action plans should contain (at least) the measures set out in Annex III of the directive. This means, at a minimum that they are to include such mandatory measures as the specification of periods when the application of fertilizers is prohibited; limits on the quantities of fertilizers applied; and a limit on the application of livestock manure per hectare to an amount containing no more than 170 kg N, or 210 kg N during the first four year action program. According to one respondent, it is hard to predict the outcomes of this directive because 'the real meat of the Directive - the action programs - comes into effect in 1996.'

Pesticides. Finally we look briefly at pesticides. Several community regulations exist that deals with pesticides. Some of these directives regulate the residues of pesticides in food. We will not describe these directives because these directives are not primarily intended to protect the environment but to protect consumers by setting limits on the amounts of pesticides on food. Apart from the residue policy the EU also regulates the pesticide market. Since 1979, the EU has prohibited the marketing and use of a number of active ingredients in the EU through the *Directive concerning the banning or the restriction in use of active substances* (79/117/EEC)[10] that are considered to cause harmful effects in human and animal health, the environment.

More important in this respect is the *Directive concerning the placing of Plant Protection Products on the market* (91/414/EEC). This directive seeks to harmonize the pesticide registration schemes of member states. In 1976 the Commission proposed a directive on this matter. It took 15 years of discussions to be able to have a proposal that was acceptable! The original proposal of the Commission contained a sort of European authorization of pesticides: the EU would authorize the pesticides, and if after two years there were no objections from the member states, the EU authorization would automatically become legal in the whole Union. This proposal though did not make it.

Directive 91/414 is intended to provide a basis for controls over pesticide composition, marketing and use so as to protect human health and the environmental. The principal aim of the directive is to introduce a community system for the authorization and placing on the market of plant protection products. On basis of the directive of 1991, authorization is implemented at two levels: (1) the EU authorizes active substances and (2) the member states authorize the pesticide products on basis of the so-called 'Uniform Principles' (UP). The use of specific pesticides can be authorized if the active substance is on the list and it complies with the UP. At European level all active substances will be evaluated within ten years.

The *Uniform Principles* which were added to this registration directive (as Annex VI) are intended to provide guidelines to enable decision making on specific pesticides in a similar way. The Uniform Principles were adopted on 27 July 1994. They are supposed to ensure protection of ground- and surface water (groundwater intended for the production of drinking water and surface water in general). The UP set standards divided into 'evaluation' and 'decision making'. Member states are required to evaluate, on the basis of data supplied by the applicant, the risk that a pesticide can pose to the environment (the soil, groundwater, surface water, air, animals) or to human health. Having evaluated the data, member states must then make decisions on the authorization according to decision making criteria. These decision making criteria are formulated in the Uniform Principles. Important is that in these criteria there is a link with the standards of the Drinking Water Directive and the Directive Surface Water for Drinking Water. In the UP it is said that:

> *... a pesticide can only be authorized by the member states when on basis of the evaluation it is clear that a pesticide will not end up in the groundwater (meant for the production of drinking water) in concentrations exceeding the Maximum*

[10] OJ L33 8.2.79.

Admissible Concentration for pesticides in drinking water formulated in the Directive 80/778/EEC.

For surface water, the UP refer to the Directive 75/440/EEC in which the standard for pesticides in surface water for the production of drinking water has been formulated.

The fact that a direct link exists between the water directives and the pesticide directive was occasion for many discussions. The development of the Uniform Principles was - like the development of the directive itself - a difficult and slow process. Especially because at the same time there was a debate going on about the water directives and the standards being used. In order to get a qualified majority in the Council for the Commission proposal, an exemption was added to the UP: it is possible that temporarily registrations will be given by member states that exceed the drinking water standard of 0.1 μg/l. These temporarily authorizations are not mutually recognizable (which in fact frustrates the objective of the directive - to harmonize pesticide registration in the EU)!

It is perhaps interesting to note that the European Parliament has started legal proceedings in the EU Court of Justice to have the Uniform Principles declared null and void. The reason for this action is that the Parliament was not consulted on the Uniform Principles.

Apart from these directives there also is a directive that deals with the labelling and packaging of pesticides, the *Directive classification, labelling, and packaging of dangerous substances* (78/631/EEC).[11] Via labelling farmers can be informed of the possible environmental impact of pesticides.

3.2.3 EU AGRICULTURAL POLICY (CAP)

Kraemer (1993) reminds us that the Drinking Water Directive is primarily an instrument to protect human health. For this purpose it laid down standards for the quality of water intended for human consumption. At the same time, the directive has the additional effect of protecting the environment, since drinking water sources must be free from contamination to allow inexpensive water treatment. In this connection the directive also has indirect links with Directive 75/440/EEC concerning the quality required for surface water intended for the abstraction of drinking water in the member states and Directive 79/869/EEC concerning the methods of measurement and frequency of sample and analysis of surface water intended for human consumption, as well as the directive 80/68/EEC on the protection of groundwater against pollution caused by certain dangerous substances. However, while there may be logical connections between these different elements of the EU's water management policy, the Drinking Water Directive can be been seen as a general environment directive for the protection of water resources. Among other things this means that if we do not want to pay expensive treatment costs for drinking water and would instead rather guarantee supply of sufficient amount of raw water of acceptable quality we will need to take action to protect the surface and groundwater sources from which this water is abstracted. It is in this regard that the Drinking Water Directive has

[11] OJ L206 29.7.78

been linked with Directive 91/414/EEC on the placing of plant protecting products in the market and Directive 91/676/EEC concerning the protection of waters against pollution caused by nitrates from agricultural sources. However, in the last analysis, the enforcement of the appropriate quality standards for these standards will require that we ensure that activities leading to excessive discharges are regulated or otherwise controlled. This means, ultimately, that the offending economic activities will have to be changed.

Agricultural activities in the EU are regulated by the Common Agricultural Policy (CAP), which was developed in the late 1950s and early 1960s. The objectives of CAP were set out in the treaty of Rome in 1957. Since then, CAP has remained central to the Community and for a long time accounted for over 80% of the total budget. Although this has now been reduced to below 60%, the CAP is still very important. The combination of an upsurge in new technology and high prices and guaranteed markets resulted in a consistent increase in farm productivity. By the late 1980s all major farm commodities (cereals, milk, meat, sugar, wine) were in surplus (Williams, 1993). The CAP has a clear impact on the use of pesticides and fertilizers by the farmers. Because of the guaranteed prices, farmers tried to produce as much as possible and therefore use more pesticides and fertilizers. According to Williams:

> *cereal prices were set at a particular high level and over time this led to relatively cheap cereal substitutes being imported to replace indigenous cereals in livestock rations. This in part explains the development of intensive livestock farming units near the ports in Belgium and the Netherlands and subsequently the problems of surplus farmyard manure and soils supersaturated with phosphate.* (Williams, 1993: 4)

Since the start of CAP various changes in the policy measures - reforms - have been implemented, each to control the continuing increase in farm output. The May 1992 'Macsharry Reform' is the most radical. Firstly, it reduces prices towards world free market levels. Secondly, it introduces income compensations to farmers (for the price cuts) and thereby cuts the direct link between farm output and farm income support. According to COPA, statistics show a decrease in the use of pesticides in the last two years as a result of the CAP reform of 1992. Alongside changes in the main commodity regimes, the Commission proposed three significant 'Accompanying Measures'. One is a new system of incentives for older farmers who retire early, another is an enhanced aid regime for afforestation of agricultural land and a third is an Agri-Environmental Action Program. The *Agri-Environmental Action Program* (2078-92) extends the existing community schemes for aiding extensification and environmentally sensitive farming systems. Also the program adds new elements, such as assistance for long-term set-aside for environmental reasons, incentives for the upkeep of abandoned farmland and woodlands and a new education and training scheme for farmers (Reus, et al., 1994: 25). Member states have to introduce agri-environmental measures which are to be based on a series of defined options which seek to encourage extensification and a reduction in chemical use. One of the respondents though doubted the impact of the measures because the EC budget is very limited.

3.2.4 PENDING POLICY DEVELOPMENTS

Drinking Water Directive. At the present time the Drinking Water Directive is under review by the Commission. Three reasons can be pointed out for reviewing the directive. Firstly, the fact that the directive leaves no scope for member states that are in non-compliance with the directive: the exceeding of the standards automatically means an illegal situation independent of the question of the member state is to blame for the situation or not. A wide expectation exists within the water sector to review this situation. Secondly, it was felt that the directive should be given a technical and scientific review since the directive stems from 1980. The third reason lies in the desire to bring the directive in line with the subsidiarity principle.

In September 1993 the Commission organized a conference at which all interested parties could express their view on the Drinking Water Directive. The event brought together some 200 scientists and other representatives from water companies, the agro-chemical sector, manufacturing industry, the World Health Organization, and governments. The discussions specifically focused on the limit values regarding pesticides: the value limit of 0.1 mg/liter for each pesticide, and a 0.5 mg/liter on the concentration of total pesticides. These values, fixed at the limit of what was detectable at the time, are so low as to be surrogate zeros. The chemical industry and some water companies (in Britain) argue that the existing limits have no scientific validity, are far tougher than is needed to protect human health, and impose unjustifiably heavy purification costs on the water industry. They (and the British government) are lobbying hard for the limits to be brought into line with new, toxicology based guidelines drawn up by WHO for 33 pesticides. For 30 of the pesticides, the WHO-guidelines are up to 1000 times less stringent than the current EC limit.

Environmentalists and other water companies (including EUREAU) argue for keeping the current strict precautionary approach in order to safeguard raw water resources to the maximum degree. They maintain that pesticides have no place in drinking water (Environmental Watch, 1993: 2). In a position paper on the revision of the Drinking Water Directive (September 1993) the European Environmental Bureau (EEB) argues that the WHO standards should not be used:

> *Adopting the WHO guideline values will result in an increase of acceptable concentrations up to a factor 1000. WHO has not taken ecological effects into account and these very high levels could be disastrous for the protection of the environment if tolerated in the environment. These changes are contrary to the integration of environmental policy in other EC policies as stated in the Fifth Environmental Action Program.... The WHO guideline values are set by using available scientific evidence. Nevertheless, there is a lot of missing information.* (EEB, 1993)

The pesticides industry (represented by ECPA) is especially concerned to get the limit value for pesticides raised because of the directive concerning the placing of pesticides on the market (91/414/EEC) according to which a pesticide can not be authorized for marketing in the Community if its concentration in groundwater may be expected to exceed the limit value for drinking water under normal conditions of use.

The Commission has been working on the directive since this conference took place. It was expected that the Commission would present its proposal for the revision of the directive at the end of 1994. On basis of (unofficial) drafts of the revised directive, it is very likely that (1) the standards (for nitrate and pesticides) will be maintained, only the 0.5 mg/l total of pesticides will be left out (this was requested by both ECPA and EUREAU!), (2) the aesthetic parameters such as taste and color, etc. (the non-health related parameters) will be dropped from the directive and left to individual member states to set themselves, (3) member states will be allowed to derogate from the values specified for the parameters in Annex I, part B (chemical parameters including pesticides) for a limited period and to a maximum value they determine themselves, and (4) reporting by the member states to inform the Commission on exceeding of the standards is more flexible (Kraemer, 1993).

Directive on Ecological Water Quality. A draft directive was presented in August of 1994 (proposal for a Directive on the Ecological Quality of Water, (COM93 680 final). The Parliament and the ESC are deliberating on the proposal at this moment. The purpose of this directive is to create the necessary framework for the member states to be able to define and apply complementary measures beyond those already envisaged under community legislation, so that their surface waters can be of good ecological quality.

In order to implement this proposal for a directive, the member states will have to (1) introduce a system of monitoring and assessing the ecological water quality of surface water, (2) and will have to define operational objectives for good ecological quality. Thereafter, (3) integrated programs will have to be implemented in order to attain those operational objectives. The measures and practices required under the integrated programs will be rendered obligatory for natural persons and legal entities, whether public or private. The directive is likely to have an impact on agriculture, since it specifically mentions diffuse pollution, nitrate and eutrophication.

Integrated Pollution Prevention Control Directive. Another related directive that is currently be debated is the proposal for an Integrated Pollution Prevention Control Directive. This directive contains a permit system for industrial installations, also for intensive livestock productions! It requires application of the best available technology.

Ground Water Action Program. As we have already noted above, one of the main subjects discussed at the ministerial seminar in The Hague In November 1991 was the unsatisfactory implementation of the groundwater directive. In the final declaration of the Seminar, the European Commission was requested to develop a ‘Ground Water Action Program’. Accordingly, this request was repeated in a Council Resolution (Council resolution of 25 February 1992 on the future community groundwater policy (92/c59/o2)). Meetings of national experts were set up by the end of 1993. So far the Commission produced a working document. The general expectation is that an official Commission document will be presented early next year (1995). According to the Working Document:

> *the GWAP should provide key elements for a future framework directive on ground water which will take a similar approach as the directive on the ecological quality of surface water.*

But, before such a framework directive can be drafted a lot of talking needs to be done. In this sense, the purpose of the action program is to stimulate discussions, in getting attention for the problem of groundwater pollution. The awareness of the vulnerability of groundwater among policy makers needs to be increased so that (1) the groundwater issue gets more involved in decision making procedures in broader contexts and (2) in discussions in context of the Common Agricultural Policy, the groundwater interest is taken into account. One of the eight themes of the working document concerns 'Agricultural practices'.

> *The considerable increase in agriculture of nitrogen fertilizer and pesticide applications is at the root of diffuse pollution of soil and water. The progressive increase in nitrate content in ground water has become alarming since the seventies, followed by pesticides showing in ground water in recent years, in certain regions particularly. This trend is on the increase.* (CEC, 1994: 27)

What is needed? According to the document:

1. Changes in agricultural practices to prevent pollution from nitrates as provided for in the nitrate directive, including the progressive reduction of fertilizer application to the level required for good agricultural practice.
2. The establishment and implementation of rules for good agricultural practice.
3. The progressive replacement, by degradable and less harmful products, of persistent and accumulative pesticides which are likely to cause deterioration of ground water quality. The implementation of the directive on the placing of plant protecting products on the market (91/414/EEC) must take place with respect to groundwater protection.
4. The progressive use limitation of plant protection products, and other pesticides for non-agricultural purposes, to prevent adverse effects on groundwater, and with the objective of ensuring, where technically feasible, that for already contaminated groundwaters, after a period of sanitation, their concentrations respect at least the values for drinking water.

3.2.5 CONCLUSIONS

The EU deals with the protection of the water resources in several ways. The EU has been regulating the quality of drinking water supplies in Europe since 1980 by setting standards for maximum concentrations of nitrate and pesticides in drinking water. Via the Groundwater Directive and the Directive on Dangerous Substances in Water the EU is regulating certain dangerous substances among which pesticides. Because of a lack of concrete measures, the impact of these directives on agricultural practices (and thereby the protection of the water resources) is limited.

Protection of the water against nitrate pollution should follow from the implementation by the member states of the Nitrate Directive. Applying Good Agricultural Practices by the farmers should prevent nitrate from ending up in water. More can be expected from the action programs to be developed by the member states before the end of 1995.

Pesticides are being regulated by the EU through several regulations. Since 1979, the EU has prohibited a number of active ingredients from use in the EU. Until 1991, the regulation of specific pesticides was left to the member states. In 1991, the Directive on the Placing of Pesticides on the Market was adopted in order to harmonize this situation. Protection of the water from pesticides should flow from the Uniform Principles (annex VI of the directive), recently adopted by the Council. The Uniform Principles use the drinking water standards for pesticides and nitrate as criteria for maximum allowable concentrations in the ground- and surface water. Although one might expect that with the UP the protection of ground and surface water is guaranteed, several aspects undermine this. First, it will take years before effects might be expected because (1) the member states need to implement these UP in their national legislation which will take some time, (2) a transition period exists for those pesticides that do not comply with the UP. Moreover, it is possible that pesticides currently used in agriculture will show up in groundwater in 20 or 30 years. Second, on basis of the Directive (91/414/EEC) it is possible for member states to temporarily authorize pesticides that do not comply with the UP.

We conclude that the EU developed legislation to protect the water against agricultural pollution (although sometimes it is difficult to point out the impact of the directives on agricultural practices). Moreover, several developments like the Ground Water Action Program and the Proposal concerning the Directive on Ecological Water Quality, indicate an even more active role of the EU in the future. Also, the changes of the position and contents of the common agricultural policy (CAP) should be taken into consideration. It is very likely that pesticide use and fertilizer use will decrease because of the developments within CAP.

3.3 Policy network

This section will provide an overview of the actors involved with the policy issue of preventing agricultural water pollution. It is assumed that the focal network consists of all actors who have a stake in the policy issue. The study of Richardson on the EU water policy identifies the actors involved with the water policy (Richardson, 1995). The actors involved are mainly organizations, governmental agencies and interest organizations. Two sub-networks will be distinguished: the drinking water sector and the agricultural sector. After describing these sub-networks, the focus will be on the interactions between the two sectors concerning the issue of agricultural water pollution.

3.3.1 GENERAL INSTITUTIONAL CONTEXT OF THE EU

Before looking more closely at these two sub-networks, a few general comments are in order to set the general institutional context within which these actors operate. The EU policy process has been described as:

> *a multi-national, neo-federal system, extremely open to lobbying by a wide variety of organizations [...], with an unpredictable agenda setting process [...], creating*

an unstable and multi-dimensional policy making environment. (Richardson, 1995: 140)

In such a situation it is, according to Richardson, difficult or impossible to speak on 'the' water policy network. Instead, what we find are different configurations with regard to different sectors, time and stages of policy process. Again Richardson notes that:

[i]n some sectors of EU water policy, as for example with groundwater, it is virtually impossible to identify any coherent network of actors as such, simply because the range of actors is so diverse and unstructured. (Richardson, 1995: 140)

Furthermore, the large range of possible actors - also quite varied - makes it difficult to 'scale the influence of participants' (1995: 140).

Nevertheless, there is little doubt that the Commission is the pivotal actor in the water policy process - at least as far as the initial formulation of policy proposals is concerned. In subsequent phases of the EU policy process, it shares center stage with other institutional actors. The key role of the Commission is the result of its formal power of policy initiation and its pivotal position - and skill - as 'broker of interests' and bourse for exchange of policy ideas (Richardson, 1995: 141). Sub-networks, concerned with various aspects of water management policy are, for these reasons, centered around the Commission.

The long, complex multi-layered nature of the community processes provides many points of access for interests, and so many opportunities for them to keep informed and to press their cases. Groups organized along both national and Europe-wide lines are actively involved with putting forth whatever collective views there are on issue under consideration; disseminating information and exchanging views both with the Commission and their associated members; and, in general, trying to get their interests and views incorporated into community policy. Given the fact that the Council of Ministers can, in most cases, only be 'reached' indirectly, by trying to influence the positions taken by national governments on issues up for decision, the Commission is the main target for most interests, in view of its importance with regard policy initiation and formulation as well as the important role it plays in the development of secondary legislation, filling in the details of Council decisions.[12] Moreover, the Commission is also involved in the fate of a policy measure throughout its whole life cycle. Equally important is the fact that the Commission is known to be highly approachable and permeable.

Of course the Commission reaps its advantages from these contacts with various interests. It this way Commission officials, responsible for drafting legislative proposals, gain access to specialized information on the matter at hand. This exchange of ideas with and inputs to the draft proposals from the ultimate target groups or otherwise involved sectoral interests can also strengthen the Commission's hand vis-à-vis the Council insofar as the Commission can appear to be speaking for and supported by important interests in member states. At the same time, the responsible officials of the Commission can use such exchanges to gauge the technical soundness and the political feasibility of proposals so as to reduce the risk that proposals will be shot down in the Council of Ministers. Of course,

[12] As a consequence of its growing role and influence, European Parliament has also become more interesting for lobbyists. Groups increasingly seek contacts with individual members, political groups and committees.

working with Eurogroups has the advantage of providing aggregated view instead of approaching so many organizations individually. (Although it cannot be assumed that such organizations are always able to speak with one clear voice!) The organizational structure of these interest groups is usually extremely loose; the affiliated national organizations themselves are quite autonomous in most respects and are not subject to central discipline. Nevertheless such Eurogroups can be used to mobilize support for Commission proposals. On the other hand, it is usually the affiliated national associations that are involved, in one way or another, together with national governmental and other sectoral actor, with the implementation of EU policy.

Once it has been decided to produce a proposal - a decision usually taken at a senior level within the most relevant DG - a text is prepared. Normally, a junior official in the appropriate DG writes the initial draft. This draft is then passed upwards through her superiors; as it is passed upwards, it is discussed with all DG's and specialized commission services which have an interest. When all directly-involved commission interests have given their approval, the draft is sent to the cabinet of the Commissioner responsible for the subject. The cabinet may re-work the draft before submitting to the Commissioner for her approval. When the Commissioner is satisfied, she asks the Secretary General of the Commission to submit the draft to the College of Commissioners. Prior to this, or in preparation for this meeting, the draft will be scrutinized and possible amended by the 'chefs de cabinet' at their weekly meeting. If the draft is judged non-controversial, the so-called written procedure for approval can be used. If it is controversial, the Commission may, after debate, decide to accept it, reject it, amend it or send it back to the responsible DG.

Not surprisingly, while such a draft text is being prepared, the responsible officials usually find themselves the focus of attention from many directions. The Commission's thinking is probably most open and flexible at this preliminary stage. Once the text has been formalized, it is more difficult to change. Therefore, interested parties use whatever means at their disposal to press their views as early as possible. They normally do not have to fear an unresponsive audience since traditionally the Commission has organized extensive consultations during this pre-proposal stage. The Commission wants to incorporate national and sectional views (from the outset) in order to broaden the basis of support and to ensure the feasibility of the measure; or it may need the technical knowledge and outside expertise.

There are several ways in which external views can be brought to the attention of those drafting a text or a report: the Commission itself may request a report on such views; interest groups may submit briefing documents; professional lobbyists, politicians as well as officials from the COREPER may press their preferences in informal meetings; EP committees and ECOSOC sections may be sounded out; or use may be made of the extensive advisory committee system. There is no standard consultative pattern or procedure. The Commission - officials responsible for drafting (and then higher levels to further sound out and gain support) - engages in an extensive sounding and listening process, that precedes and accompanies policy development by the Commission. An important part in these processes, especially at the pre-proposal stage (i.e., before the Commission has formally presented a proposal to the Council) is played by a vast network of advisory committees that have been established over the years. Different DG's will have different practices and styles of consultation. DGXI is 'one of the most advanced DG's within the Commission in terms of identifying and mobilizing the relevant constituency of groups and individuals,' (Richardson, 1995: 151)

The contacts and communications between the Commission and interested parties are numerous and are both formal and informal in nature:

1. First of all, there exists an extensive advisory committee system which is intended to provide regularized channels through which interests can make their views known to commission authorities. There are two main types of these committees: [13] First of all, there are the so-called expert committees consisting of officials, experts and specialists of various sorts nominated by national governments. The second type are the consultative committees, which are composed of representatives of the sectoral interests and are organized and funded by the Commission without reference to national governments. Normally, the members of the consultative committees are appointed by the Commission from persons nominated by representative European level organizations.
2. Representatives of the Commission also receive numerous delegations from interests and discuss with them pending and prospective matters.
3. Commission officials often travel to member states to meet interests, and to attend conferences or other kinds of meetings organized by these interests.
4. A few large Eurogroups invite commission representatives to some working parties and committee meetings.
5. And, of course, there are the numerous informal meetings and telephone calls, as well as the various written documents, information, and other policy statements submitted by the different organizations to the relevant DG's of the Commission.

The case of COPA illustrates nicely the extensive set of links and access channels between an interest organization and the Commission. Approximately every 4-6 weeks the presidium of COPA meets the Commissioner for Agriculture; likewise, the Secretary General of COPA and the Director General of DGVI meet and phone regularly. At all levels, staff members of COPA are in constant touch with counterparts in DGVI, and also with officials from other relevant DG's. COPA is strongly represented on all agricultural advisory committees and also sits on certain other leading advisory committees not dealing specifically with agriculture matters (Nugent, 1994: 202). Of course, not all Eurogroups enjoy the same degree of access to the higher levels of the Commission. Some are better able to represent and defend the interests of their members than others.

---

13 There is also a number of so-called management and regulatory committees that are involved in monitoring and, adjusting on-going EU programs. Both are chaired and serviced by the Commission (director or head of a division) Committee members are national representatives, with middle ranking officials from the appropriate ministries attending on behalf of each country. Most management committees are concerned with agricultural matters; regulatory committees tend to concerned with issues of harmonization and vary greatly in sectoral interest. There exist regulatory committees (although they may 'go under' another designation) for the Quality of Water Intended for Human Consumption and for Fertilizers.

3.3.2 WATER SUPPLY SECTOR

At the national level we find close relationships between groups and governmental departments. One of the common features of the policy processes in these countries is the development of relatively stable communities and networks of policy actors. At the level of the European Commission there has also been a '...trend towards the development of close relationships between groups and the Commission.' (Lodge, 1993: 40) However, unlike the rather stable communities of actors and patterns of relationships which have grown up at the national level, the situation remains rather more fluid at the EU level.

With regard to water policy, Richardson notes that:

> *over time, the number and range of actors has increased, moving the system from a relatively private and limited range of participants to a more open, conflictual and public system of decision making.* (Richardson, 1995: 142)

Among all these groups, however, there exists a 'core group of actors', that is those who will need to be drawn into the process more closely before proposals are submitted to the Council of Ministers - indeed, even much earlier than this, while the issue is still being defined. According to Richardson, there continues to be a sort of inherent 'logic', followed within the Commission, which leads it to identify and nourish certain key actors (inner and outer circle of consultation). Why is this so? There is a number of quite simple and understandable reasons: commission officials get to know over time who matters and whom they 'need' - with regard to formulation those that 'matter' are those that define the problem and how to process it; or those that can block such efforts.

The core actors constituting the water sector, with regard to drinking water quality, includes:

- the water supply companies in Europe;
- interest groups at national/ European level (EUREAU);
- the ministries of the member states (and the German *Länder*); and
- the European Commission: DG XI.

Generally, there are three inputs that can be drawn from these actors for the formulation of directives: the scientific input, the industry input and the political input. The scientific input includes independent and community-sponsored research institutes. The recently established European Institute for Water is intended to play a co-ordinating role in this field. The industry input includes water suppliers and any other industry which is influenced by the provisions of water directives. Industry can play a useful consulting role in terms of technical and cost/practicality issues. The political input stems from the approval mechanisms that exist for environmental legislation. Water directives are the subject of unanimity in the Council of Ministers and in order of attaining an agreement between each Member State's objectives in many cases the legislation deviates from its initial purpose and is watered-down. For future legislative proposals in the field of water, the co-operation procedure is envisaged and the EP has a very important role to play in the formulation of the final legislation. The political input is subject to influence through lobbying by the industry and non-governmental organizations, and all these factors must be taken into account when the final legislative outcome is assessed (Kallis, 1995: 32-3). Through these procedures and mechanism a balance is sought, which incorporates all of the above inputs. The result should be standards based on scientific evidence, taking account of their socio-economic implications and respective the views of the public and the opinion of industry.

At the center or core of the non-governmental actors today is the water industry itself, which has come to appreciate the value or importance of being involved early in preparation of EU legislation and actions. The water supply companies in Europe vary much according to their status: they can be private or public. The water supply companies are represented by EUREAU, the Union of National Associations of Water Suppliers. The Union's principle aim is to represent the common interests of its members to the community organizations responsible for dealing with the development of community directives and European standards in the field of water. EUREAU's remit is two-fold: to review and discuss prospective legislation and standards in order to be able to give to community organizations the common view of the water sector and to analyze existing legislation so that, at the time of a revision, a sensible view can be put forward, balancing the politically desirable with the technically achievable, based on the experience of people actively involved in water supply.[14] Besides EUREAU also national organizations lobby in Brussels. Especially the British and the German organizations are very active in this respect, often having different views on proposed EU policy. So, within the sector there is not always a common view. EUREAU has a hard time getting the sector on one line. The revision of the Drinking Water Directive, especially the issue of the pesticide parameter, illustrates this. On the one hand a group exists that is in favor of a science-based parameter (a lower standard than the current standard that is based on a precautionary approach). On the other hand there is a group that is in favor of the current standard and rejects a change of the parameter. EUREAU's position corresponds with the last group. In practice this means that individual water companies lobby against EUREAU! At the Conference on the revision of the Drinking Water Directive (September 1993) the conflict within the drinking water sector was obvious.

Nonetheless, EUREAU nowadays plays an important role in Brussels. There is a close contact between DG XI and EUREAU. According to F. Rillaerts, Secretary General of EUREAU, this relationship has changed a lot during the last two years. The European Commission used to be a very closed organization and for EUREAU, trying to get access to them was frustrating. The last few years though, EUREAU is invited to the DG XI on an regular basis, in particular to talk about the revision of the Drinking Water Directive. According to Rillaerts, information/technical knowledge of its members is the most important source of EUREAU. It is much easier for DG XI to ask EUREAU for information than to contact the national ministries which would take more time. Rillaerts expects EUREAU to keep being involved with water issues in the future more than they used to do. He also explains EUREAU's lack of involvement with the development of the Nitrate Directive (developed by DG XI) from this historical perspective:

> *If the Nitrate Directive was being developed these days, EUREAU because of its close relationship with DG XI (as a result from the process of revising the Drinking Water Directive) would be involved.*

---

14 Information brochure EUREAU.

Although more at home in the agricultural network, the agro-chemical industries and agriculture itself have been very active in the deliberations concerning the revision of the drinking water directive and other related water quality measures.

3.3.3 AGRICULTURAL SECTOR

The actors of this sub-network:

- farming unions at local/national/european level (COPA);
- interest organizations representing the chemical industry (national/European), CEPA;
- Ministers of Agriculture of the member states; and
- European Commission: DG VI.

The agricultural sector traditionally plays an important role in Brussels. Several reasons can be pointed out. First, the EU carries, via the Common Agricultural Policy (CAP), major policy-making and decision-making responsibilities for agriculture. Second, as the major recipient of EU funds, agriculture is central to EU financial deliberations and serves to restrict policy development in non-agricultural spheres. And third, there is a greater institutional presence and activity in the agricultural field than in any other: the agriculture ministers meet more frequently than all other councils except for the foreign ministers; uniquely, Agriculture Councils are prepared not by COREPER but by a special body, the Special Committee on Agriculture (SCA); and DG VI (Agriculture) is the second largest DG of the Commission (Nugent, 1994: 361).

Although CAP has caused major difficulties and has been under reform several times, the agricultural sector is still very important. The environmental issue is getting more and more important and according to several respondents the attitude within the sector is changing slowly. According to one respondent, the agricultural sector moves from pure opposition against any environmental requirement to the acceptance of environmental measures. On the other hand it is argued that within the agricultural sector environmental measures are not taken out of a concern for the environment, but are used as a means to solve the problem of overproduction.

The agricultural lobby operates at two levels, the national and the European level (and probably in most countries also at the local level). The lobby at the national level is important because here the national positions are being decided which determines the discussions within the Council of Ministers. At the European level many agricultural groups have been formed. The most important of these groups is COPA, which is an umbrella organization representing all types of farmers on the basis of affiliation of national groups. COPA is the best resourced interest organization in Brussels, with a full-time staff of around 50. Nonetheless, all respondents indicated that COPA nowadays is not as powerful as it used to be. Regarding environmental issues, the North-South contradiction within the agricultural sector dominates the discussion. The North-South conflict consist of several issues. First, in the northern member states agriculture is more efficient and causes more environmental problems than in the southern states. Moreover, the 'green' pressure in the Northern states is bigger than in the southern states. Second, the environmental technology in the northern states is much more advanced than in the southern states which means that the southern states have to buy the technology from the northern states. And third, the southern states are less developed than the northern states and feel frustrated that they are being restricted in their development while the northern states were able to develop themselves without these restrictions. It is in this context that COPA operates. In practice this means that COPA position papers will be (too) general. Nonetheless, COPA of course tries to influence policy making in Brussels using its contacts at DG VI.

Regular meetings take place between COPA and DG VI at different levels of the organization. COPA also maintain relationships with the chemical industry. In the context of this research, three organizations are of interest: the European Fertilizer Manufacturers Association (EFMA), The European Crop Protection Association (ECPA) and the European Chemical Industry Council (CEFIC). Although the chemical industry in this report has been put under the agricultural sector, this doesn't mean that they (COPA and the chemical industry) always share the same view. Even more, one respondent argued that:

> *the traditional lines between agriculture and the chemical industry tend to fall apart. COPA more and more shifts the responsibility for environmental problems to the chemical industry and argue that the chemical industry should solve the problem.*

3.3.4 OTHERS

Finally, there is a group of actors that have a stake in the issue and can not be categorized into one of the two sub-policy sectors. First there is the environmental movement. The environmental organizations are not as powerful as the agricultural and chemical industry because they have less resources. On the other hand, the environmental groups do have the support of the general public. In Brussels several organizations are active. One of them is the European Environmental Bureau. EEB's strategy is two-fold. On the one hand they try to influence policy making in Brussels by talking to the Commission, writing position papers, attending conferences, etc. On the other hand they make use of their members, the national organizations. These national organizations have to convince their national ministries of their positions with an issue.

Besides these interest organizations, the World Health Organization (WHO) is an important player in the process of revising the Drinking Water Directive. This organization has developed guidelines for a number of parameters. In most cases, the WHO standards are more flexible than the EU standards. Richardson also mentions the consumers organizations having a stake in water issues (Richardson, 1995: 154;161-162). Several organizations were present at the Conference on the revision of the Drinking Water Directive. According to our information though, they are not very active in the field of agricultural water pollution.

3.3.5 NETWORK DYNAMICS: LINKAGES AND DEPENDENCIES

The contacts between the two sectors will be studied in the context of their involvement with two directives: (1) the revision of the Drinking Water Directive, (2) the development of the Uniform Principles (Annex VI of the Directive concerning the placing of plant protection products on the market). The descriptions of the relationships between the actors and the processes of development of the directives are not complete. In particular, information is lacking on processes in the past because people in Brussels (especially within the European Commission) shift working places (projects) very often.

Revision of the Drinking Water Directive. Like most policy development processes in Brussels, the revision of the Drinking Water Directive started in the so-called working groups. In these working groups experts from the member states participate. The working groups prepared a conference on the revision of the Drinking Water Directive (organized by the Commission) which took place on 23 and 24 September, 1993. The function of the conference was to hear all interested parties. This way the conference served as an important source of information for the Commission. Also it was argued by several respondents that the conference served as a kind of 'alibi' for the Commission. In any case, more than 200 people attended the conference. Richardson in his paper on EU water policy points out which interest groups were present at the Conference: (1) the water suppliers, (2) the polluting industries, (3) the environmental groups, (4) consumer organizations, (5) scientific experts, and (4) others like the WHO, DG's from the Commission, etc. (Richardson, 1995: 151-155). At the conference, the participants presented their position papers and discussed these in various panels. After the conference the Commission had an overview of the different views on the revision of the Drinking Water Directive.

Besides their presence at the conference, the drinking water sector - represented by EUREAU - has had an important input in the revision of the directive by writing a proposal for the revision. This proposal formed an important source of information for the employer at DG XI who had to write the revision (in general, all respondents stressed the importance of being present at an early stage of policy making). Moreover, EUREAU had meetings at DG XI about ten times to talk about the revision of the Drinking Water Directive. This way EUREAU was able to influence the process a lot. Also, respondents from the chemical industry indicated that EUREAU played a dominant role in the process of revising the directive. EUREAU did not form coalitions with the environmental organizations, like EEB. Although they both argued for maintaining the precautionary based pesticide parameter, they stood against each other on the issue of the parameter concerning the total of pesticides in water: EUREAU wanted that parameter out of the directive, while the EEB lobbied for the maintenance of this parameter.

Once the Commission has presented its final draft to the Council of Ministers, and the other European institutions, it is obvious that the lobby of the water sector (and the lobby of the other interest organizations) will move to the Parliament, the Economic and Social Committee and the national ministries. EUREAU will request her members, the national associations, to influence national positions.

Input from the agricultural sector comes from DG VI (Agriculture) under the process of 'inter-service consultation'. Because the revision of the Drinking Water Directive is a commission proposal, DG VI is co-responsible and, therefore, has a say whenever agricultural issues are at stake. According to a respondent of DG XI: 'DG XI and DG VI had some big fights over the Uniform Principles, this has eased the discussions on the revision of the Drinking Water Directive.'

COPA representing the farmers in Europe, has not been very active in the process. The chemical industry on the other hand has and still is lobbying hard for their position. They produce position papers, based on a lot of research. ECPA has been at DG XI several times to talk about the pesticide parameter. (ECPA asks for a meeting with DG XI, while EUREAU, on the other hand, is invited to DG XI!). According to Vlahodimos of ECPA, 'they talk to the authorities, EUREAU, COPA, EFMA - to a dense network of people.' Besides the lobby in Brussels, the national organizations put a lot of time and money in influencing the national governments.

Development of the Uniform Principles. The Uniform Principles (UP) were drafted by DG VI.[15] The UP were prepared in working groups made up of experts from the member states. Also DG XI participated in these working groups. Especially, the link in the UP to the parameters of the Drinking Water Directive caused a lot of conflicts and delays in the development of the UP. EUREAU was not directly involved with the UP, because they do not have access to DG VI. Via other canals, EUREAU tried to influence decision making on the UP. They talked to DG XI and some of the member states.

The agricultural sector was directly involved with the Uniform Principles: COPA through her contacts at DG VI, and ECPA because of her direct interest in the issue. In general, respondents indicate that ECPA has a very dominant position in the network because the organization possesses a lot of resources, money and personnel. Their strength mainly results from their lobby capacity: they have many lobbyists active in Brussels and also in the member states. Moreover, the industry organizes receptions, diners, seminars, etc. Also, they have a huge research capacity. According to one respondent, the chemical industry has a 'battery of scientists' who can work on an issue.

This situation is in sharp contrast with the environmental groups, who lack resources. Still, as one respondent formulated, the environmental groups can rely on the general support of the public and press. Especially concerning pesticides there has been a tremendous debate in the press in some member states which impacts policy making. The chemical industry on the other hand is not supported by public or press. In fact, their intentions are being distrusted. The chemical industry lacks trust (which also can be considered as a resource) because their (economical) interest in the issue is very big. Apart from the conflicts between the various interests (drinking water, environmental, industrial and agriculture) with the UP, the conflicts between member states were even bigger. At one point a proposal was presented by the Commission. Most member states were in favor of this proposal, but there was not a qualified majority because Denmark, the Netherlands, and Germany were against. They demanded higher standards.

---

15 For more information on the Uniform Principles, see section 3.2.2 above.

3.4 Conclusions

Interactions between the drinking water sector and the agricultural sector takes place within different networks and during different stages of policy making. Within the formal network we saw that several interactions take place. (1) Interactions take place via initiatives of the Commission, like the conference on the revision of the Drinking Water Directive. (2) DG XI and DG VI are required to consult each other on relevant issues, so-called inter-service consultation. (3) The sectors meet through committees or working groups that prepare commission proposals. (4) The actors of the drinking water and the agricultural interest interact in the member states through the process of deciding on national positions on EU policy proposals.

Besides all that, interactions take place in the (more) informal circuit. (1) DG XI talks to the interest groups that represent the agricultural sector (COPA, ECPA, EFMA). It seems that DG XI is more 'open' in this respect than DG VI. The intention of DG XI is to stimulate discussions on environmental issues within the agricultural sector. (2) Between the interest groups relationships exists. The chemical industry, especially ECPA, is very active in establishing relationships with everybody involved with an issue. EUREAU also maintains relationships with individual chemical companies. Rillaerts explained that within the chemical industry some companies are in favor of high standards for competition reasons. In any case, every respondent stressed that coalitions between interest organizations are always flexible and vary dependent on the (sub)issue.

References

Brouwer, F.M., I.J. Terluin, F.E. Godeschalk (1994) *Pesticides in the EC.* The Hague: Agricultural Economics Research Institute.

Commission of the European Communities [CEC] (1993) *Fifth Environmental Action Programme. Towards Sustainability.* Brussels: Commission of the European Communities (*PB C 138, 17.5.93*).

Commission of the European Communities [CEC] (1994) *Working paper concerning the Ground Water Action Programme.* Brussels: Commission of the European Communities, August 1994.

European Environmental Bureau [EEB] (1993, September) *Position paper: comments on the possible revision of the EC Drinking Water Directive.* Brussels: European Environmental Bureau.

Environmental Watch (1993) *Environmental Watch Western Europe, 2* (19).

Haigh, N. (1987) *EEC Environmental Policy and Britain.* Harlow (Essex, UK): Longman.

Haigh, N. (1992) *Manual of Environmental Policy: the EC and Britain.* Harlow (Essex, UK): Longman.

Hanf, K.I. (1992) European Community Policy and Water Management at the Level of the Member States. In: J.Th.A. Bressers and L.J. O'Toole, Jr. (eds.) (1992) *International comparative policy research,* pp: 85-98. Enschede: University of Twente, Center for Clean Technology and Environmental Policy.

Johnson, S.P. (1983) *The pollution control policy of the European Communities.* London: Graham & Trotman.

Kraemer, R.A. (1993) *Water Policy Report.* Bonn: Institut für Europäische Umweltpolitik.

Kraemer, R.A. (1994) *Water Policy Feature. Opinion on 80/778/EEC.* Bonn: Institut für Europäische Umweltpolitik.

Krämer, L. (1990) *EEC Treaty and Environmental Protection.* London: Sweet & Maxwell.

Kallis, G. (1995) *Assessment of EU water quality policy,* Luxembourg: European Parliament, the STOA Programme (EP 165.512).

Lodge, J. (ed.) (1993, 2nd ed.) *The European Community and the Challenge of the Future.* London: Pinter.

Nugent, N. (1994, 3rd ed.) *The Government and Politics of the European Union.* Basingstoke: Macmillan.

Reus, J.A., H.J. Weckseler, and G.A. Pak (1994) *Towards a future EC Pesticide Policy: An Inventory of Risks of Pesticide Use, Possible Solutions and Policy Instruments.* Utrecht: Centre for Agriculture and Environment.

Richardson, J.J. (1995) EU Water Policy: Uncertain Agendas, Shifting Networks and Complex Coalitions. In: J.Th.A. Bressers, L.J. O'Toole, Jr., and J.J. Richardson (eds.) *Networks for water policy. A comparative perspective,* pp: 139-167. London: Frank Cass.

Williams, A.J. (1993) The CAP and the general environmental policies concerned with agriculture in the EC and their implications for fertilizer consumption. *Paper prepared for the Environment Northern Seas Foundation International Conference and Exhibition.* Norway: Stavanger.

4. CONTROL CAPACITY - EUROPEAN UNION

ELLIS L.M. BRAND
Center for Clean Technology and Environmental Policy, University of Twente, Enschede, The Netherlands.

KENNETH I. HANF
Erasmus University Rotterdam, Department of Public Administration, Rotterdam, the Netherlands.

JOS LIPMAN
Bureau for Contract Research, Rotterdam, the Netherlands.

4.1 Introduction

In this section the focus is on the control capacity of the water supply sector and specifically on the role of the EU in strengthening this capacity. What do we mean by control capacity? By using the term 'control capacity' we refer to the capacity of the water suppliers to influence behavior of the farmers. Several indicators for this concept can be distinguished: capital, human resources, information, time, money, (legal) rights, trust. The EU, for example, could support water suppliers in Europe by increasing one of these resources of the water suppliers. In very simple and quite general terms we might say that there are two options for EU action: (1) The EU commits the member states to do 'something' via a directive or (2) the EU only facilitates, there is no duty for the member states. As we will see, perhaps the most important contribution that the EU can make to improving the capacity to deal with the problem of agricultural pollution of drinking water supplies lies in the area of promoting and assisting the building of networks of actors around this issue at the different levels of European, national and regional or local decision making. It is in this combination of regulatory and networking activities that the EU contribution to increased control capacity is to be sought.

In the first section we describe current EU practice: in what ways is the EU strengthening the control capacity of the water suppliers or in what ways is the EU stimulating water suppliers to act as environmental agencies? In the second section we then

Geerten J.I. Schrama (ed.), Drinking Water Supply and Agricultural Pollution, 73-91.

focus on the possibilities for the EU to strengthen this control capacity of the water supply sector. First we present the responses of our respondents to this issue.

4.2 Actual control capacity

In the previous chapter EU policy regarding the protection of the water resources against agricultural pollution is described. There is no policy of the EU to stimulate the water suppliers in Europe to act as environmental agencies. One could only argue that by setting high standards for the quality of drinking water, the EU indirectly stimulates the water suppliers to play a more active role (besides their role in a more technical way) in the prevention of agricultural water pollution.

Traditionally, the EU has been notorious regarding two things: its lack of capacity for policy co-ordination and its (relatively) poor implementation record. The highly sectoral nature of Commission has inhibited efforts to produce any kind of routine interdepartmental coordination. However, as we will examine in more detail below, increasing criticism of this situation has recently provoked more determined attempts to secure better co-ordination within the Commission and to integrate more effectively environmental policy within rest of Commission's policy making. With regard to implementation, the problem lies in the formal division of labor between the Commission and the member states. Front line administration in the EU has been delegated to the appropriate agencies within the member states. Consequently, although the Commission is formally charged with the enforcement of EU laws, it is in a very difficult position when it comes to trying to supervise the actual execution of policy in the member states. Here too, increasing pressure for more effective implementation has lead the Commission to explore ways of reaching those actors that ultimately 'count' in the hope that they will be able to work against the 'policy erosion' that now often occurs during implementation. For example, the perceived seriousness of the problems surrounding the implementation EU environmental policy has elicited the following observation made to the House of Lords Select Committee on the European Communities:

> *We have now reached the point in the EC where if we do not tackle implementation and enforcement properly, there seems very little point in producing new environmental law.* (Hildebrand, 1995: 18)

Together with the increasing attention paid to the applying the 'subsidiarity principle', this has meant that the Commission tends to prefer an approach that is based on general, union-wide quality objectives, defining the goals of the policy, which are to be implemented by the member states in accordance with national and regional conditions and needs. Effective monitoring and control of the results achieved, in the first instance by mobilizing the citizens in the different countries to exert pressure 'from below' for effective goal achievement is a crucial element of this division of labor between the Commission and the member states.

Before exploring possible ways for the EU to strengthen the environmental role to be played by the water suppliers, we should ask ourselves if this is necessary. Clearly, several key actors within this field indicate that this is a non-issue. Maybe the EU should concentrate on the implementation of existing policy. The reactions to this issue of the respondents (of the European Commission, and various interest groups) are presented.

Water suppliers. F. Rillaerts (Union of National Associations of Water Suppliers, EUREAU):

> *The drinking water companies are in the first place drinking water producers. They do not have the task to act as environmental agencies, which is more a task of the government. They can have an alarm function: they can inform the government of offenses. Giving public powers to water supply companies is not a real possibility for several reasons: (1) Water supply companies in Europe are a mixture of public and private companies. (2) This would put the water suppliers in a very powerful position vis-à-vis the farmers which certainly will not stimulate co-operation and communication between the two.*
>
> *The payment of compensations by the water suppliers to the farmers is in conflict with the 'polluter-pays-principle' and EUREAU does not support this. There could be a role for the water supply companies in the educational programs together with local partners such as farmers organizations, extension services. In this respect the EU might stimulate (financial) such information or educational campaigns.*

The polluter-pays-principle is an important element of most of the answers of the respondents. Several respondents agree with Rillaerts in the sense that paying compensations to the farmers in order to prevent groundwater contamination is in conflict with the polluter-pays-principle. This is one of the basic principles of the environmental policy of the EU (set out in Article 130r of the Treaty).

European Commission - DG XI (Environment). According to Gammeltoft (DG XI) compensations to the farmers as an instrument to change their behavior is especially difficult because the agricultural sector is already heavily subsidized. The same tax payers then would have to pay for the consequences of this policy (because in the end, of course, the consumers of drinking water pay for the compensations).

Not surprisingly, from the agricultural side (COPA), there is support for the use of compensations payments to the farmers. Moreover according to COPA, this should be a condition for restricting farming practices.

European Commission - DG VI (Agriculture). According to M. Scheele, working within the environmental unit of DG VI, supporting the water suppliers in preventing agricultural water pollution, is not necessary because the EU already regulates the use of pesticides, considering their (possible) impact on water. According to Scheele, the 'regulation gap'[1] does not exist: pesticide regulation is linked to the drinking water standards. Scheele argued that there will always be a last part in the chain that is not regulated. Because of the subsidiarity principle, the EU should not try to get involved at that policy level.

---

1 The assumed regulation gap between the regulation of water supply authorities and the regulation of farmers.

The subsidiarity principle is one of the principles of EU environmental policy. It limits the scope of community action to the extent that environmental policy objectives can be better attained at community level rather than national level. In general almost everybody agreed that a solution to the problem - agricultural water pollution - should be looked for at the regional and local level.

Chemical industry. Mr. Vlahodimos of the European Crop Protection Association (ECPA) formulates this idea as follows:

> *At the European level there is rather comprehensive legislation concerning pesticides: it defines what products will be on the market and how they will be used. However a gray area exists - the distribution system. The distribution system could be used to promote Integrated Crop Management. Policy solutions at the European level are no use because local conditions are important and should be taken into account when applying pesticides. The solutions for water pollution should be looked for at the local level. The local players are important: the farmers organization, the extension service, the local industry representative and... the water supply companies might co-operate in this local network. The network has to educate the farmer in integrated crop management.*

Farming union. Mrs. D. Dejonckheere (COPA)[2] worked out the idea of local initiative and also indicates what role could be left for the EU. Starting point in her argument are the so-called protection zones to be established by the national governments. Within these zones agricultural activities are to be restricted in order to prevent water pollution. According to Dejonckheere it is imaginable that the protection zones will be required on basis of EU regulation. This could very well be one of the recommendations of the Groundwater Action Program which is being drafted by the Commission at the moment. At a later stage the protection zones could be a requirement on the basis of the revised Groundwater Directive. In concrete terms, the EU could require the member states to take the following steps (in line with the Directive on Ecological Water Quality): (1) member states have to measure the quality of the groundwater; (2) on basis of the results, the member states have to define protection zones; and (3) member states have to develop and implement action programs for these areas. The EU could require that the action programs contain (1) local or regional measures to improve or maintain water quality; (2) compensations for the farmers (the agri-environmental measures which contain compensations for farmers could be linked to the protection zones); and (3) the set up of a platform of local parties involved in order to stimulate communication among them.[3]

Conclusions. In general, the reactions of the respondents regarding the role of the EU, were rather reserved. Important aspects are two principles of EU environmental policy, the polluter-pays-principle and the subsidiarity principle. One sees possibilities for the EU in

---

2 Committee of Professional Agricultural Organizations in the EC.

3 These ideas are presented on a personal basis and do not reflect COPA's position on this issue.

(financially) supporting the development of educational programs for farmers. The purpose of these programs then would be: (1) to make farmers aware of the problem of agricultural water pollution, and (2) to educate farmers on ways to prevent water pollution with nitrate and pesticides.

4.3 Prospects of increasing the control capacity

The problem posed in the research design for this project was how the water supply network could 'penetrate' the agriculture-chemical network so as to create the conditions - political and organizational - for co-ordinated and integrated approach to dealing with the problem of the pollution of drinking water sources by agricultural activities.

The main point to keep in mind in this regard is that there is very little what the EU can do directly in dealing with a problem that, in the last analysis, has to be solved the local or regional level on the basis of national programs that, to be sure, have been developed to put into effect union-wide regulatory schemes. Having said this, it does seem that the EU has a very important role to play in generating resources and action opportunities for other actors: in the form of legislation and regulations that define the quality of objectives for water quality and set the parameters within which various groups of actors must operate. Within this common regulatory framework, the different member states are given the responsibility for working out the most appropriate measures for translating the general program into specific constraints and possibilities upon the action leeway of local and regional actors. In some limited cases the Commission can provide financial assistance or information as well as guarantee the regulatory flexibility needed to adjust general regulations to the conditions in a particular locality. In this way, the EU can create the framework within which regionally-specific solutions to the problem are to be sought and can work to facilitate the joint efforts of water supply and agricultural actors.

In what follows, no attempt has been made critically to evaluate the substance of measures being contemplated or taken by the EU to deal with the problem of water pollution through agricultural activities. Instead we have considered what types of actions could be taken by the EU to bring about an integration of water resources management and agricultural restructuring so as to solve the problem of deteriorating quality of drinking water resources.

A further qualifying observation is in order at this point. We assume that whatever the strategy chosen to integrate concern with protection of water resources and the restructuring of agriculture in order to change behavior of farming community of relevance to quality of freshwater resources, it can be assumed that it will be shaped by and consistent with certain basic principles on which present policy is supposed to be based:

- the general strategy of the Commission which places emphasis on the formulation of general, EU-wide quality standards, leaving to the member states the specification of the ways in which these are to be achieved and relying of the mobilization of pressure from the general and organized public to ensure that member states make an honest effort to achieve the quality objectives;

- respect for subsidiarity principle, which means allowing for regional and local variation in adapting general EU-wide policy objectives to the specific conditions within individual member states;
- making use of a more varied set of policy instruments in implementing the legislated quality objectives; and
- developing mechanisms for the effective application of the idea of partnership between European, national and sub-national levels of decision making in sharing responsibility for achieving joint policy objectives.

In the following sections we will look more closely at the types of actions that the EU might undertake to make a variety of resources available for the joint development of policies and programs designed to deal with the different manifestations of the general problem in the member states under investigation.

Many commentators have noted that the EU already has an adequate regulatory system in place (and is in the process of improving it with the recent proposals for revised and new water directives) but that the problem lies in the effective implementation of measures already on the books. As we have noted above there is actually little that the EU (acting through the Commission) itself can do directly to ensure that the necessary and appropriate actions taken at the local levels. At the same time, it is important to keep in mind that the 'implementation' of water quality protection policy actually involves the 'making' (and, once again, subsequently implementing) policy affecting other policy sectors in order to change the behavior of the 'polluting' actors. Once the quality standards and emission limits have been set, 'enforcement' involves ensuring that polluting substances are not introduced into the environment in amounts or in forms that would directly or in the course of time lead to the deterioration of the water source in question. In this respect, the 'substance' of the appropriate EU policy strategy is clear: steps need to be taken to bring about structural changes in agriculture in Europe so as to modify significantly present farming practices so as to bring about environmentally friendly farming. The question is then how these objectives can be achieved and how networks of actors drawn from the area of water supply and agriculture can be brought to co-operate in seeking this kind of solution to the problem.

It should be kept in mind that the expansion of the control capacity of the next work of actors takes place both simultaneously and sequentially at the different levels of decision making. There is a number of separate but linked games going on at the same time involving varying configurations of different aggregates of actors (individual farmers, regional and national farmers organizations, and the COPA at the European level) involved directly, and through representative organizations.

4.3.1 AUTHORITY AND LEGITIMACY

Regulatory framework for water quality management. The first prerequisite for effecting the necessary changes in agricultural activities is the presence of effective regulatory pressure in the form, in the first instance, of quality standards for the European Union as a whole, which define the objectives to be met and specify the procedures and criteria to be followed in developing national programs for meeting these goals. The decision processes with regard to both the formulation and implementation of these EU regulations create the occasions as well as the arenas around which functionally specific networks of actors take shape or can be formed.

We are assuming that the EU, on the basis of proposals coming from the Commission, provides the regulatory framework, within which other actors then subsequently must operate, once these proposals have been accepted and passed as official EU legislation. This policy framework - in our case, the relevant water directives - lays down the quality objectives that are to be achieved or the substances the emissions of which are to be limited. It would appear to follow from the discussion thus far that if stricter controls are to be imposed on quality of water for human consumption, than more effective protection will have to be provided for the water sources from which the raw water for this product is abstracted. This, in turn, means that the Drinking Water Directive must form part of a general comprehensive regulatory system for the integrated management of water quality (and quantity) within the EU. Such a policy requires that the Drinking Water Directive be complemented by measures to protect the quality of surface and groundwater under the broader umbrella of legislation to protect the ecological quality of these waters. As we have seen, four directives, that could make up such a policy, have been presented by the Commission for consideration and critical examination. At the same time, questions have been raised by various commentators on these proposals - see for example the STOA report for the European Parliament (Kallis, 1995) - as to whether these measures constitute the 'coherent and comprehensive legislative framework for the European Union's policy on he protection and quality of water' that is needed.

Furthermore, it is equally important that other EU policies also be in conformity with the Drinking Water Directive. For example, in the opinion of EEB, the objectives of the Drinking Water Directive needed to more effectively integrated into agricultural policy and into the policy concerning pesticides and the placing of pesticides on the market. For example, the EEB notes that '[w]hereas there exists legislation to reduce nitrate input, nothing comparable exists for pesticides' (EEB, 1995: 15). Only EU-wide legislation on reduction of pesticide input will give the polluter the right signal. Otherwise, according to the EEB, polluters will continue their damaging agricultural practices relying on the fact that it is not possible to 'tackle' them as polluters (EEB, 1995: 15).

In the past, EU policy has been characterized by the use of regulations as prime mode of policy action. The preference for this regulatory instrument can be understood by recalling that - with the exception of the CAP (and other structural funds) - the EU has few tangible resources to distribute. Nevertheless, the present Fifth Environmental Action Program lays out a strategy that intends to make use of a more varied mix of policy instruments, including the promotion of joint decision making and shared responsibilities among societal and governmental actors. EU legislation will, then, continue to be needed to provide the 'push' for behavioral change coming from binding general quality standards. Furthermore, EU-wide regulation is necessary to protect the integrity of the internal market and, to this end, to define the common substantive and procedural rules of the game. Ultimately, however, the solution to the problem is not be found in more regulation, in the traditional sense, of union-wide rules and regulation specifying what must be done; an effective solution involves more the marshaling or mobilization of the resources of various actors in order to bring about these changes in the behavior of the ultimate target group, the individual farmers in the different member states.

The ultimate objective of these regulations is to guarantee the quality of drinking water. Within the limits of the technological development of a given moment, this goal can be achieved by water treatment procedures to remove unwanted substances from the water. The possibilities for doing so are limited by the technical possibilities and the economic considerations of additional costs, with the resulting higher prices for users or lower 'profits' for the water company. (It is, of course, also possible to mix water from different sources to arrive at a quality that is acceptable under the requirements.)

Another way of trying to deal with the problem of drinking water quality is to guarantee that the water bodies from which the raw water for the production of drinking water is abstracted - surface and groundwater sources - is of such a quality that expensive treatment to achieve the prescribed product quality is not necessary. In this regard, we have noted the links between the Drinking Water Directive and those directives aimed at protecting the quality of surface and groundwater resources. This means applying measures to prevent contamination from occurring in the first place, and in this sense, involves implementing effectively the directives that already exist or are currently under consideration.

The 'implementation' of these directives actually involves 'making and enforcing' policy with regard to the economic activities, in this case, farming, which cause the pollution of these bodies of water. It is in this sense, that the regulation of water quality or the achievement of the quality objectives for surface and groundwater - by limiting the emissions of certain substances into the water - can only be implemented by changing the behavior of the offending activities. And here is where the nub of the problem has been. Lack of effective implementation of the relevant directives - especially the Nitrate Directive - has been the result of the unwillingness and the inability of member states (and the CAP) to intervene in the traditional growth objectives and associated farming practices of intensive agriculture in Europe. Here, in protecting the integrity of its bailiwick, is where the agro-chemical establishment has been strong.

Agriculture and water resources protection. The reduction of groundwater pollution by nitrates and pesticides as well as other dangerous agro-chemicals remains one of the important objectives of water resources protection policy. In areas with intensive agriculture use there are water bodies with high levels of pollution from these substances. In the recent past, countries have increasingly been taking measures to change agricultural practices and in particular to reduce amounts of fertilizers applied.

An institutional imperative was underlined as an important principle at the Water Meeting in The Hague in 1991, where the EC environment ministers stressed:

> *that water management policy should be integrated within the wider environmental framework as well as with other policies dealing with human activities, such as agriculture, industry, energy, transport and tourism.* (Kraemer, 1993)

The idea here is to further sustainable development and environmental protection through changes in other policies. Specifically, with regard to nitrates and pesticides, the environmental ministers agreed that actions aiming at sustainable management and protection of freshwater resources should include:

> *changes in agricultural practices to prevent pollution, e.g. from nitrates as provided for in the Nitrate Directive. This objective could be achieved by the 'progressive reduction of fertilizer application to levels required for good agricultural practice.... Water management policy should also aim at 'the progressive replacement of persistent and accumulating pesticides which are likely to cause deterioration of groundwater quality... The ministers also suggested a 'progressive limitation of plant protection products and other pesticides for non-agricultural purposes' so as to prevent adverse effects on groundwater...* (Kraemer, 1993)

In short, the line of reasoning here is that less intensive farming practices will be less environmentally damaging. A number of actions along these lines have been included in the Fifth Environmental Action Program for achieving the objective of sustainable agriculture and the protection of specific water resources. These include:

- strict application of the Nitrate Directive by member states working closely together with agriculture sector;
- setting of regional emission standards for new live stock;
- programs for reducing the use of phosphates;
- registration of the sales of pesticides;
- control of sales and use of pesticides;
- promotion of integrated control programs (training programs) and practices of bio-agriculture;
- setting up of protection zones for agriculture and environment; and
- training farmers in environmental friendly practices and exchange of information on these matters.

Each of these measures would involve different sets of actors at the level of the individual member states, in the interaction between the member states, and at the European Union level. Some would involve taking measures to enforce and implement already existing EU and national legislation; while others would involve the passage of new EU directive to set Union-wide standards (CEC, 1993: 37).

According to research, carried out by IEEP, integrating environmental objectives into CAP will involve, in practice, an ability to match agricultural policy more closely to local conditions. Therefore, it is not a question of coming up with new Europe-wide measures, but creating for Europe as whole conditions under which member states can tailor measures to regional/local conditions within their countries. Regulation 2078/92 encourages member states to adopt locally appropriate schemes. The broader programmatic context, within which particular measures could build upon and draw from the possibilities contained in this regulation, would be based on voluntary agri-environmental schemes developed by actors in the member states - and for which it will be necessary to develop institutional arrangements and roles for national, regional and local authorities as well as non-governmental actors at different levels.

As a consequence of reforms of the CAP in 1992 land can now be set aside and left fallow. In addition land can be taken out of production as a result of a down-turn in agricultural activity. EC Regulation 2078/92 concerning environmentally compatible agriculture, which was conceived as a part of the so-called accompanying measures of the CAP reform, opens up the possibility for promoting agricultural production compatible with water resource protection requirements. There already exist national compensation schemes under Art 19 of EC Regulation 797/85 which encourage farmers in environmentally sensitive areas, for example in nitrate vulnerable zones, to adopt, under voluntary management agreements, methods of crop production that are less damaging to the environment. The systematic evaluation and discussion of relative merits of these measures, together with what is needed to make them work, by the parties involved would provide a sound basis for provide a sound basis for the development of agricultural practices which would pay sufficient attention to the environmental impacts of these activities. If and where necessary, some kind of facilitating legislation, defining concepts and procedures, and laying down common 'rules of the game' for the EU as a whole would be in order. Both the need for and the contents of such legislation would itself be a product of the consultations and negotiations among network members, under the general direction of those EC officials jointly responsible for this problem area.

An important element of structural reforms of European agriculture is some kind of compensation or other form of financial assistance to farmers who are willing to move toward environmentally friendly practices. It should be pointed out that such schemes are not to be interpreted as payments for compliance with existing laws, something that would be in conflict with the polluter-pays-principle. On the contrary, these payments involve compensation and support for structural change and adjustments as part of the broader policy of agricultural reform. In this connection, and also there where other kinds of economic instruments can be used to promote agricultural change, such as water charges, the EU can function as a facilitator of change by providing the flexibility needed to interpret and apply European law in a manner to encourage creative solutions to the problem at issue. Moreover, where required, European legislation can be used to provide the common 'rules of the game' within which member states develop national programs for meeting these objectives.

4.3.2 ORGANIZATIONAL CAPACITY

Effectively integrating environmental considerations into agricultural policy is not primarily a question of trying to regulate present practices, but rather of attempting to influence long-term structural changes as well. This would involve, in the long run, more than special protection zones (a short term measure) and would be aimed at or form part of broader effort of restructuring agricultural policy at both the European and national levels. This involves extending the present issue definition from the narrow sectoral focus of agricultural reform or water quality to the development of environmentally sensitive farming as the precondition for sustainable agriculture. In turn, this means actively creating an extended policy community by bringing together the experts and 'policy' people from various associations in water management and agricultural areas to discuss the 'contribution of agriculture to the protection of water resources. A convincing argument can be made, we think, for the existence of a new strategic context for farmers (and their friends) as a consequence of increasing pressures for serious reform of CAP and, consequently, the agricultural practices on which this policy is based. At the same time, there is also recognition of the need for the agricultural community to take effective action in order to mitigate the negative environmental impacts of farming, particularly with regard to the quality of water resources. These two sets of pressures are important considerations in determining why the various actors in the agricultural network should be interested in engaging in such discussions and working together in the development of joint programs for dealing with the water quality issue.

The 'easy' answer to our problem of effective management of (drinking) water quality - and, thereby, for improving the poor implementation record with regard to present water quality legislation of the EU - is, then, to change present farming practices in the direction of environmentally sensitive agriculture. As we have noted above many of the elements of the substantive strategy for putting such a solution into practice are already in place or presently being discussed. Realizing such structural changes presupposes extensive institutional changes at all levels of decision making (some of which are presently taking place at the local and regional levels). In this sense, a program of water quality management through structural reform of agricultural practice requires important changes in organization and management in order to create the institutional prerequisites for the formulation and implementation of such joint policies. In particular the Commission will need to improve its own capacity for co-ordinated policy development as well as to contribute to strengthening those capabilities of other actors by actively helping to create and support the networks bringing together the necessary inputs from the water quality and agricultural communities. In this sense networking skills will have to be employed, at different levels of action, in order to create occasions and arenas for mobilizing the problem solving resources of different interrelated actors.

Internal reorganization of the commission. Where policy issues cut across the Commission's administrative divisions, it is essential that provision be made for liaison and co-ordination. When we talk about integrating environmental policy considerations into other Community policy areas, we are asking the Commission to develop a capacity for cutting across these functionally specialized lines of responsibility and to developing integrated policy proposals.

As we have seen (discussion of policy networks), the Commission - as the policy initiator in the sense of being responsible for preparing and submitting policy proposals for Council decisions - has tended to act in a segmented and disjointed fashion. Traditionally it has tended to be organized and to operate along sectoral lines and even within these areas, along highly specialized functional lines. Similarly, its contacts with other interested governmental and socio-economic interests have tended to be functionally specialized around a particular problem area or issue. While there are mechanisms for a certain amount of co-ordination, this tends to be of the kind that Scharpf (1973: 85-89) refers to as 'negative co-ordination', i.e. giving potentially affected interests a chance to make sure that proposed actions do not encroach on their own territory or affect the interests they represent in a negative manner.

Still, various commentators have detected a gradual shift toward more co-ordinated approach, something which has important implications for an analysis of EU water policy making. There are a number of signs of attempts to co-ordinate environmental policy (including water) with other sectors, such as regional and agricultural policies. And the more this has in fact taken place, the more the process has shifted from narrowly based set of actors to a more extended network of actors.

A number of procedures and mechanisms exist which are supposed to provide for the necessary co-ordination in such cases. For example, the President of the Commission has an, admittedly ill-defined, but generally expected co-ordinating responsibility. How effective it ultimately is, depends on how forceful the personality of the incumbent is and if he enjoys sufficient prestige among his fellow Commissioners to be able to 'forge measures of collective identity.' The College of Commissioners is, in theory, also in a strong position to co-ordinate activity and to take the broad view of things. In the last analysis it would seem in any case, that the effective initiative and political support for policy co-ordination and integration within the Commission would need to come from this more 'political' level and be passed on to the responsible DG's.

Furthermore, a number of practices and forums have emerged as settings for more co-ordination by providing the occasions and structures for bringing together officials lower down in the Commission hierarchy. For example, standing and ad hoc committees are used, as are task forces and project groups to promote better co-ordination between different officials in DG's dealing with related issues. At a somewhat higher levels, the affected Directors General meet. The Secretariat General also has a role of play - once the decision has been taken elsewhere - in making sure that these co-ordination mechanisms function well.

Networking and the creation of an advocacy coalition. These kinds of developments, while not completely overcoming the strong tradition of segmented work patterns, are steps in the right direction. However, there is also an external dimension of co-ordination and joint problem solving that needs to be developed. This involves the bringing together of representatives from the presently separate water quality and agricultural communities. And here the Commission also has an important role to play. By working to redefine the relevant problem to focus on the environmental dimensions (especially with regard to water quality) of agricultural restructuring - within more general and broader issue of sustainable agriculture - the range of potentially interested parties is enlarged. In this way, the Commission can work to create the raison d'etre as well as the institutional basis for the emergence of a new 'policy community' or, perhaps more correctly, for an 'advocacy coalition' organized around this issue. Both the regulatory framework and the associated decision processes serve as occasions and focal points for interactions among sets of actors drawn from this broader coalition at the different levels of deliberation and decision making. Apparently there already exists an informal working group advising DG XI on environmental aspects of CAP. This group has been considering such things as cross-compliance, i.e. the possibility for attaching environmental conditions to agricultural support payments, and the use of voluntary environmental management schemes involving incentive payments for farmers. Such an advisory group, while integrating external interests, still remains within the traditional sectoral division of labor inside the Commission.

In this sense, the Commission can build further on its active tradition of consultation, conferences and advisory groups. The difference would lie in the opportunities provided in this context for an important element of 'positive co-ordination' in the sense that agriculture and environmental interests jointly work on problem definition and solutions, rather than one group 'checking' the initial work of the other to see whether or not it can live with what has been put together. Here we need to build on the acknowledged experience and reputation of the Commission as a 'skilled broker of interests'. In more contemporary jargon, the Commission must function as a network manager to bring together affected and interested parties, not in the first instance as suppliers of information or for the mobilization of support for Commission proposals, but rather for the purpose of jointly defining the problem and laying the foundations for collaboration among them. In this sense, the enterprise, even while requiring a certain amount of facilitation and direction from the Commission, is quite different from manipulating or building strategic relations with some actors to generate inputs to or to gain legitimacy and political support for positions of one DG or the other.

Networking at the national level. Ultimately, most actions to modify agricultural practice will have to be organized and taken regionally or locally. It will be necessary to convince individual farmers of the nature and extent of the problem and how it affects their long term survival interest. At each level - national, regional and local, depending of the particular conditions and arrangements in a given country - it will be necessary to bring together the relevant actors in the combination that is appropriate for that level of action and point in the decision process. At each level monitoring will need to be organized and carried out, and information gathered and exchanged. Even if it is subsequently processed at a higher, even European level, it will ultimately have to be desegregated to be used in specific situations.

In this sense then, there will need to be a series of interrelated networks of actors, involved in a two-way exchange of information and decision premises. The EU can play an important role in directly 'creating' the European-level networks, and provides the regulatory and facilitating framework for the development of equivalent structures at other levels.

4.3.3 LEGITIMACY (POLITICAL SUPPORT)

Redefining and extending the policy issue to one that combines agricultural restructuring and water quality management will, therefore, change the nature of the policy process and the actors involved. The network structures sketched above provide the context for coming up with shared problem definitions, consensual knowledge regarding claims and proposed solutions, and opportunities for the joint development of mutually acceptable solutions.

They also can serve to generate the political pressures which, ultimately, are necessary to generate interest and support for dealing with the problem and making the institutional adjustments that are needed to handle it effectively.

In trying to develop the kind of integrated policy strategy discussed above, we confront a sort of 'chicken and egg' problem. Effectively combating the pollution of drinking water from agricultural activities requires both policy and institutional changes that will not be forthcoming unless the issue itself achieves sufficient political visibility and urgency to lead to changes in traditional modes of operating that, at present, impede the development of an integrated approach. Yet, in order to generate these pressure, the Commission would have to take an active lead in promoting this kind of strategic reorientation. As we have noted above, the recent moves toward co-ordinated action within the Commission are encouraging but unlikely to gain a significant foothold vis-à-vis the sectoral 'dossiers' approach unless and until there is pressure and support from the College of Commissioners itself to encourage the officials within the DG's to develop closer working relations with one another. For example, agreements between affected Commissioners (such as Environment and Agriculture) could create the 'political basis' for activities on their behalf by their chefs de cabinet in pushing and monitoring co-operation between the relevant DG's. At the moment, neither Commissioners nor their chefs de cabinet normally get involved until commission proposals have been launched by the DG's, although earlier consultation does sometimes take place.

Here is where input from the 'political' side is needed, although it would be possible for initiative to bubble up from the DG level, through chefs de cabinet, to the Commissioners. Still, the response from this highest level-would be heavily influenced by the political relevance or priority of the issue. However, without the pressure and monitoring from above in organization, appropriate initiatives are unlikely to come forth from the level of the DG's themselves. At least until they see the advantage and payoff in political support to make internal co-ordination 'work' and to lend credibility to their 'threats' of stricter regulatory action enough credibility to elicit credible enough to elicit co-operation from diverse interests affected.

What is necessary is effective political pressures on and from the Commission and from the outside through the European Parliament, the member states, and the relevant social and economic interests. EU policy making is, normally, an intergovernmental affair (even with the changes in voting rules and procedural changes regarding input and role of European Parliament are changing the decision making dynamics and, therefore, the relationships among the different EU institutions). Political commitment is needed from the Commission - reflecting its collective reading of the political tea leaves - to generate and sustain pressure for more integrated and co-ordinated policy developments within the commission bureaucracy. The Commission, working through its DG's, can make the problem visible by providing the fora and other mechanisms to bring together the different interested parties to generate substantive inputs to the problem definition together with the social and political support to carry the search for solutions. In this sense, the Commission is not a completely passive actor, merely responding to demands from outside; it, too, can take actions to help generate the pressure and support needed to create the political context favorable to the development of integrated policy in this area. An equally important question in this regard is why various sectoral interests are likely to be susceptible to these appeals for support and interested in considering collaborating in the search for joint solutions.

For one thing, the agricultural community is no longer the impregnable fortress it once was, and together with its associated interests, should be more responsive to read the handwriting on the wall from market changes and environmental concern. Richardson has described the European farmers as:

> *perhaps the most effective industry in terms of EU lobbying generally, securing enormous financial support via CAP. Moreover, they have their own DG (VI) which has in the past been adept at defending its clients.* (Richardson, 1995: 160)

However, as he point out, there is evidence of a gradual erosion of this enormous power. The objective of sustainable agriculture has become a real policy and behavioral option and it may be that COPA's fall back position might be to press for some form of compensatory payments for farmers. In any case, there exists at the moment a perhaps unique 'window of opportunity' for redefining the issue to bring together a new configuration of interests and to search for common ground. The farm lobby is normally too strong; now it is under attack and, even more importantly, itself engaged in a critical re-evaluation of accepted practices. Consequently there is an opening for policy change and the inclusion of environmental considerations in the process of restructuring. The same is true of some of its traditional allies.

According to Richardson the fertilizer industry is in serious economic trouble and has become preoccupied with the danger of irreversible economic decline. It, too, has come to realize that the industry itself must develop solutions to nitrate problem. Although not plagued with the same economic problems, the pesticide industry also faces a number of threats in the water field.

The point is, that there are developments that indicate a potential for a more co-operative attitude on the part of traditional 'adversaries'. Collaboration between DGXI and DGVI within the Commission could build on these changes in attitude and potential convergence of interest in mobilizing a joint search for appropriate measures to reduce agricultural pollution of water supplies.

4.3.4 FINANCIAL CAPACITY

The general strategy that the Commission presently favors - combining general EU-wide quality objectives with implementation by member states - looks, at first glance, a lot like the regulatory federalism approach in the United States. There the EPA offers the individual states the chance to administer the national regulatory programs within their own jurisdictions. To do so, the states must submit an implementation plan that is acceptable to the EPA. If the plan is approved, the national government will provide financial assistance to the states for the costs incurred in implementing a national program. If the state is not interested in taking on the executive tasks, or if the implementation plan submitted does not receive approval, the EPA retains responsibility for the implementation of its programs.

In the case of the EU, there is no EU competence for implementing its own policies directly, through a European administration; implementation is in the hands of the member states, with the Commission reduced to a the role of enforcer with limited capacities for doing so. Nor does the EU possess substantial financial resources with which to entice reluctant member states into taking on executive responsibilities. There are, to be sure, limited resources available through the structural funds and the newer Cohesion Fund, which can be used to promote environmental policy objectives. Furthermore, in our case, their is CAP money that has been earmarked for accompanying measures in connection with agricultural reforms. As we have noted above, these can be used for compensating farmers for restrictions on their farming activities. There are also limited funds for research and development, and demonstration projects that could be used in this connection.

The EEB, among others, had proposed that access to EU funds, limited as they may be, be made conditional on the development of effective programs for managing water quality in the member states. In this sense financial support from EU sources would be tied to environmental protection measures.

There are, in addition, different kinds of subsidies to agriculture for structural change. There appear to be interesting possibilities for making use of these fund to support and assist farmers in changing farming practices and promoting environmental objectives, without contravening the polluter-pays-principle with subsidies for 'complying with law'. As we have already noted, there are already various national compensation schemes to encourage farmers to adopt less environmentally damaging farming methods.

4.3.5 INFORMATION

If we define 'information' in the broadest sense, there are a number of supportive measures that could be taken by the Commission. The most obvious are in connection with its role in gather, processing and disseminating information on the different aspects of the problem. This information, which would have to be provided by national agencies or organizations and analyzed and published through the European Environmental Agency would document the measures taken in the different member states and the results achieved in dealing with the problem. What kinds of information is required, in what form and for what purposes, could be decided upon by the parties themselves. Agreement on data needs, and the methods for gathering and analyzing them, would go a long way in creating a body of both of information that is acknowledged as being both legitimate and reliable for decision making with regard to the problem. One of the critical factors in a guaranteeing the effectiveness of regional and local implementation activities will be a trustworthy and effective monitoring and control of the results of these measures. Here is where EU-wide standards and procedures can play an important role in ensuring both a level playing field and the transparency of the national efforts.

Again most of these activities would be carried out through co-operative efforts of different actors in the member states themselves. The Commission could play an important role in providing opportunities for EU-wide exchanges of results and opportunities for consultation and deliberation on different courses of action. In short, the European level would be the vehicle for generalizing national experiences, evaluating programs and drawing conclusions regarding necessary facilitating or even regulatory measures to support measures being taken.

The idea would be to develop through co-operation at the EU and Member State levels the regulatory, institutional and financial conditions to encourage joint research and development programs - between industry and government - on new farming practices (ecological farming, for example), or new products (biological or other non-chemical substances) in place of environmentally damaging pesticides.

An important supporting measures in this connection would be the development - on the basis of EU-wide standards - of 'agricultural extension services' to provide information and advice to farmers in connection with both everyday farming operations and more fundamental structural changes. These services could be organized in various ways as co-operative undertakings involving agricultural ministries, farming organizations, agricultural schools and, even, agro-chemical organizations. The idea would be to provide mutually recognized expert input into programmatic and operational decision making. These extension services could also be involved in developing training programs for farmers on integrated crop treatment and other more environmentally sound practices with regard to the use of fertilizers and pesticides. There is general agreement among the different parties that better training of farmers would lead to significant improvements in farming practice. There is a willingness on the part of both the farming organizations and the agro-chemical industry to co-operate in the development of better agricultural practice. The Commission could cash in on this willingness by creating the opportunities for bringing together both agriculture and environmental people to discuss what needs to be done in these regards.

These various co-operative - and, in the last analysis, voluntary - schemes could be both supported and 'encouraged' by more formal regulatory actions. For example, there could be provisions for licensing and certifying machinery and equipment for applying or even selling pesticides. If the industry itself proves reluctant to join in the search for replacements of present products, taxes and other charges could be introduced on chemical substances used in agriculture.

4.3.6 CONFIDENCE AND TRUST

Above we have noted how the Commission can act - as a networking organization and facilitator - to general social and political support for joint programs for dealing with the problem of drinking water quality and agricultural pollution. At the same time, and insofar as these efforts bear fruit, the experiences with these kinds of collaborative efforts and joint decision making will contribute to the building a mutual trust and confidence among the parties involved. As long as DG's 'represent' and defend their own constituencies, they will not succeed in getting out of confrontational mode. To the extent that the Commission relies on formal regulation, it will remained locked in continuing conflicts with the member states regarding the implementation and enforcement of its rules. In this sense, previous approaches to environmental management have tended to divide the interests involved. By stressing both the need and opportunities for joint efforts and shared responsibilities - and by creating the networks of actors through which these activities can be organized - chances are created for becoming familiar with the needs and positions of others and for developing a shared history of co-operative action. These are the kinds of experiences that lead to greater trust in the motives and objectives of one's partners. It does not mean that there will be no conflicts or differences of opinion, or that everyone will always be willing to go along and agree to a joint course of action. But it would appear to be feasible to construct of set of institutional arrangements within which to construct an advocacy coalition for sustainable agriculture and water quality, and through which these actors can participate in formulation and implementation of appropriate policy measures at the different levels of collective action.

4.4 Conclusions

The main point, with regard to strengthening the 'control potential' of the drinking water industry is to think in terms of contributions that can be made, by the EU - in the first instance, by the European Commission - in stimulating and facilitating co-operative action among these different sets of actor within the water quality and agriculture policy areas. In this sense, the water supply industry needs to be embedded in a broader policy network focused explicitly on the extended policy issue: 'agricultural restructuring and water quality management.' We have noted that there is a number of roles that the Commission could play in this connection.

There will, to some degree, be a need for additional legislation to provide the common regulatory framework and sets of quality objectives to be pursued within the member states. These common rules of the game would also protect integrity of internal market, prevent emergence of non-tariff trade barriers or contravention of other important policy principles. However, more important is the combination of pressure and opportunities for creating a new advocacy coalition to carry an effective policy commitment to the solution of the problem of agricultural pollution of water resources. The Commission of the EU has an important role to play in creating the conditions under which this new configuration of actors can emerge and work. In so doing, it can contribute to 'rescuing' the problem of water quality from the confines of technical and scientific debate on quality parameters and exposing it once again to the broader political debate on general principles and strategy.

In all of this, the central task of the Commission is that of creating and managing the networks of actors through which this advocacy coalition takes shapes and operates. The Commission is 'in the business' of inter-relating existing and creating new networks, of providing the raison d'être (policy issue) and mechanisms for bringing together the array of actors and interests needed to deal effectively with the problem at hand.

References

Commission of the European Communities [CEC] (1993) *Fifth Environmental Action Programme. Towards Sustainabilit'* Brussels: Commission of the European Communities (*PB C 138, 17.5.93*).

European Environmental Bureau [EEB] (1995) *EEB Comments on EU Water Policy.* Brussels: European Environmental Bureau.

Hildebrand, P. (1994) *Compliance in International Environmental Politics. The Case of the European Union.* Oxford: Ph.D. Thesis Oxford University.

Kallis, G. (1995) *Assessment of EU water quality policy,* Luxembourg: European Parliament, the STOA Programme (EP 165.512).

Kraemer, R.A. (1993) *Water Policy Report.* Bonn: Institut für Europäische Umweltpolitik.

Richardson, J.J. (1995) EU Water Policy: Uncertain Agendas, Shifting Networks and Complex Coalitions. In: J.Th.A. Bressers, L.J. O'Toole, Jr., and J.J. Richardson (eds.) *Networks for water policy. A comparative perspective,* pp: 139-167. London: Frank Cass.

Scharpf, F.W. (1973) *Planung als politischer Prozess: Aufsätze zur Theorie der planenden Demokratie.* Frankfurt am Main: Suhrkamp.

Interviews

Mrs. D. Dejonckheere, Committee of Agricultural Organizations in the EU (COPA);

P. Gammeltoft, European Commission, DG XI (National Expert Water Protection, Coastal Zones, Environment and Tourism);

Mr. Geffarth, European Chemical Industry Council (CEFIC);

P. Godin, European Commission, DG XI (Water Protection Unit);

F. Marien, European Commission, DG XI (Nitrates/pesticides in water);

Mr. Oosthuizen, European Crop Protection Association (CPA);

F. Rillaerts, Union of National Associations of Water Suppliers (EUREAU), Secretary General;

M. Scheele, European Commission, DG VI;

I. Steen, European Fertilizer Manufacturers Association;

A. Stokkers, Member of the Economic and Social Committee and Representative of the Agricultural Organizations in Brussels;

Mrs. K. Taschner, European Environmental Bureau (EEB), Scientific Advisor;

Mr. Vlahodimos, European Crop Protection Association (ECPA).

5. POLICY FRAMEWORK - GERMANY

RALPH PIOTROWSKI AND R. ANDREAS KRAEMER
Ecologic, Centre for International and European Research, Berlin, Germany.

5.1 Introduction[1]

Germans regularly refer to drinking water as *Lebensmittel*, meaning literally 'a means of life', also encompassing all food staples, such as bread and butter. The term indicates that water is essential for human life, and assigns it the same degree of importance as products of agriculture and animal husbandry. Furthermore, water resources are understood to be part of *natürliche Lebensgrundlagen*, the fundamental life support systems provided by the natural environment, and the natural preconditions for any life to exist. The use of this term is an indication of the significance attached in Germany to the protection and management of water as a natural resource, which is referred to as *Wasserwirtschaft* or 'water management'.

For the English word 'agriculture', the German term *Landwirtschaft* is used, a compound noun similar to *Wasserwirtschaft*. This literally means 'land management' and is much broader and has richer connotations than the rather narrow translation of agriculture as *Ackerbau*. *Landwirtschaft* is about maximizing the long-term carrying capacity of the natural environment and, like water resource protection and management, it is perceived as an integral part of landscape development and management. As a result, the role of the farmer, or *Landwirt*, in landscape management is traditionally very complex and places food production in the wider context of managing the natural environment and enhancing its function of sustaining life.

This traditional perception of farming as the management of rural areas has faded somewhat with increasing urbanization and industrialization. Nevertheless, it still has an

---

1 This case study on water policy and agricultural policy in Germany builds on previous work carried out at the Institute for European Environmental Policy in Bonn (Rüdig and Kraemer, 1994; Ganzert, 1994). A literature survey revealed a dearth of relevant publications (the most comprehensive source is Conrad, 1992). Accordingly, materials published by all relevant actors were evaluated, and interviews were carried out at the federal and *Länder* levels.

Geerten J.I. Schrama (ed.), Drinking Water Supply and Agricultural Pollution, 93-122.

influence on expectations, preferences and priorities, and therefore also on political decisions. Water resource protection and management as well as farming are activities traditionally regarded as one and the same, and until recently this was reflected by the fact that the ministries of agriculture had legal control of water management. This also changed in the modern era, when the protection of water resources against pollution became a dominant issue (see below). At the local level, the importance of joint management of water and land is still underlined by the existence of many *Wasser- und Bodenverbände*. These are associations of landowners (and others) created for the purpose of 'water and soil management'. Their function is also called *Kulturbau*, literally 'culture construction', highlighting the impact of bio-regional conditions on social institutions, and the importance of *Kultur*, as distinct from the urbanized *culture* in French, in maintaining the customs connected with the long-term management or sustainable development of the landscape. It follows that the bio-regional variations within Germany are reflected in the makeup of its federal state institutions which, as explained below, provide the political framework for water resource protection and the regulation of agricultural activities.

5.1.1 ENVIRONMENTAL AWARENESS

In Germany, water pollution was not always perceived as a striking problem requiring immediate political action. In the 1950s, public environmental awareness was still underdeveloped and the reconstruction of post-war Germany was given highest priority. Although scientists had already begun to discuss problems of agricultural water pollution, it took nearly twenty years until the predicament of water resources was recognized outside the scientific community. The general public became increasingly aware of environmental problems during the seventies (Conrad, 1990: 57). In the following decade, groundwater pollution by agriculture eventually became an issue of political salience, especially with regard to drinking water pollution. Public attention focused particularly on high concentrations of nitrates present in both drinking water and food (Conrad, 1992: 64). Today, agricultural water pollution is a widely acknowledged environmental problem.

The intense press coverage of the planned revision of Council Directive 80/778/EEC concerning drinking water demonstrated how carefully the issue of drinking water quality is treated in Germany.[2] The fear of lower quality standards caused a public uproar to which the European Commission reacted with astonishment. Greenpeace succeeded in mobilizing thousands of citizens to send signed postcards to the German Ministry of Agriculture and to the European Commission. Nearly every proposal for revision was discussed and commented on in the German press. But not only environmental concerns were articulated in the debate. The agro-chemical industry criticized the Drinking Water Directive for its rigidity and its unscientific basis. They were referring to standards developed by the WHO, and demanded that individual standards be set for each type of pesticide. Environmental organizations were accused of

[2] Official Journal of the European Communities L 299 from August 30, 1980.

profiting from a 'business with fear' by giving the public the impression that drinking water in Germany was of poor quality.

Today, agricultural water pollution is a controversial and widely discussed political issue. The complexity of these issues involves a broad range of actors, ranging from government officials, industries, and interest organizations, to NGOs and, ultimately, the consumer.

5.1.2 GERMAN POLITICAL FRAMEWORK

Political participation, access to decision-making bodies, and the establishment of networks are closely related to the formal political system of a member state. Due in part to the late establishment of a parliamentary democracy in Germany, its political system differs in many ways from the Dutch or British systems. The most outstanding features are the concepts of *Rechtsstaat* ('rule of law' or 'legalism') and federalism.

The notion of a *Rechtsstaat* is based on the idea that the only state is entitled to enforce laws, which are passed to regulate the relations between the state and citizens (Hesse and Ellwein, 1992: 380). The state is able to impose duties on its citizens only if it is so empowered by law. Policy making is not mere formulation and implementation, but must always conform to existing legal regulations. As a consequence, actors have a much more limited scope of action at their disposal. Their interrelationship is governed by laws and other legally binding rules. New policies frequently fail at the policy formulation stage, not because they are politically unacceptable or do not pass an economic cost-benefit test, but because they are incompatible with established legal structures. Over the years an enormous body of laws, rules, and regulations relating to water has been built up (Rüdig and Kraemer, 1994: 54). A collection of only the most important legal texts relating to water easily fills six large loose-leaf binders (Wüsthoff and Kumpf, 1994).

In Germany, legislative powers are divided among the federal parliament (*Bundestag*), the second federal chamber (*Bundesrat*) and the parliaments of the sixteen states (*Länder*). Executive and enforcement powers are divided between the federal government and the governments and authorities of the sixteen Länder. The federal constitution, known as the Basic Law (*Grundgesetz*) grants all legislative powers not already granted to the federal level to the Länder. That means that the Federal Republic of Germany, the *Bund*, is allowed to legislate only in the fields explicitly named in the federal constitution. The federal constitution distinguishes among three main types of legislative competencies:

- Exclusive competence (article 71): in this fields only the *Bund* enjoys legislative competence. The Länder are allowed to pass legislation only if they are so empowered under federal law.
- Concurrent legislation (article 72): the Länder are only allowed to pass legislation as long as the *Bund* did not do so. The same article lays down certain requirements which have to be met by the *Bund* to prevent it from taking over all legislation in these fields.

- Framework legislation (article 75): the framework legislation is a special case of concurrent legislation. The *Bund* has the right to pass framework legislation with regard to the requirements for concurrent legislation.

The constant struggle between the *Länder* and the *Bund* is one of the most important characteristics of German politics. The *Länder* passionately strive to maintain their competencies and regard Federal and European legislation as a threat to their autonomy and statehood. Foreign relations falls into the exclusive competence of the *Bund;* the *Länder* are not supposed to exercise any influence. Since European legislation strongly affects domestic policy, the *Länder* were able to force the *Bund* to consult them on issues of EU policy (Müller-Brandeck-Bocquet, 1992: 162).

The constitutions of the sixteen *Länder* as well as the federal constitution (article 28-2) also guarantee local authorities (cities, towns, and rural districts), collectively referred to as municipalities (*Kommunen*), the right to self-government within the confines of the law. Municipalities are constituent parts of the *Länder*, rather than the Federal Republic of Germany, the *Bund.* Municipal autonomy includes the responsibility for the local environment and providing vital services on their territory. Water supply and sewerage services are part of the services that must be available to satisfy the basic needs of society (*Daseinsvorsorge*).

The *Bund* has made extensive use of its competence to pass concurrent and framework legislation. Two main arguments are usually brought forward to shift competencies from the state to the federal level. First, to make policies more efficient, and secondly, to ensure comparable standards of living throughout Germany (Böhret, Jann and Kronenwett, 1988: 98). Cooperation among the *Länder* and the cooperation between the *Länder* and the *Bund* are often referred to as the 'third' and the 'fourth level' of the political system respectively. The *Länder* usually establish joint working groups or sign state treaties to regulate policy areas that still fall into their competence (e.g. water policy, educational policy). Similar cooperation and coordination takes place between the *Länder* and the *Bund* especially of the correspondent state and federal ministries. In the literature these developments are described as 'unitarian federal state' or 'cooperative federalism'. Fritz Scharpf describes the interlocking decision-making process in Germany as 'horizontal and vertical *Politikverflechtung*' (Scharpf *et alii* 1976; 1977). The horizontal and vertical *Politikverflechtung* follow the same logic. Decisions that have consequences beyond state boundaries should not simply be made by majority vote at the higher level. Instead they should (as in the vertical *Politikverflechtung*) be decided on a higher level only with the consent of all subunits, or (as in the horizontal *Politikverflechtung*) by the subunits themselves (Scharpf, 1992: 12).

Networks usually expand into these two dimensions, the vertical and the horizontal. Actors have to be identified on the state as well as on the federal level. The networks might differ considerably, depending on the level at which they concentrate. Networks of state-level organizations might pursue different interests than would corresponding networks of federal groups. When analyzing the networks, it must be taken into account that the policy of an interest organization is not automatically defined by the federal umbrella organization. Conflicts often arise between federal umbrella organizations and their subunits, especially if the former promote solutions which are sensible from a federal point of view, but are perceived as disadvantageous by the states.

5.1.3 DIFFERENCES IN WATER AND AGRICULTURAL POLICY

Water policy and water law, notably policy on water quality, as well as landscape planning and nature conservation (Kraemer and Jäger, 1997) are the responsibility of the *Länder*. In relation to water management in general, the Basic Law empowers the *Bund* merely to adopt framework legislation (Kraemer and Jäger, 1997: 22ff). Because agricultural policy is a traditional domain of European legislation, member states can only implement directives and enforce regulations. Consequently there are very few instruments of agriculture policy-making that can be used to influence farming practices at the member state level. This seems to be a major reason why financial compensation for farmers is the most commonly applied instrument in Germany for improving farming practices. But buying the 'good behavior' of farmers is turning the 'polluter-pays-principle' on its head.

The issue of compensating farmers for costs due to restrictions imposed for groundwater resource protection purposes, whether in water catchment zones or more generally, is suspended at the border of environmental and agricultural policy. In Germany, responsibility for groundwater protection measures and many other aspects of environmental policy is given to the *Länder*. The *Länder* also have competence in the administrative implementation of agricultural policy, and the German Farmers Association (*Deutscher Bauernverband e.V.*) has devolved responsibility for the issue of compensation payments in water protection zones to its regional subunits, associations within the *Länder*. The otherwise very detailed Report of the Federal Government on the State of the Agricultural Sector 1996 remains silent on the issue of farming in water catchment areas and compensation payments received by farmers facing restrictions on agricultural practices (BML, 1996).

Furthermore, because public drinking water supply organizations are the intended beneficiaries of groundwater resource protection, the water industry plays a vital role, at the local level in a practical sense and at federal or *Land* (provincial) level politically. The water industry in Germany is organized at the municipal level and includes a large number of individual suppliers (currently about 6,300 in the western *Länder* and about 2,300 and rising in the eastern *Länder*). Therefore there is a wide variety in the legal framework, administrative implementation, and even in the content of groundwater protection policy and agricultural policy.

Furthermore, there are significant differences in attitude as well as approaches to protecting water resources among the various *Länder*. Obviously, therefore, an investigation of networks at this level is necessary. Because it would have been impractical to study all sixteen *Länder*, after initial interviews it was decided to focus on just two, Bavaria and North Rhine-Westphalia. A second round of interviews was carried out in these *Länder*; the results for Bavaria remain inconclusive, however. The public policy of a third *Land*, Baden-Württemberg, which pioneered the payment of compensation to farmers in water protection zones for restrictions on otherwise normal farming practices (see Kraemer, 1993), is fairly well documented in the literature, although not in relation to policy networks.

North Rhine-Westphalia is the most populous and economically powerful *Land*. The federal government is currently located there, as are most of the actors who operate at the federal level. The nature of the interconnections between North Rhine-Westphalia and the *Bund* make it impossible to understand networks at federal level without reference to North Rhine-Westphalia. Bavaria was chosen because it provides a 'counter-model' to the approach taken by North Rhine-Westphalia; indeed there is a traditional and friendly rivalry between these two *Länder* when it comes to developing water policies and influencing federal policy formulation. Furthermore, Bavaria also has the means and the will to influence European policy directly, i.e. without necessarily involving the *Bund*. So far, it is the only *Land* to do so.

In this chapter, we have described the policy framework in Germany. Relations were classified among actors involved in public debate and consciously or unconsciously influencing policy outcome; coalitions of actors were identified and changes in the various networks observed. In the following section (5.2), we go on to examine the causes and the extent of agricultural pollution of drinking water resources. An outline of the present regulatory framework is given in section (5.3). Section (5.4) contains a description of the relevant actors and the policy networks. In addition to the federal level, the state and the local level are described by case studies of North Rhine-Westphalia and Bavaria. In the last section of this chapter, the discussion is summarized in concluding remarks. In the chapter below, the control capacity of the water supply sector is outlined and possible alternatives models for improvement are evaluated.

5.2 Agricultural pollution of drinking water resources[3]

In Germany, around 70% of all drinking water is extracted from groundwater. About 55% of the total area of Germany is used for agriculture(BMU, 1994: 32). Most of the water catchment areas are situated in agricultural areas. Pollution of these resources would constitute a major threat to the population's drinking water supply. This demonstrates the particular need for comprehensive groundwater protection strategies (BMU, 1994: 130). The reduction of nitrate and pesticide pollution of groundwater resources is seen by the German water policy community, including probably all water supply organizations, as the most difficult and most pressing issue in water pollution control. Until recently, the adverse trends of the past seemed to continue unabated and there was a fear that the level of nitrate pollution resulting from the excessive use of fertilizer had yet to reach its peak.[4] Legal regulations on agriculture and forestry,

---

3 This sections draws from a paper prepared by Frank Jäger at the Institute of European Environmental Policy in Bonn.

4 This fear was expressed by the President of the influential German Association for Water Resource Protection (*Vereinigung Deutscher Gewässerschutz e.V.*), Dieter Flinspach, on the occasion of the Association's general assembly in Bonn on September 3, 1992.

including the obligation to use fertilizers according to 'good agricultural practices', have not been able to significantly reduce the pollution load (Kraemer and Jäger, 1997: 93ff). In many cases the groundwater is negatively affected by intensive farming practices; partial amounts of fertilizers and pesticides or their metabolites are leached into the groundwater.

The precise amount of groundwater pollution is difficult to asses. As already mentioned, water management in Germany is primarily the responsibility of the *Länder*. In 1956, a joint *Länder* working group (*Länderarbeitsgemeinschaft Wasser* - LAWA) was founded to unify land water law and to discuss all matters of common interest (Kraemer and Jäger, 1997: 45ff). Since 1983, the LAWA has tried to develop a groundwater monitoring program for the purpose of generating comparable data for all of Germany. However, after more than ten years, there was still no comparable data on the state of groundwater pollution available (UBA, 1994: 438). In 1990, the groundwater monitoring program entitled 'The development of a unitary survey-system for groundwater quality'[5] for the eastern *Länder*[6] was initiated. During recent years, more sampling stations were installed to comply with the duty of Germany to report to the European Union. The 245 sampling stations (as of 1992) are not yet sufficient to supply representative data on all hydrological landscapes and areas (UBA, 1994: 445). Since 1990, the Federal Environment Agency (*Umweltbundesamt*) has collected data on the appearance, occurrence, and spreading of pesticides in groundwater. Thus the Federal Environment Agency is dependent on voluntary actions of the *Länder,* since they are not legally obliged to supply such data. Recent overviews do not seem to be available, however.

The Federal Ministry of Environment identifies agriculture as one of the main water polluters. They point out that the highest input form agriculture into water are caused by:

- the erosion and washing away of soils contaminated by phosphate and pesticides;
- leaching of soluble fertilizers (nitrate) and pesticides; and
- the release (and subsequent deposition) of ammonia gases from livestock holding areas and the spreading of slurry by means of splash-plate spreaders (BMU 1994: 41).

The following two sections compile the available data on nutrients (i.e. fertilizers) and pesticides. The data are given here as rough indications only and should not be taken to define actual levels of groundwater pollution.

5.2.1 NUTRIENTS

Unpolluted drinking water usually contains less than 10 mg/l of nitrates. Higher concentrations are usually the result of human activities (UBA, 1994: 444). In recent decades a continuous rise in the nitrate concentration of drinking water, especially in intensive farming areas, has been observed (see table 5.1). In nearly all regions of

5 *Entwicklung eines einheitlichen Grundwasserbeschaffenheitsmeßsystem.*

6 These are the *Länder* in the area of the former German Democratic Republic.

Germany, some drinking water sources had to be closed down because nitrate concentrations exceeded drinking water standards (BMU, 1994: 131).

TABLE 5.1: Concentration of Nitrates in Drinking Water in the Western *Länder*.

	1915	1926	1959	1975	1983	1989
Total number of samples	*278*	*909*	*984*	*5043*	*4462*	*7360*
under 20 mg/l[%]	98	88	81	75.5	74.5	74
20 to 50 mg/l [%]	2	9	15	21	22	23
50 to 90 mg/l [%]	-	2	2.5	3	3.5	3
above 90 mg/l [%]	-	1	1.5	0.5	0	0

Source: UBA (1994: 444).

Table 5.1 provides an overview of the rise of nitrate concentrations. The pollution load in the category 20 to 50 mg/l was continuously rising, not only in absolute terms but also by percentage. The share of the category 50 to 90 mg/l has stagnated or even decreased over the last twenty years. But taking into account that the total amount of samples has risen considerably, one has to conclude that drinking water quality standards were more frequently exceeded.

The potential nitrate load of groundwater is dependent on three factors:

- the potential nitrate load of the percolating water;
- the denitrification process in the water-bearing layer (aquifer); and
- total amount of time that the groundwater remains in the aquifer.

Groundwater nitrate contamination is the result of various kinds of human activities, not only agricultural operations. It is therefore important to asses the share of agriculturally-generated pollution. Distinctions should be made between:

- point sources of nitrate pollution: waste disposal, leaking sewage systems, small scale sewage plants; and
- diffuse sources of nitrate pollution: agricultural manuring, natural mineralization processes of organic substances, (*Deutsche Bauernkorrespondenz*, 1988).

Only rough estimates of the share of each pollution source exist. However, these estimates allow an assessment of the scale of single pollution sources and to determine where measures have to be taken most urgently (BMU, 1993: 3). Diffuse sources account for about 61% of total contamination, of which agriculture is responsible for about 90%, or 55% of the total share (UBA, 1994: 428).

The most striking aspect of the distribution of nitrate concentrations is the fact that concentrations near the limit value of 50 mg/l are continuously rising. A gradual increase in nitrate concentrations can be expected over many years even if nitrate pollution is halted. Because of this, a sizeable proportion of the groundwater resources currently used for drinking water production will have to be abandoned and their water mixed and diluted or subjected to extensive treatment. Drinking water supply and treatment costs might then rise, especially in areas where high levels of nitrate pollution coincide with high population density or a scarcity of suitable water resources. One notable area would be the *Land* of North Rhine-Westphalia and parts of Western Lower Saxony.

The pollution of groundwater resources by the agricultural sector has become a central issue of the water policy debate in the last decade or so, even though the causality of the pollution has been known for a longer time. Legislators and administrators thus continue to confront a new challenge, leading to various instances of policy innovation, as will be seen in the following sections.

5.2.2 PESTICIDES

In Germany, 200 agricultural pesticides and about 900 chemical substances that are dangerous for the aquatic environment are currently licensed. The negative impact of pesticides on the environment and human health is beyond doubt and according to the Federal Association of Gas and Water Industries, the use of these pesticides implies a potential health hazard (BGW, 1993: 12). The goal of the official policies of Germany and the European Union is to keep our drinking and groundwater free of pesticides. The allowable concentration of pesticides in drinking water was set at the level of the lowest detectable concentration. Recently this conviction was challenged by the agricultural lobby. Agricultural interest organizations promote the idea of setting quality standards for each pesticide type individually, according to alleged human health risks.

By the end of 1990, test results on samples were collected, ca. 14,000 provided by water suppliers, and ca. 35,000 samples from *Länder* authorities. In about 10% of all cases, biocides were found to be present.[7] A breakdown of positive results reveals the importance of the triazine family of agents, which also tends to be at the center of the political controversy about agricultural pesticides and the environment.[8] Table 5.2 gives an overview:

According to the groundwater monitoring program of the UBA, pesticides and their metabolites were detected in about 10% (about 19,000) of the 195,000 samples taken between 1986 and 1992. The quality standards set by the Drinking Water Directive were exceeded by 3% (about 6,300) of the samples (UBA, 1994:439). Due to a lack of information, the data could not be arranged by year or cumulatively indexed. The number of samples in which pesticides are found is continuously rising in absolute terms. However, as the total number of samples taken rises, the relative amount of samples with pesticides shows a slight relative decrease. This might be an initial sign of the success of the cooperation between agricultural and water industry, intensified

7 An independent study carried out by the Federal Health Authority (*Bundesgesundheitsamt*) found atrazine to be present in 14.5 per cent of samples, simazine in 17.3 per cent and one or more of a further 35 biocides were found in 8.2 per cent of samples (see Schleyer and Kerndorff, 1992: 213-215).

8 In the sense that pesticides were found to be present in a sample.

TABLE 5.2: Biocides in groundwater samples in (West) Germany 1990.

	Information provided by:	
Substance	*Länder* authorities	Water suppliers
Atrazine	35.3%	31.1%
Desethylatrazine[9]	27.4%	17.4%
Desisopropylatrazine	1.8%	1.9%
Simazine	11.1%	15.8%
Other triazines	3.3%	9.2%
Other active substances[10]	21.0%	24.7%
Number of Cases	3,800	1,394
of which more than 0.1 µg/l	1,700	438
of which less than 0.1 µg/l	2,100	958

Source: UBA (1992: 349).

efforts to monitor groundwater, and growing environmental awareness among pesticide users (UBA, 1994:441).

In 1995, Greenpeace conducted a survey on pesticides in the groundwater sources in Germany that resulted in a less promising conclusion. They sent a questionnaire to the about 500 municipal health agencies (*Gesundheitsämter*) asking for evidence of concentrations of pesticide in the groundwater from public and private wells between 1989 and 1994. The results of this survey demonstrate that pesticides are present in most of the groundwater sources that are used to obtain drinking water:

- Pesticides were detected in 260 districts *(Landkreise)* (comprising 61% of the total responses of the Lankkreise);
- about 103 health agencies reported that no pesticides were found (24% of the total responses of the health agencies); and
- 93 agencies did not supply any information (Greenpeace, 1995: I).[11]

Taking into account that water catchment areas are usually surrounded by groundwater protection areas, it is likely that the actual level of pollution is even higher. Greenpeace published two maps, one showing the level of pesticide contamination in untreated water from 1989 to 1994 (Greenpeace, 1994a), and the other showing the number of samples per district where pesticides were detected in raw water from 1989 to 1994 (Greenpeace, 1994b).[12] The Federal Association of German Gas and Water Industries (*Bundesverband der deutschen Gas- und Wasserwirtschaft* - BGW) organized a one day

9 Desethylatrazine is a metabolite of atrazine.

10 Organic chemical substances, including their main toxic metabolites.

11 In the *Landkreis* of Vechta, which is known for its intensive farming practices, the *Amt* replied that the material Greenpeace was interested in is typically incinerated.

12 See also: *Der Spiegel*, 2/1995, p.33.

discussion forum on the revision of the drinking water directive (BGW, 1994). During the discussion the methodology of the survey was repeatedly criticized (BGW, 1994: 86, 89, 94f). It was noted that Greenpeace did not use comparable data; they mixed data from local authorities with data on private wells, which they received at random from private individuals. Furthermore, the maps suggest that the groundwater of states which have established a functioning monitoring program like Bavaria is more polluted than the groundwater of the remainder of Germany. In any case, in the absence of a functioning monitoring program for the *Länder*, the study offers a preliminary indication of the present levels of pesticide contamination in groundwater.

5.3 Regulatory framework

This chapter discusses national pieces of legislation in the field of water, specifically concerning water quality in agriculture. European law is not discussed in detail (see Chapter 3 of this volume). This overview is devoted to a formal description of the laws, rather than an explanation of the relevant policy instruments that are made available by the legislation.

The current German water resource protection legislation has its roots in the late 1960s and early 1970s, even though the individual laws and ordinances have been subject to numerous amendments. In 1957, central water management law, the Federal Water Management Act (*Gesetz zur Ordnung des Wasserhaushaltes - Wasserhaushaltsgesetz*)[13] was adopted. The Act provides for many elements of qualitative and quantitative water resource management, like licensing of abstractions and effluent discharges. Secondary legislation based on the Federal Water Management Act adopted nation-wide sets uniform standards. These relate to processes (limit values for the amount of pollutants in industrial and municipal effluents), products (regulations on detergents, for instance) and a fiscal instrument, the *Abwasserabgabe*, or 'effluent charge' (Bongaerts and Kraemer, 1989; Kraemer and Bongaerts, 1991). The Federal Water Management Act also provides the basis for establishing water resource protection zones.

Most aspects of water policy, legislation, and administration are nevertheless situated in laws passed at the *Länder* level, transposing federal legislation where necessary. *Land* water laws stipulate rules of procedure and establish the organizational and institutional structure of water administration. Other *Land*-wide water legislation establishes rules concerning enforcement and water rights. There is great variation in the water legislation of the various *Länder*. This variation reflects differences in local tradition and specific water management issues stemming from specific climatic, geological, geographical and hydrological conditions. Attempts by the Federal Government to 'federalize' water policy - either directly, by attempting to amend the

[13] Of 27 July 1957 in the version of 23 September 1986 (BGBl. I, p. 1529 ff; corrected in BGBl. I, p. 1654 ff) as amended by Act of 12 February 1990 (BGBl. I, p. 205 ff).

federal constitution, or indirectly, by attempting to impose EC legislation - were fought vigorously by the *Länder* and largely failed.[14]

Of importance to agricultural causes of groundwater pollution is the fact that as early as 1957, the obligation to establish licensing procedures for all types of land use that pose a threat to soil and groundwater quality was contained in the Federal Water Management Act (Salzwedel, 1991: 196-197). However, this provision was not applied, even in cases of severe damage to groundwater resources through nitrates used in agriculture. In effect, the German water authorities have ignored a potentially powerful instrument for protecting groundwater.

In contrast, individual *Länder* have adopted what amounts to radically new approaches in dealing with water resource protection and consumption. Perhaps the most famous - or notorious - of these is the *Wasserpfennig*, or 'water penny', developed and adopted for the first time in the *Land* of Baden-Württemberg. This new concept involves a resource tax on water extraction with a conditional subsidy to farmers. It has stimulated the development of similar as well as different policies in other *Länder*, all aiming to provide for more effective water resource protection. Some of them are described in more detail in the sections on Baden-Württemberg, Bavaria and North Rhine-Westphalia.

More recently, the controversy opposing the agricultural sector to the goal of protecting water resources and drinking water has thus understandably intensified. The agricultural sector is now recognized as the single most important source of surface and groundwater pollution. The political salience of this issue is reinforced by the reality that many water suppliers in the new *Länder* of the former German Democratic Republic confront problems of nitrate pollution. The future of water services in these new *Länder* thus becomes a second key issue for German water policy.

5.3.1 WATER PROTECTION ZONES

Germany has developed a zoning approach to water resource protection which is increasingly being recognized as a model solution. The Water Management Act provides for the establishment of water protection zones (*Wasserschutzgebiete*) in order to:

- protect water bodies from harm, in the interest of the public water supply;
- recharge groundwater aquifers; and
- prevent pollution from run-off containing soil components, fertilizers and biocides.

---

14 One recent controversy involved Bavaria. The Bavarian government did not agree with the federal government's position that all of Germany is to be treated as a sensitive area in relation to the Municipal Waste Water Directive (91/271/EEC). Against the wishes of the Federal Ministry for the Environment, Bavaria excluded the entire Danube river basin from the sensitive area. Müller-Brandeck-Bocquet draws a different conclusion, pointing out that framework legislation, especially the federal water management act, does not leave the *Länder* enough administrative freedom to pursue their own interests (Müller-Brandeck-Bocquet, 1992: 168). Schodder goes so far to call framework legislation a sub-case of concurrent legislation (Schodder, 1989: 37).

Water protection zones must be identified in water registries. Activities may be restricted or even banned in water protection zones and property owners are obliged to tolerate measures connected with water protection, such as monitoring. These restrictions can be applied seasonally or all year round. Specific land uses may be imposed according to provisions in *Land* water legislation.[15] Precise restrictions are laid down for each protection zone through individual *Länder* legislation in the form of ordinances.

On the basis of water resource protection applying to ground surface area (*flächendeckender Gewässerschutz*) protection zones are usually divided into several zones around the water body or catchment facility:

Extended Protection Zone. Zone III should extend to the limit of the underground catchment area of a water catchment facility (i.e. the area from which water will eventually flow into the facility) and provide protection against long-distance pollution by persistent or radioactive substances. Not acceptable to water resource protection are such things as wastewater discharge, waste deposits, and agricultural activity (unless in compliance with the precautionary principle of groundwater protection).

Narrow Protection Zone. Zone II should extend from the catchment facility to the line marking the distance from which water takes 50 days to arrive at the facility via underground routes. This should secure protection against pollution by pathogen micro-organisms, which could be dangerous if the underground route was shorter. In addition to the installations, activities and measures regarded as dangerous in zone III, the application of dung, slurry are unacceptable in zone II.

Catchment Locality. Zone I includes the immediate environment around the catchment facility and is to assure its protection against any pollution or impact. In addition to agriculture, forestry and horticulture and the application of fertilizer, pesticides and other agro-chemicals are not tolerated.

Generally, water protection zones should cover the whole water catchment area. In accordance with *Land* law, adequate compensation must be paid for any economic losses resulting from restrictions on normal agricultural or forestry use of property (according to the law - *ordnungsgemäße Land- und forstwirtschaftliche Nutzung*).

In addition to the joint guidelines of LAWA and DVGW, LAWA has adopted Guidelines for Protection Zones for Medicinal Sources (*Richtlinien für Heilquellenschutzgebiete*), which have been enacted in ministerial circles in North Rhine-Westphalia and Schleswig-Holstein. These guidelines no doubt influence decision-making in other *Länder* as well.

---

15 See for instance Article 35 of the Bavarian Water Act (*Bayerisches Wassergesetz*) as adopted on June 1, 1994.

5.3.2 INFLUENCING FARMING PRACTICES IN WATER PROTECTION ZONES

Agricultural policy is one of the traditional domains of the European Union; very little legal control has been retained by the member states. Compensation payments are one of the few remaining instruments that can be applied at the state level and therefore developed to the most important instrument to influence farmers behavior. In the absence of the ability to pass and enforce economic legislation, the good behavior of farmers must be bought.

Obligations are almost always imposed by law, such as federal legislation on chemical use or *Land* ordinances establishing water protection zones. But contractual agreements on good practices can also be created between a farmer and a water supplier or private industrial water user. In some cases, the two approaches are combined, as is shown below.

Compensation payments can be made either by the *Land* government, a water supplier, or a private industrial user. The payment amounts can reflect the higher cost of farming actually incurred and justified. They can be based on general estimates or banding, or they can be made a lump sum payments. Again, the different methods of calculating and paying compensation may be combined. Finally, water suppliers have the option of buying agricultural land in protection zones around in their catchment areas. Agricultural land leases can also impose obligations and restrictions or even stipulate specific agricultural practices such as organic farming. The pluralistic structure of *Länder* legislation and of the water suppliers have produced a wide variety of arrangements for compensation.

Table 5.3 gives an overview of the diversity of arrangements that exists in the western German *Länder*. The second column indicates who is liable to pay compensation to farmers. In most cases it is the 'beneficiary' of land use restrictions, i.e., the water suppliers. Where there is no beneficiary, for instance where a protection zone is established to preserve a groundwater resource for future public water supply, the *Land* pays compensation. In most cases, compensation is calculated by estimating the loss of revenue or the additional costs incurred by farmers due to land use restrictions. One interesting point is that Bavaria, for instance, provides for compensation for restrictions imposed by a federal ordinance concerning the application of plant protection agents *(Pflanzenschutzmittel-Anwendungsverordnung),* whilst other *Länder* explicitly exclude such payments. In general, compensation is not available for restrictions imposed outside of water protection zones.

Administrative responsibility is distributed among various government authorities or agencies. The 'upper water authorities' are ministries of the environment or district governments; 'lower water authorities' are local authorities or district governments. When calculating or estimating the additional cost or loss of revenue a farmer is likely to incur in a water protection zone, one must have a point of reference. Logically, this would be the cost structure and revenue level of a farmer operating without specific obligations and restrictions but within the law, at least as far as environmental protection is concerned. Some *Länder* do provide a definition of good agricultural practices, even

TABLE 5.3: Arrangements for Compensation Payments in Western *Länder.*

Baden-Württemberg	The *Land*	310 DM/ha/y or compensation according to a detailed costing.	Upper water authority for restrictions, agricultural authority for compensation.	General definition of good practices. Limit of 45 kg N/ha at the end of growing season. Positive list of permitted substances.
Bavaria	Water supplier	Compensation for higher costs, compensation for restrictions imposed by federal legislation.	Local authorities (*Kreisbehörden*)	General definition of good agricultural practices.
Berlin	Water supplier	Contract between water supplier and farmer, otherwise by order of water authorities.	Water authority	None
Bremen	Water supplier	Compensation for higher costs.	Water authority	None
Hamburg	Water supplier	Compensation for higher costs, no compensation for restrictions imposed by federal legislation.	Water authority	None
Hesse	Water supplier or other user; or the *Land.*	Compensation for higher costs.	Upper water authorities	Specific restrictions on fertilizer use.
Lower Saxony	The *Land*	Compensation for higher costs.	Upper water authorities (District Government)	Definition will be agreed between Ministries.
North Rhine-Westphalia	Water supplier; or the *Land.*	Compensation for higher costs; lump sum compensation in exceptional cases.	District Government	None
Rhineland-Palatinate	Water supplier; or the *Land.*	Compensation for higher costs; lump sum compensation possible.	District Government, consultation of agricultural or forestry authorities.	Exists
Saarland	Water supplier; or the *Land.*	Compensation for higher costs; mediation possible.	Lower water authority.	None
Schleswig-Holstein	Water supplier; or the *Land.*	Compensation for higher costs.	Water authority (Ministry for the Environment)	None

Source: Information provided by the *Länder*arbeitsgemeinschaft Wasser (LAWA): LAWA-AG 'Grundwasserschutz'.

though these are in the minority. It would appear from this preliminary overview that not all *Länder* do so.

The size of payments to farmers falls into a considerable range (from 80 DM up to 300 DM per hectare and year). Furthermore, in some cases payments are subject to regular reviews and thus exhibit a degree of variation over time. A complete overview of the sums actually paid is not currently available. The method of calculating compensation payments varies among the *Länder*, depending on differing approaches and forms of payment. In particular, the economic effects of any restrictions imposed on farmers are assessed differently. To what extent these differences reflect unequal natural conditions cannot be ascertained in this investigation. Such other aspects as the distribution of political power between farmers and water suppliers, or among the political parties within the *Länder*, might also play a rôle.

5.3.3 ABSTRACTION CHARGES AND ADVICE TO FARMERS

An increasing number of German *Länder* have adopted or are adopting legislation imposing water abstraction charges (charges, levies or fees) which usually involves an amendment of the *Land* water law. A common feature in this legislation is that the revenue is reserved for specific purposes, including giving advice to farmers on agricultural practices that promote groundwater resource protection.[16] By the end of 1992, five Western *Länder*, Baden-Württemberg, Hamburg, Berlin, Hesse and Lower Saxony, had adopted such legislation. Hesse and Lower Saxony explicitly made part of the revenue available for providing advice to farmers, and this approach is also likely to be followed in many eastern *Länder* who are currently adopting new water legislation. No details of the implementation in Hesse and Lower Saxony are available at this time; in both cases the relevant legislation was adopted in the middle of 1992.

5.4 Policy network

The policy networks interfacing water policy and agriculture policy in Germany are described here, first exploring water policy and agriculture policy separately. Their relationship is then investigated and recent and possible future developments in the network structures are identified.

As an explanatory tools, actors are identified by their position on a two-dimensional matrix (Figure 5.1). The first distinguishing feature is their origin or current position in relation to water supply and agriculture. Some actors clearly belong to the 'water industry' and can be expected to defend its interests,[17] such as the Federal Association of Gas and Water Industries (*Bundesverband der deutschen Gas- und Wasserwirtschaft* - BGW). Other actors are clearly agricultural, such as the German Farmers Association (*Deutscher Bauernverband* - DBV); many actors fall in between.

16 An overview of the use of abstraction charges is given by Kraemer (1992).

17 The term 'water industry' is not used in Germany, where there are currently no dominant water and construction firms, as in France or the United Kingdom, who widely use the term. Instead, the term water 'economy' or 'management' (*Wirtschaft*) is used in Germany in relation to the operators of water supply systems.

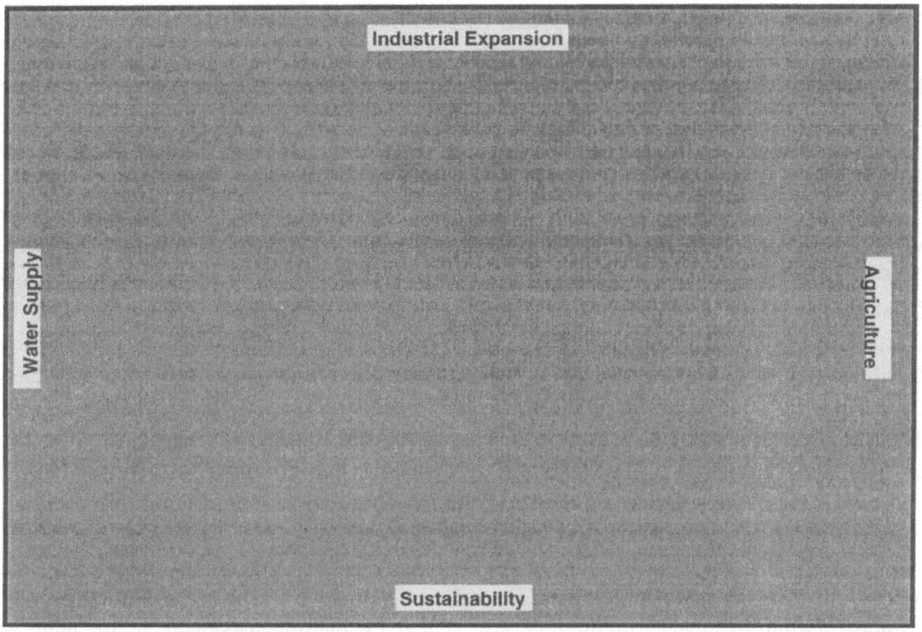

FIGURE 5.1: Two-Dimensional Matrix

As a second distinguishing feature, actors are located according to whether they tend to support ecological approaches ('sustainability') or defend industrial interests ('industrial expansion'). Ecological approaches tend to favor changing behavior rather than seeking technical fixes or integrating anthropogenic and natural material flows. The industrial approach is to separate these flows by technical means, and thus to increase the material and energy intensity of technical systems.[18]

In the water industry, there are such actors as equipment manufacturers who have an interest in increasing the material intensity of water systems. Opposed to them are actors interested in the ecological function of water in protecting wetland habitats who do not favor engineering solutions. In agriculture, for instance, the chemical industry, who seeks to shield the market for environmentally dangerous substances from regulation, stands in opposition to farming associations who promote organic farming as an 'ecological way of life'. Clearly, however, there are no parameters for a cardinal axis by which to place an actor in this matrix; we have used our own judgment to locate them. Differentiating actors according to their industrial or ecological approach is nevertheless a useful way to explain many features of the networks observed.

[18] Refers to a method of creating barriers characterized by high intrinsic values of the materials used, high energy and high pressure levels.

5.4.1 WATER POLICY NETWORKS RELATING TO AGRICULTURE

This analysis of water policy networks in Germany explores those segments that relate to environmental policy and, most importantly, agriculture.[19] It focuses on the protection of water resources for drinking water supply (in-stream or 'in-ground') and water supply (off-stream and 'in-pipe') itself, without ignoring other segments of the water cycle.

The most influential water policy actor in Germany is the joint *Länder* Water Working Group (*Länderarbeitsgemeinschaft Wasser* - LAWA) established in 1956. LAWA consists of the highest water authorities (Ministries of the Environment of each *Land*). This is specifically aimed at unifying *Land* water laws, but also intended to discuss all matters of common interest and to prepare joint reports. Within LAWA, the *Länder* agree on common objectives and monitor their implementation. They discuss issues that fall into their jurisdiction, develop solutions, and issue recommendations. They advise the regular Conference of Environmental Ministers bringing together the environmental ministers of the *Bund* and the *Länder* (*Umweltministerkonferenz* - UMK) and the Conference of Department Heads (*Amtchefskonferenz* - ACK) and execute their orders. The LAWA is thus a network of actors in itself.

Although it cannot issue binding regulations, one strength of LAWA lies in the power of its members resulting from their control over the development and implementation of water policy and law. LAWA's central weakness is its lack of permanent structures which make it difficult to establish and carry out long-term strategies. The presidency and the administration of LAWA rotate every three years. LAWA is not a very public actor in the sense that its actions are not normally communicated to or by the media. LAWA must build consensus among its members, or at least avoid a single veto. Its position thus is generally neutral, situated somewhere between 'industrial expansion' and 'sustainability'. It is neither an instrument for furthering purely industrial policy objectives, nor does it promote ecological policies in particular, because it must consider all water uses. Therefore, LAWA does not pursue the interests of water suppliers, but aims to integrate them with other users. However, the Federal Water Management Act (*Wasserhaushaltsgesetz*) makes the protection of water resources for drinking water supply a priority objective of water management.

Through its members, LAWA is represented on the boards, councils and executive committees of the three standard-setting associations in water management.[20] The first is the German Water Pollution Control Association (*Abwassertechnischer Vereinigung* - ATV), with more than 12,000 individual, industrial, and municipal members. ATV is active in sewage and waste management, sewage sludge and compost disposal, which brings it into contact with agriculture. ATV, whose members include many equipment suppliers, tends to promote engineering solutions to water pollution problems. It has

---

[19] For a broader description and analysis see Rüdig and Kraemer (1995). Rüdig and Kraemer do not discuss the internal structure of key actors in the field of water policy.

[20] The Water Standards Committee (*Normenausschuss Wasserwesen* - NAW) within the German Standards Institute (*DIN Deutsches Institut für Normung*) elaborates and publishes a wide variety of standards in all areas of water management. There appear to be no links, however, to agricultural policy.

However, ATV belongs neither to the side of water supply nor agriculture, although many of its municipal members are also involved in water supply.

LAWA is also linked to the German Association of Gas and Water Experts (*Deutscher Verein des Gas- und Wasserfaches* - DVGW), to which belong about 5,800 individual, corporative and municipal members. In relation to agriculture, DVGW's interests lie primarily in the protection of drinking water resources from agricultural pollution. Together with LAWA, DVGW has drawn up an information paper (*Informationsblatt*) 'Influence of land use and fertilizer application in water protection zones on the nitrate pollution of ground water'.[21] This paper serves as an explanatory memorandum to Standard W 101 on groundwater protection.[22] DVGW is also a member of the Würzburg Circle (*Würzburger Gesprächskreis*) involving water suppliers and water management authorities (see below). Together, DVGW and the Federal Association of Gas and Water Industries (*Bundesverband der deutschen Gas- und Wasserwirtschaft* - BGW) closely cooperate in many areas and have reciprocally appointed several members or guests to their respective boards. In order to further strengthen co-operation, DVGW will soon to relocate to Bonn and will share a building with the BGW (shown as a broad black line linking these two actors in Figure 5.2).

The third standard-setting association linked to LAWA is the German Association for Water Management and Landscape Development (*Deutscher Verband für Wasserwirtschaft und Kulturbau* - DVWK), which was originally an agricultural association.[23] With a membership of about 2,500 individuals, companies, scientific institutions, and authorities, DVWK has maintained an agricultural profile through its involvement in irrigation and drainage. To some, it is purely a water resource management association, while others view it as a specialized agricultural association. But DVWK, which has few industrial members, tends to promote ecological rather than technical approaches, particularly in relation to the development of water courses.

All three standard-setting associations have a number of characteristics in common. Together, they not only elaborate most of the relevant standards but also publish, in some cases jointly, the most important journals covering water resource protection and management (*Wasser + Boden*; *Wasserwirtschaft*), water supply (*gwf - Das Gas- und Wasserfach, Ausgabe Wasser* and *Abwasser*), and sewerage (*gwf - Das Gas- und Wasserfach, Ausgabe Wasser* and *Abwasser*; *Korrespondenz Abwasser*). They thus dispose of a major instrument for influencing the debate among experts and thus help shape the agenda. More importantly perhaps, the journals act as 'gatekeepers'.[24] Anyone who 'has been asked to contribute' to one of these journals will benefit from name recognition and the authority bestowed on someone who is thus accepted as an expert ('I

21 *DVGW - Wasser-Information no. 35* of 1993.

22 *DVGW - Regelwerk - Arbeitsblatt W 101* of 1975; amended in 1994.

23 And still listed as an association of agrarian sciences in the standard German manual of public life (Oeckl, 1993/94).

24 'Gatekeeper' not 'bouncer'; no one has the power to deny an individual recognition or access to all journals. Thus their role is to welcome experts into the community but not to expel them.

know him from the literature' is a common form of acknowledgment; and it is mostly a 'him').

Even mere membership offers similar benefits. It is not uncommon for water engineers to refer to one another as 'colleagues', in recognition of shared membership in ATV, DVGW, or DVWK (but mostly in relation to ATV). They may do so despite working for different towns or types of organizations, or having attended different universities. In this sense, individually and even collectively, the standards-setting associations form communities, though not necessarily policy communities. A common feature of all three associations and their journals is that they provide a bridge between the two main professions involved in water policy and management in Germany: lawyers and engineers of various vocations.

Internally, however, the associations are structured as formal networks. The executive directors of the associations have some influence but no power of the associations' positions and policies. The expert work on rules, guidelines and standards, the main influence on water management in practice, is delegated to numerous committees and sub-committees; the ATV has about 200. Whether permanent, temporary, or ad-hoc, each committee is supervised by a chairman who is usually an eminent expert with practical experience or a scientific background in the field in question. Other committee members usually contribute without pay and appear to be partly co-opted, partly delegated by interested parties, partly based on the personal interests. A balance of interests is sought in each case. A member of a *Land* Ministry of Environment, representing LAWA interests, normally serves in each committee developing standards that would affect the application of the law.

The associations and their staff are best described as permanent secretariats to these committees, facilitating their work without too much interference. Although typically weak in such decentralized structures, power is exerted by the associations' boards and councils in which the members' interests as well as government are represented. Influence is scattered over a large number and a wide variety of experts, notably those actively serving on the various committees and sub-committees.

LAWA and the above three associations, together with the Federal Association of Gas and Water Industries (*Bundesverband der deutschen Gas- und Wasserwirtschaft* - BGW), form the 'hard core' of the water policy network in Germany.[25] BGW's members consist of about 1,200 enterprises in gas and water supply, which include private and publicly owned companies active at the municipal, regional or national level. In contrast to the associations, BGW is a public entity and, as even its opponents acknowledge, its strength lies in its relations with the media as well as its understanding of the policies and politics of the European Union.

---

[25] A fourth standard-setting association is the Expert Group on Water Chemistry (*Fachgruppe Wasserchemie* - FW) in the Association of German Chemists (*Gesellschaft Deutscher Chemiker* - GDCh) with 1,400 individual members working in water chemistry and limnology

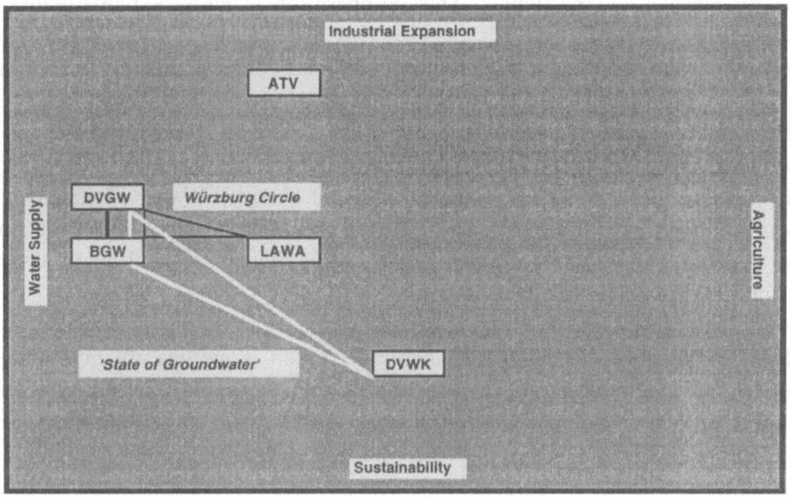

FIGURE 5.2: The 'Hard Core' Water Policy Network.
Note: Abbreviations and linkages as explained in text.

LAWA, BGW, and DVGW form the Würzburg Circle (*Würzburger Gesprächskreis*) between the water supply sector (BGW and DVGW) and the authorities responsible for water management and drinking water resource protection (LAWA). The Würzburg Circle has adopted a joint declaration outlining principles of cooperation which are directed at water suppliers and water management authorities and agencies. The aims are:

- to make relevant data and information available to both sides as needed;
- to improve implementation of legislation on water protection zones;
- to expand water protection zones to cover total water catchment areas;
- to impose restrictions on polluting activities, notably agriculture and commercial horticulture; and
- to restore polluted water resources (Würzburger Gesprächskreis, 1994)

Through either the ordinances establishing water protection zones (*Wasserschutzgebietsverordnungen*) or legislation concerning all farmers in a *Land*, the *Land* authorities are charged with forcing farmers to reduce farming intensity (i.e., use of agro-chemicals) wherever necessary, including, but not only in water protection zones. Water suppliers are encouraged to consider land purchases and subsequent leases to farmers with land use restrictions attached. The Würzburg Circle is shown in Figure 5.2 by thin black lines.

DVWK, DVGW and BGW co-operated in drawing up a guidelines on assessing the state of groundwater resources, BGW mainly through comments provided by its members (DVGW 1995). The paper, although formally neutral, constitutes a threat to the agricultural sector, as it contributes to publicizing the extent of groundwater pollution by nitrates and pesticides. This co-operation is identified in Figure 5.2 by broad white lines.

Other 'core' actors (not shown above) include the Federal Ministry of Environment, Nature Conservation and Nuclear Safety (*Bundesministerium für Umwelt, Naturschutz und Reaktorsicherheit* - BMU) and the Federal Environment Agency (*Umweltbundesamt* - UBA). The BMU deals with water resources management as a part of environmental policy; it has control over the federal Water Management Act (*Wasserhaushaltsgesetz*). It supervises UBA and the Institute for Water, Soil and Air Hygiene (*Institut für Wasser-, Boden- und Lufthygiene* - WaBoLu) in charge of monitoring drinking water quality.

Around this core group of actors, there are a large number of other actors, some of which form stable networks. One such network, which is gaining in importance as a result of the public attention paid in Germany by the media and the public to drinking water quality and environmental pollution by agriculture, and its willingness to expand beyond Germany, is an 'NGO Water Policy Network' co-ordinated by the conservationist German Association for Nature Conservation (*Deutscher Naturschutzring* - DNR) and shown in Figure 5.3. The activities of the network mainly involve discussion and co-ordination - where desired - of positions on policy issues of interest to all participants, notably European directives such as the proposed amendments to the drinking water directive, the directive on licensing and marketing of pesticides and the proposed directive on the ecological quality of surface groundwaters.

Indicated below are only those members most active and interested in water resource protection from agricultural pollution. Whether or not associations are active depends primarily on the interests of individuals. Only one core actor, the BGW, is involved part-time. Members otherwise fall into three categories. The first category includes environmental and nature conservation associations such as the Association for Environmental and Nature Conservation in Germany (*Bund für Umwelt und Naturschutz Deutschland* - BUND).[26] Another is the *Land* Association of Citizens' Initiatives for the Environment of Lower Saxony (*Landesverband Bürgerinitiativen Umweltschutz Niedersachsen* - LBU), or the political-ecologist Federal Association of Citizens' Initiatives for the Environment (*Bundesverband Bürgerinitiativen Umweltschutz Niedersachsen* - BBU).

All of these groups are opinion leaders on water policy issues in the environmental movement who contribute actively to the discussion and drafting of joint statements (indicated by broad white lines). Both Greenpeace and WWF of Germany take part in the network, but they do not always support joint statements (indicated by a thin white line).

---

[26] BUND and DVWK have reciprocal membership, indicated by a thin black line linking the two.

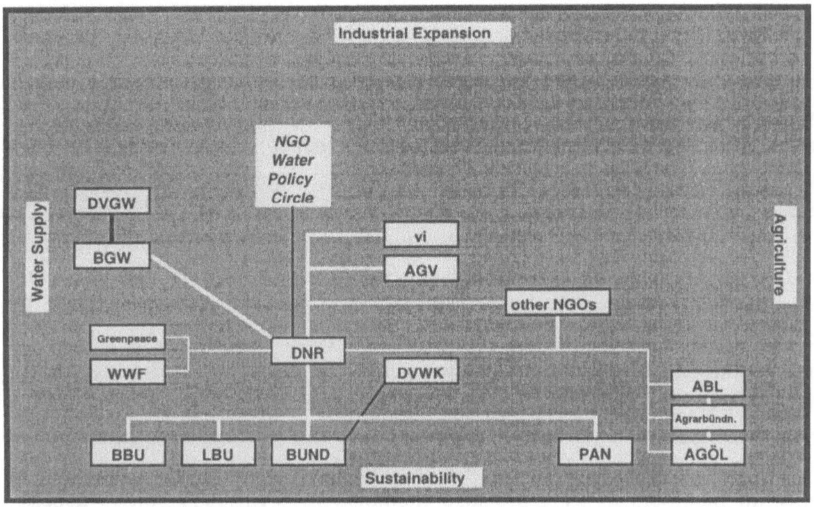

FIGURE 5.3: The NGO Water Policy Network
Note: Abbreviations and linkages are explained in text. Positions of some actors are determined by graphical clarity, and do not necessarily indicate other characteristics such as positions on policy issues.

The two most important consumer protection associations in Germany are a second part of the network. These are the Consumers' Initiative (*Verbraucherinitiative* - vi) and the Task Force Consumer Protection (*Arbeitsgemeinschaft Verbraucherschutz* - AGV). A third category comprises links to associations promoting organic farming, such as the Task Force for Organic Agriculture (*Arbeitsgemeinschaft Ökologischer Landbau* - AGÖL), an umbrella organization of producers' associations and marketing organizations. Also included is the Task Force for Traditional Farming (*Arbeitsgemeinschaft bäuerliche Landwirtschaft* - ABL). ABL issues the annual 'Critical Report on Agricultural Policy' as a response to the official annual report by the Federal Ministry for Food, Agriculture and Forestry (*Bundesministerium für Ernährung, Landwirtschaft und Forsten* - BML), and the Agrarian Federation (*Agrarbündnis*), itself a network of associations promoting environmentally friendly farming.

In the future, the involvement of these groups is likely to grow (indicated by thin white lines). In addition, there is the environmentalist Pesticides Action Network (PAN) which is mainly active in agricultural policy. Beyond the two described here, there are many more networks in water policy, which deal with a range of issues such as taxation, but none attempt to influence agricultural policy.

5.4.2 AGRICULTURE POLICY NETWORKS RELATING TO WATER MANAGEMENT

Compared to the water policy networks described in the previous section, the agriculture policy networks appear relatively simple, and consist of a few powerful actors. The most important is the German Farmers Association (*Deutscher Bauernverband* - DBV), the umbrella organization of regional and *Land* farming organizations in which almost all farmers in Germany are organized. Because of the dominance of the European Common Agricultural Policy (CAP), the role of the *Bund* and the *Länder* in policy formulation is much smaller. There is no agricultural equivalent to LAWA and the most important government wing is the Federal Ministry for Food, Agriculture and Forestry (*Bundesministerium für Ernährung, Landwirtschaft und Forsten* - BML).

One of the most prominent 'public' actors in terms of media exposure (often negative) is the Agro-Industrial Federation (*Industrieverband Agrar* - IVA). IVA is the lobbying organization of pesticides manufacturers and operates from the same address as the Federal of Chemical Industries (*Verband der Chemischen Industrie* - VCI). Probably all members of IVA are also members of VCI. IVA enjoys access at the highest political level (the City of Ludwigshafen, home of the chemical company BASF, is the constituency of Chancellor Helmut Kohl). It appears to operate aggressively behind the scenes - so aggressively in fact that some in the chemical industry have privately expressed reservations about IVA's methods. IVA and DBV jointly support the Federation for the Promotion of Integrated Plant Protection (*Förderverband Integrierter Pflanzenschutz* - FIP) which develops and promotes 'low-chemistry' agricultural practices.

On most issues, agreement among all these actors is such that it is often impossible to distinguish one from the other (indicated by broad black lines). They appear to co-operate very closely, exchange board members, and rely on the same experts.[27] This network structure is reproduced both at the European level, where many functionaries hold equivalent positions in respective European Associations, and at the *Land* level, most notably in North Rhine-Westphalia, home of both the president of DBV and its chief environmental policy expert.

Strictly speaking, BML and DBV should be neutral towards both industrial expansion and sustainability. In fact, they are not. Chemical-intensive farming methods are still widely favored in the agricultural sector, as the close link between DBV and the chemical industry shows. So far, this does not seem to erode the membership base of DBV, even though there appears to be a trend towards dual membership in one of the more environmentally-conscious farming associations.

[27] The president of the DBV is on the supervisory board of the chemical company Bayer.

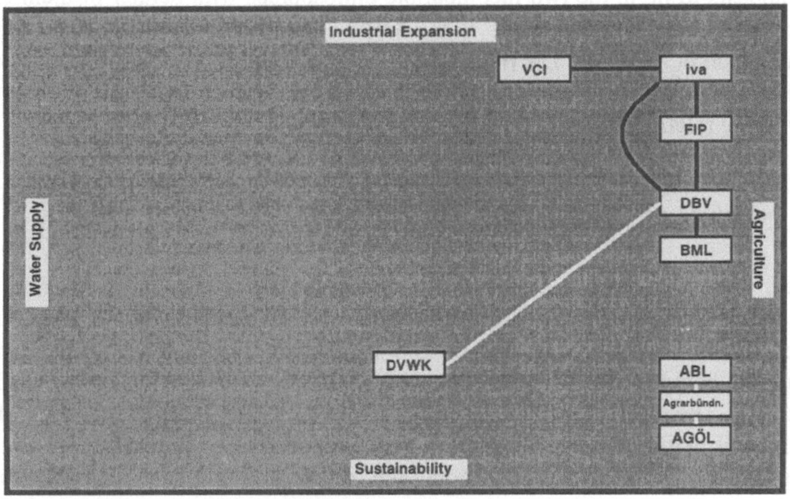

FIGURE 5.4: The Agriculture Policy Network
Abbreviations and linkages as explained in text.

DBV co-operates on specific issues with the German Association for Water Management and Landscape Development (*Deutscher Verband für Wasserwirtschaft und Kulturbau* - DVWK) as indicated by a broad white line. Areas of co-operation include the maintenance of river and stream banks by farmers in accordance with water resource management requirements, an area encompassing water management, nature conservation, forestry, and farming. There is more significant co-operation in developing guidelines for pesticide use.

There appears to be no communication between the actors represented in the top right corner of Figure 5.4 and the associations promoting organic or traditional farming shown in the bottom right corner. The former regard the organic farming associations as marginal, in an economic as well as in a political sense. This assessment is borne out by comparative statistics of cultivated acreage and by the obvious inability so far of those promoting organic farming to influence agricultural policy. The organic farming associations complain about difficulties in obtaining an appointment in the BML. In part, however, these associations are suffering from success: public opinion of them is excellent and their membership is growing fast. Even critics can only claim that organic farming, with its use of dung or manure as part of the nutrient cycle on the farm, is not suitable for zones I and II in water protection zones and that the ploughing of grassland as part of crop rotation can induce nitrates to be flushed into ground water. The latter problem is only acute on land that has been farmed intensively and is then converted to organic farming.

Nevertheless, there appears to be a barrier between the agricultural actors promoting industrial expansion and those favoring sustainable development. A barrier of this kind does not exist between the various actors in the water policy field. The reasons for this barrier appears to lie in the two incompatible philosophies with neither of them leaving much room for the other to exist. It is unclear, however, why there is no apparent communication between the DVWK and the organic farming associations, other than a reluctance of DVWK to become involved in political conflicts and thus to threaten its good working relationship with the DBV.

5.4.3 INTERFACES AND OVERLAPS

It is clear from interviews that there is no interconnected network covering both water policy and agricultural policy. From the perspective of the DBV there is an 'unholy alliance' joining LAWA, BGW, and DVGW (Würzburg Circle) and even more sinister co-operation between BGW and Greenpeace (BGW often invites some critics to conferences). The discussions and co-ordination in the NGO water policy network only feeds this suspicion. Some members of the environmental movement see collusion between BGW and DBV in the promotion of integrated plant protection and in supporting the removal of the sum parameter of 0.5 µg pesticides per liter of water from the European drinking water directive.[28]

The two preceding sections demonstrated that the DVWK enjoys a pivotal role as a member of both the hard core of the water policy network and, albeit much less central, the agriculture policy network. Through its successful co-operation with DBV, the DVWK has built a bridge where other have failed. In the 1980s, DBV worked with ATV on sewage sludge disposal on agricultural land, but this failed, largely because no agreement was reached on the issue of liability of environmental damage, notably in case of changes in the regulatory regime. DBV also worked with DVGW and LAWA on the use of fertilizers in water protection zones, but this co-operation also ended without results.[29] By contrast, the co-operation with DVWK on the relatively uncontroversial issue of the maintenance of riverbanks has apparently encouraged both associations to jointly undertake the much more difficult task of drawing up guide-lines for the use of fertilizers.

In effect, DVWK is transporting the experience of standard-setting, which is covered itself by a standard guaranteeing public participation, from the area of water management to the area of agriculture. In the agricultural field, the sanctity of private property was always important,[30] and it appears that this might be an important reason for the inability of agricultural associations so far to draw up and mutually enforce

---

28 On 22 December 1994, BGW and DBV issued a joint declaration supporting the limit value for all pesticides of 0.1 µg/l as long as there was no scientific basis for individual limit values (*Frankfurter Rundschau*, December 23, 1994).

29 This is the reason why the information paper "Influence of land use and fertilizer application in water protection zones on the nitrate pollution of ground water" (*DVGW - Wasser-Information no. 35* of 1993) is not a standard as was originally intended but merely an explanatory memorandum to the standard W 101 on groundwater protection (*DVGW - Regelwerk - Arbeitsblatt W 101* of 1975/1994).

30 See Bruckmeier and Teherani-Krönner (1992).

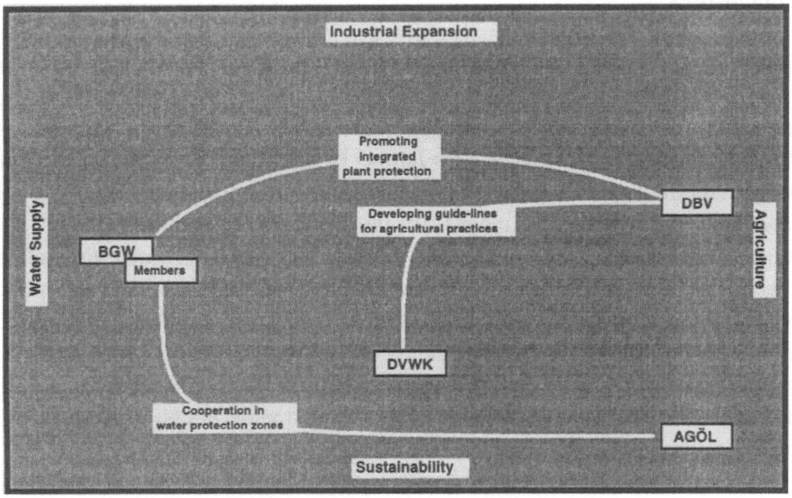

FIGURE 5.5: Cooperation Between Water Suppliers and Farmers
Note: Abbreviations and linkages explained in text.

guidelines on farming practices respecting the environment. Water resources, by contrast, have long been regarded as a resource in common ownership to be protected and managed collectively. It follows naturally, that rules and standards regarding its protection and management can and must be developed and mutually enforced.

The water management network that constitutes the DVWK has developed the instruments and mechanisms for drawing up rules and standards and thus has a social methodology not currently available to DBV. For some reason, it appears to be easier to DBV to co-operate with DVWK rather than with organic producers' associations, which have also developed mechanisms for the mutual monitoring and enforcement of agricultural practices among their members. This underlines the strength of the barrier between organic farming associations and the actors involved in chemical-intensive farming.

This barrier certainly exists at Federal and *Länder* level, albeit with some exceptions.[31] It does not exist at local level, where numerous examples of co-operation between water suppliers, farmers and organic producer associations can be found. Some have been documented (see: Hess et al., 1992; Otillinger, 1992); and *Ökologische Briefe*, 1993).

Figure 5.5 summarizes the various instances of cooperation between both sides, concentrating on those cases involving water suppliers or directly affecting agricultural practices in water protection zones.

---

31 A curious exception is the BGW, which has on occasion invited speakers from member enterprises to present their experience with organic farming in water protection zones.

5.5 Conclusions

In summary, in this section our examination of the policy framework in Germany has shown that, at least for environmental policy, the preconditions for decentralized policy experimentation do exist. As the already high levels of pollution increase, strong environmental awareness and concern for human health help create a political climate in which water suppliers and the public authorities controlling the supply and protection of water can find the (financial) means to intervene in a polluting sector such as agriculture. The sixteen *Länder* are responsible for water resource protection and management and are largely free to develop their own approach to water policy. In addition, the great number of municipally controlled water suppliers also provides a number of independent actors not dominated by unifying external forces who can also develop policy initiatives. The fact that these initiatives have only local impact initially is not a serious limitation in a member state where media attention to developments affecting the environment and human health is high.

In the area of agricultural policy, however, there is less freedom to independently develop a variety of policy concepts because of the control of European authorities over policy-making. As in the environmental field, the factors limiting experimental policy development by agricultural organizations include a comparatively higher degree of organization and the dominance of a policy paradigm which has underlined the industrialization of the farming sector during the past fifty years. Such conditions mean that local problem-solving is generally driven by water suppliers and authorities who are responsible for the water rather than the agricultural sector.

In addition to the decentralized nature of decision-making in water supply and water management, the availability of resources is a significant factor. Whereas economic instruments for surface water pollution control have been applied since the early eighties, only recently have water abstraction charges been adopted, and only in some *Länder*. Their introduction coincided with widespread revelations about the non-sustainable nature of conventional agricultural practices. As a result, the revenue from water abstraction charges is still designated in large part to controlling agricultural water pollution.

Overall, the factors described above have created a marked asymmetry in the policy framework, an imbalance which is reflected in the structure of water policy and agricultural policy networks. Just as in the broader fields of environmental protection, nature conservation, public health, or consumer protection, the highly diverse and often competitive nature of the network structure of water resource protection and water supply contrasts sharply with the much more rigid hierarchy of the structure of agricultural policy-making. It is against this background that the control capacity of the water supply sector in Germany must be assessed.

References

Bundesverband der deutschen Gas- und Wasserwirtschaft e.V. [BGW] (1993) *Die öffentliche Wasserversorgung 1992/93.* Bonn: Bundesverband der deutschen Gas- und Wasserwirtschaft.

Bundesverband der deutschen Gas- und Wasserwirtschaft e.V. [BGW](1994) *BGW-Informationstag: 'Neue Wege im Trinkwasser? - Die Novellierung der europäischen Trinkwasserrichtlinie.' In Zusammenarbeit mit dem Bundesministerium für Gesundheit. 17. November 1994.* Bonn: Bundesverband der deutschen Gas- und Wasserwirtschaft e.V.

Bundesminister für Ernährung, Landwirtschaft und Forsten [BML] (1996) *Agrarbericht der Bundesregierung 1996.* Bonn: Bundesminister für Ernährung, Landwirtschaft und Forsten.

Bundesministerium für Umwelt, Naturschutz und Reaktorsicherheit [BMU] (1994): *Wasserwirtschaft in Deutschland.* Bericht zum Tag des Wassers der Vereinten Nationen am 22. März 1994 zur Vorlage auf der Internationalen Wasserkonferenz in Noordwijk., Bonn: Bundesministerium für Umwelt, Naturschutz und Reaktorsicherheit. (Also available in English).

Böhret, Carl, Werner Jann and Eva Kronenwett (1988, 3rd edition) *Innenpolitik und Politische Theorie.* Opladen: Westdeutscher Verlag.

Bongaerts, Jan C., and R. Andreas Kraemer (1989) Permits and Effluent Charges in the Water Pollution Control Policies of France, West Germany and the Netherlands. *Environmental Monitoring and Assessment, 12*: 127-147.

Bruckmeier, Karl, and Parto Teherani-Krönner (1992) Farmers and Environmental Regulation : Experiences in the Federal Republic of Germany. *Sociologia Ruralis, 32* (1): 66-81.

Conrad, Jobst (1990) Interpretation of the German Nitrate Debate and Policy. *Zeitschrift für Umweltpolitik und Umweltrecht,* 1990 (1): 67-86.

Conrad, Jobst (1992) *Nitratpolitik im internationalen Vergleich.* Berlin: Edition Sigma.

Deutscher Verein des Gas- und Wasserfaches und Kulturbau and Deutscher Verband für Wasserwirtschaft und Kulturbau [DVGW and DVWK] (1995): *Wasserinformationen 46: Zustandsbeschreibung des Grundwassers.* Bonn: Wirtschafts- und Verlagsgesellschaft.

Deutsche Bauernkorrespondenz (1988) *Strategien zur Ermittlung der Nitratbelastung und das Festlegen von Maßnahmen zur Begrenzung des Nitrateintrages für die Landbewirtschaftung in Grundwasserschutzgebieten. Heft 10/88.*

Ganzert, Christian (1994) *Umweltgerechte Landwirtschaft - Nachhaltige Wege für Europa.* Bonn: Economica.

Greenpeace (1994a) *IX, Festgestellte Pestizidkontamination im Rohwasser (1989-1994).* Hamburg: Greenpeace.

Greenpeace (1994b) *X, Pestizidnachweis im Rohwasser (1989-1994).* Hamburg: Greenpeace.

Greenpeace (1995): Pestizide im Grundwasser. Ergebnisse einer Befragung deutscher Gesundheitsämter 1994. Hamburg: Greenpeace.

Hesse, Joachim Jens, and Thomas Ellwein (1992) *Das Regierungssystem der Bundesrepublik Deutschland.* Opladen: Westdeutscher Verlag.

Kraemer, R. Andreas (1992) *Zukunftsperspektiven der Trinkwasserversorgung. Institut für Europäische Umweltpolitik. Studienbericht für das Büro für Technologiefolgen-Abschätzung des Deutschen Bundestages.* Bonn: Institut für Europäische Umweltpolitik.

Kraemer, R. Andreas (1993) *Compensation Payments to Farmers for the Protection of Groundwater Resources.* Bonn: Institut für Europäische Umweltpolitik (manuscript).

Kraemer, R. Andreas, and Jan C. Bongaerts (1991) Experience with Waste Water Charges in France, the Netherlands and the Germanies. In: Belgische Vereniging voor Milieurecht, *De juridische en economische aspecten van de heffingen of de waterverontreiniging in Vlaanderen*. Bruxelles: Story-Scientia.

Kraemer, R. Andreas and Frank Jäger (1997):Deutschland. In: Nunes Correia, Francisco and R. Andreas Kraemer (eds.), *Eurowater I: Institutionen der Wasserwirtschaft in Deutschland - Länderberichte*, pp: 15-187. Berlin: Springer.

Müller-Brandeck-Bocquet, Gisela (1992) Europäische Integration und deutscher Föderalismus. In: Michael Kreile (ed.), *Die Integration Europas*, pp: 160-182. Opladen: Westdeutscher Verlag.

Oeckl, Albert (1993/94) *Taschenbuch des öffentlichen Lebens*. Bonn: Festland Verlag.

Ökologische Briefe (1993) Leipziger Wasserwerke fördern den Ökolandbau. *Ökologische Briefe* (7), 6-7.

Otillinger, Franz (1992) Landwirtschaftliches Sanierungskonzept der Stadtwerke Augsburg. *Wasserwirtschaft, 82* (3): 109-115.

Rüdig, Wolfgang, and R. Andreas Kraemer (1994) Networks of Cooperation: Water Policy in Germany. *Environmental Politics, 3* (4): 52-79.

Salzwedel, Jürgen, (1991): Ziele des Gewässerschutzes im geltenden Recht. In: Max G. Huber (ed.), *Umweltkrise*, pp:183-200. Darmstadt, Wissenschaftliche Buchgesellschaft.

Scharpf, Fritz, Bernd Reissert and Fritz Schnabel (1976) *Politikverflechtung: Theorie und Empirie des kooperativen Föderalismus in der* Bund*esrepublik Deutschland.* Kronberger/Ts: Scriptor-Verlag.

Scharpf, Fritz, Bernd Reissert and Fritz Schnabel (eds.) (1977) *Politikverflechtung II: Kritik und Berichte aus der Praxis*. Kronberger/Ts: Scriptor-Verlag.

Scharpf, Fritz W. (1992) Einführung: Zur Theorie von Verhandlungssystemen. In: Arthur Benz, Fritz W. Scharpf and Reinhard Zintl (eds.) *Horizontale Poltikverflechtung. Zur Theorie von Verhandlungssystemen*, pp: 11-27. Frankfurt/New York: Campus.

Schleyer, Ruprecht, and Helmut Kerndorff (1992) *Die Grundwasserqualität westdeutscher Trinkwasserressourcen.* Weinheim: VCH Verlagsgesellschaft.

Schodder, Thomas (1989) *Föderative Gewaltenteilung in der* Bund*esrepublik Deutschland. Eine Untersuchung ihrer gegenwärtigen Wirkungen und Probleme.* Frankfurt: Lang.

Umweltbundesamt [UBA] (1992) *Daten zur Umwelt 1990/1991*. Berlin: Erich Schmidt Verlag.

Umweltbundesamt [UBA] (1994) *Daten zur Umwelt 1992/1993*. Berlin, Erich Schmidt Verlag.

Würzburger Gesprächskreis, BGW-DVGW-LAWA (1994): *Kooperation zwischen der öffentlichen Wasserversorgung und der staatlichen Wasserwirtschaft.* Bonn, Wirtschafts- und Verlagsgesellschaft Gas und Wasser.

Wüsthoff, A., and W. Kumpf (1994): *Handbuch des deutschen Wasserrechts, Neues Recht des* Bund*es und der* Länder, *Loseblatt-Textsammlung und Kommentare.* Berlin: Erich Schmidt Verlag.

6. CONTROL CAPACITY - GERMANY

R. ANDREAS KRAEMER AND RALPH PIOTROWSKI
Ecologic, Centre for International and European Research, Berlin, Germany.

6.1 Introduction

After outlining the policy framework relating to agricultural water pollution in the previous chapter, we will now describe the actual control capacity of the water supply sector. The central question to be posed is: can the water supply sector influence the behavior of farmers?

The term 'water supply sector' refers not only to the actual suppliers of drinking water, but also to (semi-) governmental bodies legally charged with control over water quality management.[1] Such a broad definition is necessary to cover the diversity of legal forms of water suppliers in Germany. The term 'water supply authorities', meaning local or regional institutions with legal authority, would quite arbitrarily exclude a variety of German water suppliers. This is due to the fact that municipalities are free to choose how they secure and organize the water supply.[2]

The actual control capacity of the water supply sector depends on a variety of resources, some of which we identify below. In the following section on the actual control capacity and its application, we will assess how useful these resources are, and to what extent they help the water sector to successfully apply possible instruments.

Roughly, we distinguish among structural, political and informational resources, each comprising several factors. 'Structural capacity' refers to formal and financial aspects of the organization; 'political capacity' indicates the ability of actors to formulate and voice their interests; and 'informational capacity' involves the ability to control data and to supply or withhold information. Note that we do not claim to give a comprehensive overview on possible resources. Table 6.1 gives a simple overview of resources and summarizes the aspects addressed in this chapter. In the section below we

1 For a detailed explanation of the concepts, see previous chapter, section 5.1.

2 In chapter 5 we already explained that municipalities are free to choose nearly any legal form for the water suppliers.

Geerten J.I. Schrama (ed.), Drinking Water Supply and Agricultural Pollution, 123-147.

will not work through the table systematically. Rather, we will address individual topics while explaining the substantive instruments applied.

TABLE 6.1: Overview of Organizational Resources of Control Capacity

Structural Capacity	Political Capacity	Informational Capacity
• size of organization, manpower	• access to decision-makers	• advice to farmers
• financial	• political credibility: weight of representation	• generation of information
• organizational flexibility	• credibility in public opinion	• input/output/prevention options
• legal status / legal standing	• legal power	• information (on agriculture)
	• inclusion in networks	• supplying information
	• media	• technical standard setting

The term 'structural capacity' subsumes the formal aspects of an organization. Thus the following questions become relevant:

- how large is a particular organization?
- what is the manpower in general, or with specific regard to agricultural water pollution?
- what is its financial volume (budget)? and
- how flexible is the organization?

Assessing the legal status of water suppliers is relevant in terms of whether a water supplier is able to act independently, or is integrated in the administrative structure of a municipality. Closely connected to the legal status is the legal standing of an actor and whether he is a legal entity with access to court action.

The term 'political capacity' refers to the traditional lobbying activity of an organization. Is the organization able to protect its interests? Does it have allies at its disposal? And is it able to influence public opinion via the media?

Finally, the term 'informational capacity' addresses the power of an organization to supply and withhold information. We shall find that the question of who is able to gather information, and who holds the power to determine the use of the data gathered, is quite relevant to the outcome of decision-making. In German, the expressions *Datenhoheit* ('control over data') and *Datenherr* ('master of the data') indicates the specific notion that there is something like sovereign rights over data. Equally important is information in the form of personal competence. Setting technical standards relies heavily on the voluntary contributions of individuals in committees. An actor might easily influence technical standards by the ability to send their representatives into these committees. Last but not least, direct contact to farmers should not be neglected. For example, it can be easily imagined that there would be a difference in the advice given to farmers by an adviser paid by the chemical industry and the advice of one paid by water suppliers.

6.2 Actual control capacity and its application

The control capacity of the water suppliers is closely linked to instruments that ultimately aim at influencing farming practices. These instruments are the tools water suppliers have at their disposal to reduce groundwater pollution (see Kraemer 1997: 167ff). The value of the instrument chosen depends mainly on two factors. The first has to do with the usefulness of the instrument itself: is the instrument effective in relation to the intended result? The second factor regards the ability of an actor to apply the instrument: are the water suppliers capable of applying the instrument effectively and efficiently? Whereas the first factor is just a question of selecting the right tool, the second factor depends on the organizational resources of an actor.

We now will illustrate the interaction of instruments and corresponding organizational resources that might lead to greater control capacity of the water suppliers.

6.2.1 LEGAL INSTRUMENTS

Legislation appears to be a seductively powerful instrument in avoiding water pollution. It would seem that in order protect water resources, authorities simply have to prohibit pollution. But this notion does not take into account the particular circumstances of agricultural water pollution. On the one hand, it is costly to identify the culprit when water authorities are faced with diverse sources of pollution. On the other hand, farmers often find themselves under economic pressure and feel compelled to rely on chemicals to gain as much output from their soil as possible. But even if it is not possible to avoid water pollution by means of legislation, it is important not to condone it. At the federal level, there is an ongoing rivalry between the Ministry of Environment and the Ministry of Agriculture, with the former trying to impose restrictions on farmers, the latter trying to abolish them.

It appears possible to influence legislation at two stages. The first is the moment when laws are passed, and the second occurs when they are implemented administratively. Lobbying was formerly the domain of agricultural interest organizations, both at the level of member states and across the EU. Agricultural lobbying groups have an entire ministry as a counterpart, which gives them a significant structural advantage. The corresponding administrative counterpart in water supply is a sub-unit of the Ministry of Environment. It is obvious that the political clout of the Ministry of Agriculture in regard to agricultural legislation is tremendous, since this is where the laws are drafted. Nevertheless, the water authorities were able to chalk up considerable successes, preventing, for instance, the softening of the pesticide limit value.

Two examples illustrate the strengths and weaknesses of lobbying by water authorities and water suppliers. These are the transposition of Directive 91/676/EEC[3] on

[3] OJ L375, December 31, 1991.

the protection of water against nitrate pollution from agricultural sources (Nitrate Directive), and the revision of Annex I (especially category 4 defining the limit value for pesticides) of Directive 80/778/EEC relating to the quality of water intended for human consumption (Drinking Water Directive).

Nitrate Directive ('Düngeverordnung'). In traditional agricultural regulation, water suppliers have great difficulty in promoting provisions for water protection. The special-interest coalition joining agricultural industries and farming associations with the Ministry of Agriculture seems to function well. This coalition was reconfirmed when the *Düngeverordnung* was drafted.

The *Düngeverordnung* is a piece of national legislation that transposes the Nitrate Directive. The directive seeks to reduce or prevent pollution caused by the application and storage of inorganic fertilizers and manure on farmland, to safeguard drinking water, and to prevent wider ecological damage (Haigh, 1992: 4.14-1). It also defines the code of good agricultural practices that are to be enforced legally in vulnerable zones and on a voluntary basis in all other areas. The transposition of the Nitrate Directive proved difficult for the German government, and it was charged with non-compliance before the European Court of Justice. Nine environmental organizations also launched an official complaint to the Directorate-General XI (Environment) of the European Commission. Ritt Bjerregaard, the environmental commissioner who confirmed that Germany did violate its community obligations, brought an action to the Court of Justice. In December 1995, the ordinance received the approval of the *Bundesrat* and after a delay of over three years it was eventually signed by the Federal Minister of Agriculture in January 1996 (*Deutscher Bundesrat*, 1995).

Environmental organizations and the BGW (a federal association of gas and water industries) harshly criticized the draft of the ordinance for its soft formulations and pointed out that the ordinance even falls short of present standards. The *Länder* (provincial states) had already passed their own regulations (*Gülleverordnungen*), some of them setting rigid environmental requirements. But such ordinances adopted by the *Länder* become void when a corresponding federal ordinance takes effect.[4] Environmental organizations were convinced that the *Düngeverordnung* was specially drafted to avoid any negative impact on fertilizer sales.[5] The DNR (the environmental umbrella organization) together with the *Agrarbündnis* (the organic farming umbrella organization) published common position papers and tried to push for rigorous implementation of the EU directive. The non-governmental organizations were not left alone by political representatives.

---

4 Article 31 of the constitution of the Federal Republic of Germany (*Grundgesetz*) states: 'Federal law breaks state law.' This applies in areas where the *Bund* has legislative competence.

5 Only one chemical multi-national company, BASF, has embraced the *Düngeverordnung*. This is not so remarkable in light of the fact that the company has developed a food additive that reduces the phosphorous load of animal excretions (*Ökologische Briefe*, 50/1995: 4) that could help farmers by conforming to a rigid *Düngeverordnung*.

Two political parties also became convinced that as presented, the *Düngeverordnung* did not meet the environmental requirements of the EU directive. In the National Parliament (*Bundestag*), the Social Democrats, and the Green Party proposed a motion to tighten the environmental conditions imposed on farmers (*Deutscher Bundestag*, 1995). However, a few days after the *Düngeverordnung* was enacted, the parliamentary agricultural committee turned down the motion. A last effort to increase the effectiveness of the ordinance was made by the environmental committee of the *Bundesrat* (the second federal chamber, which represents the executive branches of the *Länder*). However, the agricultural committee of the *Länder* vigorously defended the *Düngeverordnung*.

Eventually, the agricultural lobby proved more effective than the environmental lobby in successfully promoting their interests. During a colloquium on drinking water management in Stuttgart on 22 February 1995, severe doubts were expressed about the *Düngeverordnung* meeting the requirements for water protection at all. Dietrich Ruchay, the highest official in the Federal Ministry of Environment concerned with water, stated that the agricultural lobby appeared to be too powerful to push through a more rigid definition of the 'code of good agricultural practices' in environmental terms.[6] According to Ruchay, there appears to be no lasting solution of agricultural water pollution within the present framework.[7] It is remarkable that the combined efforts of the environmental committee of the *Bundesrat*, the Ministry of Environment, nearly the entire parliamentary opposition, environmental organizations, and water suppliers were not successful in substantially changing the *Düngeverordnung*.

Drinking Water Directive. The revision of the Drinking Water Directive was the most controversial water issue discussed in Germany during the last two years. Every move of the Brussels 'well-poisoners' was scrutinized in the press. The directive stipulates that practically no pesticide residues may be found in drinking water, and it contains limit values for single substances and for the sum of substances which were intended to be 'surrogate zero' values. It is the intention of the chemical industry and, apparently, of various officials in the European Commission, that this strict regulation should be softened. But an alliance of water suppliers and environmental organizations, supported by the Green Party and the health ministers of the states, did not cease to portray pesticides as a threat that endangers water as a vital resource for human life. The purpose of the directive is to protect human health. It covers all water for human consumption, including bottled water not recognized as mineral water and water used in food industry. The Drinking Water Directive has the additional effect of protecting the environment, since the supply of drinking water must be sufficiently free from contamination to allow inexpensive water treatment (Haigh, 1992).

Closely related to the drinking water directive is Directive 91/414/EEC[8] concerning the marketing of plant protection products (the Pesticide Directive). Its main purpose is

6 *Zeitung für Kommunale Wirtschaft*, 1996 (3): 14.

7 He further pointed out that the firm pressure of the agricultural lobby was noticeable even concerning the development of the 'groundwater action program'.

8 OJ L230, August 19, 1991.

to unify the different approaches of the authorization process for pesticides by setting uniform principles for the criteria for registration and the data to be supplied for this purpose. These procedures were established by a separate Council Directive 94/31/EEC. The Pesticide Directive refers directly to the pesticide limit values in the Drinking Water Directive. No substance should be permitted which endangers the drinking water standard.

In beginning of 1994, the Commission planned to undertake minor changes in the Drinking Water Directive. In the course of this revision, a committee was set up to review the pesticide limit values, which are defined in Annex I of the directive. This committee carried out its work at a speed unknown in European bureaucratic circles. Obviously it was hoped that during the change of presidency (July 1996, when the Greek presidency ended and German tenure began) and shortly after the election of a new Parliament, the confusion could be used to tackle political explosive regulations. The distinguished business and financial newspaper, *Handelsblatt,* reported that Ioannis Paleokrassas, then European Commissioner of the Environment, would like 'to leave his mark' and to present the industry a parting present.[9]

But unexpectedly for the European Commission, the European Parliament immediately complained vigorously about what it felt to be a violation of its jurisdiction. No revision should be made without their participation. The BGW and Greenpeace launched a press campaign that covered nearly all developments in Brussels daily. They succeeded in stopping the immediate abolition of the limit values.

By summer 1996, no final decision had been taken. All revision of water directives have been postponed until a general framework directive is passed. The industrial lobby still claims that the limit of 0.1 μg/l for individual pesticides is set on a political rather than a scientific basis. A publication of the American Chamber of Commerce (who tried to influence the decision process massively) regretted that the limit values have been retained, 'regardless of the efforts made by the plant protection industry'. It was assumed that the 'debate will probably affect all discussion relating to the philosophy that should be adopted to set appropriate standard for European aquatic systems' (American Chamber of Commerce, 1996: 27). The agricultural industry pressed further for limit values based on research of the World Health Organization. It is to be expected that the outcome of the revision will strike a compromise between environmentalist and industrialists. The limit value of single pesticide substances is held valid but the joint limit value of 0,5 μg/l might be abolished. This is exactly the position supported by German water suppliers, who even changed coalitions. Together with the DBV (the German umbrella organization of the farming sector), they published joint press releases calling for the abolishment of the joint limit value.

Aware that for the farming industry, lobbying the DGXI (Environment) is much more difficult than lobbying DGXI (Agriculture), the farming lobby found a way to soften the high limit values through the Pesticide Directive. According to this directive, pesticides are to be admitted for the whole Union, nullifying the more strict German regulations. Further, it gives member states the authority to grant conditional

---

9 *Handelsblatt*, October 15-16, 1994.

authorization of a substance for a maximum period of five years, even if estimated concentrations are likely to exceed the limit values set by the Drinking Water Directive. During the decisive meeting of the Council of Ministers it was unclear which position the German minister for agriculture, Borchert, would take. Because they were misled by Borchert, who asserted that the drinking-water directive was not on the agenda, the press was already set up. Before and during the meeting intense pressure was put on the minister. Some of those interviewed reported that the 'Free State of Bavaria' sent a fax to the meeting stating that if Borchert gave up his position in favor, he need not to return to Bonn. But the minister stood firm and the environmental lobby achieved at least partial success.

The success of the water suppliers can be attributed mainly to their ability to react flexibly and immediately to developments and use their personal resources to employ lobbyists at the European level. The BGW was able to switch coalitions between the environmental organizations and the agricultural lobby and maintain professional contact to the press, which for its part, in response to public opinion, generally argued in favor of the BGW positions. The BGW was able to act on three levels: European, national, and through its members, the local level, and it gained a high reputation for competence in this field.

But no final decision has been taken and the chemical industry still tries to attack the pesticide limit value. Recently, the IVA (the Agro-Industrial lobbying Federation) claimed that residues which result from disinfecting waterworks are similar to pesticide residues in groundwater. The IVA accused the water supply industry of contaminating their own wells.[10] The BGW admitted that in some cases water pollution is unavoidable, but not so agricultural pollution. The main point of IVA's criticism is that different limit values apply for similar residues depending on whether they result from farming practices or from disinfecting of waterworks.

Federal Water Management Act. In Germany, the Federal Water Management Act (*Wasserhaushaltsgesetz* - WHG) provides an opportunity which may be the last resort to prescribe environmental farming practices. Within water protection zones, such farming practices as the use of pesticides can be restricted (Kraemer and Kahlenborn, 1998). The WHG encouraged the chairman of the BUND Baden-Württemberg to demand that the entire region of Baden-Württemberg be designated a water protection area zone (BUND, 1994: 52).

In practice, the procedures for the establishment of water resource protection zones laid down by the German Association of Gas and Water Experts (*Deutscher Verein des Gas- und Wasserfaches* - DVGW) are followed. Informational capacity in the form of expertise is thus the main resource for strengthening the environmental effectiveness of water protection zones. Whereas the location of water protection zones might be a highly political issue, the determination of what a 'water protection zone' actually means is generally left up to experts.

The DVGW has published guidelines for three different types of protection zones:

---

10. *IVA Informationen*, November 6, 1995.

- for groundwater resources (*DVGW Regelwerk Arbeitsblatt,* W 101 of 1975; amendment 1995);
- for reservoirs and dams (*DVGW Regelwerk Arbeitsblatt,* W 102 of 1975); and
- for lakes (*DVGW Regelwerk Arbeitsblatt,* W 103 of 1975).

In addition, there was a draft guideline[11] addressing land use and fertilization, but the corresponding 'yellow paper' was regarded as being outdated following the adoption of the Ordinance on the Use of Fertilizers (*Düngemittelverordnung*) implementing Council Directive 92/676/EEC.

The DVGW guidelines do not have legal jurisdiction. However, as technical regulations, they reflect the accepted expertise of the DVGW and can be drawn upon in court in order to interpret and clarify the law. The DVGW guideline W 101 was under revision, which resulted in a need to expand existing water resource protection zones. The draft 'yellow paper' is scheduled for publication in spring of 1993. In particular, protection zone III is likely to be subdivided into zones IIIa and IIIb, allowing for higher differentiation of protective restrictions. It can be assumed that restrictions will be tightened at least in the new protection zone IIIa.[12] In the literature on water policy, the need to expand water resource protection zones to cover actual catchment areas as a whole is a recurring issue.

Responsibility for Water Protection Zones. The distribution of control over and the administrative procedures for the establishment of water resource protection zones varies among the *Länder* (Kahlenborn and Kraemer, 1998: 105). In general, the zones are established by specialized water authorities who are given powers by the *Land* Ministry of Environment. The exception to this rule is Bavaria, where the general administration at district level (*Kreise*) is given control. In smaller *Länder*, the ministry itself is involved; or the *Senate* in the case of city-states, as in Berlin, Bremen, Hamburg, and the Saarland. In line with general practice in land use planning, the procedures invariably provide for consultations with the municipalities and individuals affected. The final result is given legal authority, usually by means of a *Land* ordinance, which also designates the area in question by referring to established territorial entities (e.g., the territory of a municipality), or by providing detailed maps of the zones established. The *Land* ordinances impose any land use restrictions differentiating among various degrees of protection within the zones. The restrictions are thus imposed by law and are publicized accordingly in the relevant journals.

Compensation Payments. As Rohmann reports, in the past the acceptability of the measures taken to the agricultural sector was primary, while the acceptability of compensation payments to water suppliers in Baden-Württemberg was of secondary importance (Rohmann, 1992: 25). Farming associations seem to have been more vocal - and more negative - in their attitude to the legal regulations. But it appears that the

---

11 *DVGW Regelwerk Arbeitsblatt,* W 104 published in 1991.

12 In some Land water laws, water protection zone III is also subdivided, while others provide for a fourth zone of relatively low land use restrictions (zone IV).

water supply industry did not maximize its political potential. In particular, water suppliers do not appear to have closely observed the implementation and enforcement of the ordinance and its effectiveness in reducing groundwater pollution. Only on the basis of effective supervision can a number of steps be taken against a farmer who does not comply with the regulations. Article 6 of the Ordinance on Compensation in Protection Zones provides for the following:

In individual cases, the water authority, after consultation with the agricultural authority, can issue additional orders. In particular, it can order the farmer to:

- *analyze the soil or have an analysis undertaken;*
- *maintain records on farming practices;*
- *refrain from applying nitrate fertilizer;*
- *use, or not use, specific farming practices or methods for fertilization or pest control;*
- *participate in collective measures relating to fertilization or pest control;*
- *participate in consultation, education, or information programs.*

It is likely that the mere fact that water authorities have the power to issue additional orders contributes to their bargaining capacity. Farmers generally prefer voluntary co-operation to strict regulation by water authorities.

National Pesticide Reduction Program. On 9 March 1996, the 32nd conference of the environmental ministers of the Northern *Länder* put forward a proposal to establish a national pesticide reduction program. In the summer of 1995, the DNR launched a campaign to inquire whether actual steps were taken to implement or support the decision. Table 6.2 summarizes the responses of the environmental ministries and demonstrates that no supplemental measures to further limit pesticide pollution have been undertaken at all by the *Länder*.

Table 6.2 indicates clearly that in the near future, a national pesticide reduction program initiated by the *Länder* cannot be expected. Firstly, the *Länder* do not share a common view on effective measures, and secondly, the reactions show that a program of national pesticide reduction is not perceived as an urgent measure which must be taken immediately. It is noticeable that the integrated plant protection, the measure favored by the agricultural industry is perceived as a workable solution by most of the respondents. Only two *Länder*, North Rhine-Westphalia and Hamburg, considered alternative solutions - co-operation and organic farming - whereas co-operation in North Rhine-Westphalia usually involves integrated farming methods as well. It is not likely that a pesticide reduction program which could serve as an effective instrument for water suppliers will be implemented.

Limitations of Legal Regulation. The most obvious limitation of legal regulation is that everybody is still able to break the law. When the risk of being caught is quite low and the potential earnings are high, the temptation to do so, especially for farmers under economic pressure, can become irresistible. For example, the Freiburger Water Suppliers (*Freiburger Energie und Wasserversorgungs AG* - FEW), estimate that nearly

20 % of the farmers in the water protection zone violate the co-operation agreement.[13] This might be partly the result of the Common Agricultural Policy, which still effectively promotes intensive farming practices.

TABLE 6.2: Statements of *Länder* Ministries of Environment regarding a pesticide reduction program.

Land	Preferred farming methods and reduction measures relied on
Berlin	Since 1992 limitation on the use of pesticides outside agriculturally used areas. Rely on integrated plant protection.
Baden-Württemberg	'Numerous' measures to limit the use of pesticides; measures comparable to integrated plant protection are given compensation.
Bremen	Supports the call for a 50% reduction program, prefers voluntary compliance (non-state controlled measures).
Saxony	Existing regulations should be applied, tight implementation of existing laws.
Hamburg	Acknowledges the necessity of a national reduction program; organic farming and integrated fruit farming.
Hesse	Monitoring programs in use.
Rhineland-Palatinate	Hopes for changes at the European level.
Saarland	Supports pesticide reduction program, but points out that they have only a low level of pesticide pollution.
Thuringia	Principally supports any pesticide reduction measure, but points out that under the current conditions pesticides are an important mean of production. They therefore support integrated plant protection. A 50% reduction by the year 2000 is not possible.
North Rhine-Westphalia	Supports national reduction program. Supports local co-operation between water supplier and agriculture.

Source: DNR, personal communication.

Note: The remaining *Länderministries* and the Federal Ministry of Environment have not yet responded.

The law is generally broken in three ways. First, manure and pesticides are applied during periods where they are not supposed to be used; second, farmers spray uncontrolled substances, often in large quantities; and third, such pesticides as the herbicide Atrazin and the insecticide DDT, which are banned in Germany, are imported and used illegally.

Banned substances, such as Atrazin or DDT, can nevertheless still be detected in groundwater (Zullei-Seibert, 1996: 192). Large quantities of illegal substances are

[13] *Zeitung für Kommunale Wirtschaft*, 1995 (5): 38.

brought into Germany. As a consequence of legal confusion between EU law and the laws of member states, a given substance forbidden by one state may be authorized by its neighbor. This has two effects, the first being that the groundwater catchment area of the environmentally more strict state might be contaminated by farmers of the neighbor state. Second, the open borders of the common market make it easy to illegally import otherwise banned substances.

Several solutions (legal, technical and voluntary) could help avoid the situation described:

- The most obvious solution is to maintain high environmental legal standards throughout the community.
- One technical solution might help influence the behavior of farmers acting illegally. So far it has been nearly impossible for water suppliers to identify the actual source of nitrates and pesticides. Groundwater is not polluted by one source, but by multiple sources throughout the whole catchment area. There are now more advanced, albeit expensive, techniques of identifying polluters which would increase the likelihood of farmers being caught. In the worst case the farm might be closed down by court order; at the least, high fines are levied.
- Greater co-operation between farmers and water suppliers would lead to a higher mutual understanding of the respective problems and common solutions could be found.

In the following section, this last point will be studied in more detail.

6.2.2 BEYOND LEGAL REGULATION: CO-OPERATION

For the uninitiated, the discussion of agricultural water pollution might be surprising. From a impartial point of view, one might assume that the interests of water suppliers and farmers are one and the same, rather than opposed. Both parties use natural resources and therefore have an interest in maintaining an uncontaminated environment. And both have an interest in producing water at the lowest cost possible, meaning without elaborate treatment or expensive plant protection substances.

Beyond recent quarrels between farmers and water suppliers, there may be common ground which could increase the receptiveness of both parties to co-operation and advice, and eventually lead to increased understanding.

Co-operation: Protection through Organic Farming. Some of the local water suppliers realized early on that the most efficient way to protect water resources is to avoid any kind of pollution. In the preceding chapter, a link was shown between the members of the BGW and members of AGÖL (an umbrella organization of producers' association promoting organic farming). Co-operation only takes place at the level of the member states[14]. So far, the BGW did not use its political clout to take a general stand in favor of

[14] Last year, there was an initial indication of co-operation at the member state level. The organic farming organisations organised a seminar on agricultural water pollution in which a representative of the BGW spoke.

organic farming. As a result, successful co-operation is largely dependent on the personal commitment of individuals.

One of the first water supplier to promote organic farming practices in its water catchment area was the *Stadtwerke* in Munich. Munich draws most of its drinking water from the Mangfalltal, an idyllic valley located 35 km south of the city. The idea was to cultivate the whole valley organically, avoiding the use of pesticides. About three years ago, the *Stadtwerke* launched a promotion and advisory campaign which turned out to be fairly successful. Knut Höllein, head of the Munich waterworks, reported that initially, skepticism was considerably high among local authorities and farming associations (Höllein, 1995: 28). It took consistent persuasion and frequent personal contact to convince all parties concerned. In the end, the *Stadtwerke* was also fortunate that the local representative of the *Bauernverband*, the farmers' association opposing such projects, was ready to offer practical support. In October 1992, the *Stadtwerke* together with three organical farming organizations were able to make the following offer to the farmers in the Mangfalltal:

> *In a well defined area, where farming practices are to be changed, and which is three times the size of our water protection zone, we support every farmer who with all his operations joins an organic farming association, farms according to their rules, and submits himself to independent auditing. For six years, we shall pay an initial supplement of DM 550/ha/y. In addition, we will reimburse the cost of consultations with representatives of organic farming associations that are necessary for the start-up.* (Höllein, 1995: 28; translation by the authors).

At the end of 1993, the Bavarian Ministry of Agriculture tried to sabotage the program. They argued that state subsidies had to be cut, otherwise the farmers would profit from two sources (state subsidies and compensation payments) for the same operation (organic farming). Restrictions on farming practices therefore had to be further enforced. This made state subsidies from other state resources possible, for such purposes as de-intensification, set-aside and landscape programs.

More than 90 farms were converted from conventional to organic methods, amounting to half of the water catchment area. The first results of the program already have been achieved. The nitrate load of the groundwater has diminished: against the peak nitrate load of 15 mg/l in1992, only 8 mg/l was measured in 1995.

Following the example of the *Stadtwerke* of Munich, several other towns now rely on organic farming practices for groundwater protection. In Leipzig, Augsburg, Regensburg, and elsewhere, water suppliers try to establish organic farming in their catchment areas. But not all co-operative programs proved to be as successful as that of the Mangfalltal. In Augsburg, for example, organic farming practices are implemented in only 5% of the water catchment area.

Two factors seem to be responsible for the success of promotion campaigns that aim at convincing farmers to change their practices. First, there is a structural factor. Switching to organic farming practices is much more difficult for meat and poultry producers than it is for vegetable and grain growers. This seems to account in large part for the hesitation of farmers in the Augsburg area, who rely mainly on animal husbandry. Second, motivation is an important factor. Successful co-operation depends on the personal commitment of individuals. Agriculture is a profession with its own rules and traditions. It is not hard to imagine that a large degree of sensitivity is needed to convey to a lifelong farmer the idea that his methods are now considered wrong and harmful to the environment. The likelihood that a water supplier who tries to convince farmers of this may become frustrated is thus obvious.

However, we have so far omitted a key financial aspect that might explain the success in the Mangfalltal. In Munich, farmers were more rigorously influenced by giving them the choice of receiving compensation for organic farming, or receiving no compensation at all for conventional practices. In Augsburg, farmers were able to choose from among several farming practices, each with a different level of compensation.

Co-operation with organic farming organizations brings one considerable advantage. It relieves water suppliers from instituting measures to control farming practices, a task which is left up to the organic farming organizations. And, presuming that farmers harvest according to the rules, there is no danger that harmful substances and pesticides in particular will pollute the groundwater.

The DBV, the association of German farmers, harshly criticized the Munich model. They pointed out that farmers were forced into organic farming and had to rely on compensation payments to survive.[15] The DBV feared that after the interim arrangement of compensation payment for six years, organical farmers would not be able to survive under free market conditions. In this respect, the distribution and selling of organic products may indeed prove to be a critical aspect. In Germany, the willingness of the consumer to pay higher prices for food of higher quality is generally quite high. Nevertheless, many organically farmed products have to be sold by conventional distributors, so that some farmers receive conventional, i.e. lower, prices.

The *Stadtwerke* Munich therefore took quite some effort to work out a comprehensive concept that includes not only the growth cycle of a farming product, but also the difficulties of economical distribution. The city administration plans to purchase Mangfalltal products for staff cafeterias, and items from the Mangfalltal are labeled with a drop of drinking water, indicating that the product was produced in the valley, and that its purchase represents a contribution to a clean environment.

Co-operation: Protection through Integrated Plant Protection. The water suppliers and the DBV of North Rhine-Westphalia took a substantially different approach. Although they agreed on co-operation much like the Munich case, they promote integrated farming rather than organic farming.

North Rhine-Westphalia has a unique institutional structure in which the Ministry of the Environment and the Ministry of Agriculture are combined. As a result, both the water suppliers and the agricultural organizations focus lobbying efforts on a single agency, the Ministry for Environment, Regional Planning and Agriculture (MURL). This might have contributed to co-operative solutions, where interests of all parties involved may be considered.

---

15 Interview with Reiner Latten.

In 1989, the Chamber of Agriculture (Landwirtschaftskammern), the agricultural and the water supplier interest organization, and the MURL signed an agreement involving a 12-point program which outlines the principles of co-operation. The main aims of the program are to secure groundwater protection for the entire territory (*flächendeckend*), and to ensure compensation for farmers in water protection areas. The idea behind the program is that decentralized co-operation is more effective and efficient than are central administrative measures on the *Land* level. The main aims of the water suppliers in implementing the 12-points-program were (BGW and DVGW, 1993: 4):

- to obtain a comprehensive overview of agricultural causes of water pollution;
- to secure environmentally-friendly farming practices throughout the whole territory by offering advice to farmers individually;
- to avoid water treatment as a result of agricultural water pollution in the interim period and over the long term; and
- to provide no service without receiving service in return.

The program also provides for the exchange of information, further research on agricultural water pollution (mainly paid by the water suppliers and the *Land*), and for compensation payments and de-intensification programs administered by the *Land*. Advice to farmers is generally given in accordance with the principles for integrated plant protection, an approach favored by the agro-chemical industry. The latter's position remains that economical farming without the use of chemicals is impossible. Their concept is to spray as few chemicals as possible, assuming that all chemicals are absorbed by plants and all residues are washed down into groundwater resources.

Compliance with the program was not totally voluntary. The development of the program can best be described with what Jänicke and Weidner (1995: 21) call 'prospective intervention'. By announcing intervention in the form of regulation, for instance, a government may create uncertainty among a target group. Such measures may still arouse resistance, but some innovators, preferring a secure perspective, are likely to choose co-operation over regulation. They will co-operate voluntarily and behave as if there were legislation in order to avoid being subjected to enforcement measures ('enforced voluntary co-operation'). According to a representative of the DBV, the agreement was signed at the end of a long meeting. Representatives of the water and the agricultural organizations met together with Klaus Matthiesen, the Minister for Environment and Agriculture in North Rhine-Westphalia. The water suppliers were not ready to sign the paper until Matthiesen lost his temper and threatened that this industry would be strictly regulated if no agreement were reached that night. After a few more hours, the agreement was sealed.

So far, the success of the program is difficult to assess. The BGW and DVGW evaluate the program very positively and regard co-operation as the most efficient and inexpensive measure (BGW and DVGW, 1993: 16). However, so far no actual data are available which can verify the success of the program.

Advice to Farmers. Basically, it is not in the interest of farmers to apply huge amounts of costly chemicals which may eventually disrupt the ecological equilibrium and affect the quality of their soil. There are two possible explanations for prevailing farming practices, however: economic pressure exerted on them by the Common Agricultural Policy, and misleading advice from their own special interest organizations, who might rather serve the needs of the chemical industry rather than the needs of the farmers.

Today farming must meet requirements that are different than they were thirty years ago. Yet most farmers still apply conventional farming techniques involving the massive use of pesticides and fertilizers. Advice to farmers was once the domain of the chemical industry, whose main interest was to sell as much as possible. It has now become common practice for water suppliers to finance agricultural advisers. In North Rhine-Westphalia, the chamber of agriculture was not able to supply enough advisers to reach all farmers directly or to ensure individual attention. Within the framework of the 12-point program, water suppliers now finance advisers, who still are employed at the agricultural chamber, but water suppliers co-determine the hiring of advisers. Advice remains the means of directly reaching the farmers, who are the potential polluters of groundwater resources. The BGW in North Rhine-Westphalia states that, the experience of the groundwater pollution program shows that agricultural advice is an instrument of outstanding importance.

Water suppliers pay advisers not only for integrated plant protection, but for organic farming as well. It was initially planned to include provisions for organic farming advisers within the 12-point program, but this idea was defeated by the resistance from the DBV. Several reports on the advantages and disadvantages of organic farming had been written, each offering different conclusions on the benefits of organic farming methods. Outside the 12-point program area, some organic farming advisers are employed, by the MURL and the agricultural agency in North Rhine-Westphalia. In Munich, organic farming advisers are a more regular phenomenon, since there is direct co-operation between organic farming organizations and water suppliers. In general, most organic farming advisers are employed by organic farming associations.

Financial capacity is the main resource applied by water suppliers for advice to farmers. Because water suppliers do not have in-house agricultural expertise at their disposal, they must purchase the information from outside consultants. This requires either additional capital or the authority to increase water prices to cover the extra expense. On the other hand, advise to farmers could be a profitable investment, yielding a savings on water treatment costs.

6.2.3 INFLUENCING VALUES: CO-ORDINATION

Co-ordination differs from co-operation with regard to the motivation of an actor. Actors might only agree to co-operate in the belief that it will serve their special interest (e.g. to receive or to spend money), regardless of their motivation to contribute to a clean environment. Co-operation is the active alignment of interests and behavior. Co-ordination, however, requires the internalization of common values. In a situation where all actors share common interests, it is not necessary to align them. In this case it is more precise to speak of co-ordination.[16] Co-ordination is a form of co-operation designed to avoid any undesirable side effects that might occur despite common interests. Actors agree on common rules and do not need to rely on enforcement mechanisms to insure compliance (Young, 1994: 58; Keohane, 1984).

If both sides - agricultural producers and water suppliers - internalized the desire for an unpolluted environment, co-ordination of their actions alone, rather than co-operation, would be necessary. This view presupposes that actors truly share a single notion of , for instance, the definition of a clean environment. Today, no one would deny the necessity of preserving our natural habitat. But whereas the chemical industry and the DBV regard integrated plant protection as environmentally friendly, the environmental organizations regard this measure as a cause of pollution. Co-ordination requires that the actors involved not only have a compatible or corresponding understanding of general concepts, but also agree on specific notions and their implications. But once such a level of understanding is reached, co-ordination requires much less mutual control and is much more cost-efficient than mere co-operation.

Water suppliers may use several means to win over farmers to their side. The common ground for action is the normative viewpoint that an optimal environment remains unpolluted. This approach stems from the idea that both water suppliers and farmers depend on natural resources and thus benefit from their protection.[17] If water suppliers have internalized this environmental awareness, they are more likely to employ specific instruments to increase the level of awareness within the farming community. These instruments comprise the supply of information about causal chains of agricultural water pollution, the pursuit of press coverage of environmental problems, and direct contact between ecological advisers and farmers. Co-operation based on organic farming is an necessary step to make co-ordination possible. The intention is that farmers will not only comply with organic standards in order to receive financial compensation, but that they will adopt these methods out of personal conviction.

Today farming practices are already more environmentally friendly than they were several years ago. According to Reiner Latten, president of the Rhenish branch of the DBV and the environmental spokesperson of the COPA, young farmers are much more aware of environmental problems than are their fathers, and they thus spray less pesticides and fertilizers.

---

16 The distinction among co-ordination, co-operation and collaboration is commonly used in political-science discussions on international regimes.

17 However, farmers and water suppliers use these natural resources in a somehow different way. Whereas the 'product' of the water suppliers is already clean water as such, the farmers need soil as a mean of production that might get polluted. This view does not take into account two relevant points: first, farming methods that lead to pollution of the soil and groundwater will also affect the quality of the farming product, for instance by leaving residues of pesticides; and secondly, soil contamination will affect the growth of future crops on the same land.

6.2.4 NO INSTRUMENT: FLEEING FROM POLLUTION

A final and frequently applied means for water suppliers to maintain high drinking water standards is to give up water sources, or to seek 'end of pipe solutions'. Since pollution continues to eradicate natural water resources, this presents no real solution, and is clearly the worst alternative. At the same time, such measures require considerable expenditures and drain the finances of water suppliers.

Steadily increasing pollution loads have urged water suppliers to:

- give up critically contaminated wells or individual waterworks;
- shift to deeper layers of less-polluted groundwater;
- combine polluted with clean water; and
- utilize water treatment plants extensively.

These measures can only be related to control capacity in the sense that in order to rely on them, water suppliers must have considerable financial resources at their disposal. Normally, such costs are eventually regained by increasing consumer prices. But as the last link in the chain, consumers must then pay a higher price for water whose quality is worse. In addition to compensation payments, these measures are the second instrument that subverts the 'polluter-pays-principle'.

6.3 Prospects for increasing control capacity

If we look at the actual control capacity of water suppliers, we will find that agricultural water pollution has decreased in the last years. Most legal regulations remained intact according to the wishes of the water suppliers and were not changed to their disadvantage. Nevertheless, agricultural pollution has by no means ceased and still continues to occur.

The success of the water industry can be attributed to several factors. Their financial capacity allows them to employ people in fields where needed (lobbying, press coverage, agricultural advice). In addition, they are aided by their ability to generate and supply information (technical standard setting, the proof of agricultural pollution sources), and their flexibility and elaborate network management (changing coalitions).

On the other hand, no final decisions on all legal regulations (e.g. the revision of the pesticide limit values in the drinking water directive) have yet been taken. And we should not forget that water suppliers have nearly no possibility at all to influence the most relevant field of legislation, the Common Agricultural Policy. In the absence of the long-expected reform of the agricultural policy, water suppliers must rely on indirect means of influencing farming practices. The most obvious method was to establish water protection zones, which allows water legislation to regulate water used for farming.

6.3.1 WATER PROTECTION ZONES

The number of water protection zones either already in existence or being established has increased considerably over the past years. By 1989 a total of 94,00 water resource protection zones had been formally established. Together they covered an area of about 19,000 hectares, representing about 13% of the total surface area of the Federal Republic of Germany before unification. Since then, the total number of zones in unified Germany has risen to 19,818 (11,073 for the Western *Länder*) with a total surface area of 37,082 km^2 (23,553 for the Western *Länder*). However, these averages hide a considerable degree of variation among the *Länder*. Table 6.3 gives an overview.

TABLE 6.3: Water Resource Protection Zones in Germany.

	size of *Land*	established		in progress		total			avg. size	data of
Land	[km2]	No.	[km2]	No.	[km2]	No.	[km2]	in %	[km2]	
Baden-Württemberg	35752	2560	5646	590	3964	3150	9610	26.9	3.05	5/92
Bavaria	70553	3730	2150	370	400	4100	2550	3.6	0.62	4/92
Berlin	883	12	224	8	101	20	325	36.8	16.25	2/92
Brandenburg	29060	988	1790	–	–	988	1790	6.1	1.81	12/91
Bremen	404	4	29	1	6	5	35	8.7	7.00	12/91
Hamburg	755	1	16	5	130	6	146	19.3	24.33	12/91
Hesse	21112	1504	6262	490	1693	1994	7955	37.7	3.99	12/91
Mecklenburg-Pommerania	23838	1070	2230	85	150	1155	2380	10.0	2.06	6/91
Lower Saxony	47438	313	3260	552	3650	865	6910	14.6	7.99	4/92
North Rhine-Westphalia	34071	364	3700	450	2700	814	6400	18.5	7.84	12/91
Rhineland-Palatinate	19848	1587	1683	460	821	2047	2504	12.6	1.22	12/91
Saarland	2570	45	412	69	333	114	745	29.0	6.54	12/91
Saxony	18337	2348	2715	–	–	2348	2715	14.8	1.16	9/91
Saxony-Anhalt	20445	953	1702	–	–	953	1702	8.3	1.78	1/91
Schleswig-Holstein	15727	16	171	74	934	90	1105	7.0	12.28	1/92
Thuringia	16339	4323	5092	82	358	4405	5450	33.4	1.24	12/91
Sum	**357132**	**19818**	**37082**	**3236**	**15240**	**23054**	**52322**	**14.7**	**2.31**	

Source: Information provided by the *Länderarbeitsgemeinschaft Wasser* (LAWA): LAWA-AG *Grundwasserschutz*.

Table 6.3 contains the number and surface areas of water resource protection zones in the various *Länder*, based on the latest available data and according to their current status (established or in progress). The total surface area is also expressed as a percentage of the total area of the respective *Land*. The percentage of agricultural area protected against pollution varies widely, ranging from 3.6% in Bavaria to 37.7% in Hesse. With the exception of Thuringia, the total area covered by water protection zones in the Eastern *Länder* is below the national average. Apart from this, no clear pattern (north to south, population density, governmental structure, age of data, or the like) is discernible.

The average size of water protection zones also varies, from as little as 0.62 km^2 in Bavaria up to 12.28 km^2 in Schleswig-Holstein. The city-states of Hamburg and Berlin are exceptions, with 16.25 and 24.33 km^2 per protection zone respectively. Here it is obvious that the average size of protection zones in the eastern *Länder* is considerably smaller than that of the western *Länder*. One exception is Thuringia, which has a relatively large number of impounding reservoirs and used to provide water for the areas around Halle and Leipzig, where both groundwater and surface water sources are of poor quality. This may explain the relatively large amount of land that is given to water protection zones. Given the current process of restructuring and the need to develop additional water sources in the Eastern *Länder*, the number of zones and the total area covered are likely to increase as a result of the adoption of new water laws.

6.3.2 ORGANIC FARMING: HIGH POTENTIAL BUT NOT POPULAR

Organic farming could be a worthwhile alternative solution for sustainable agriculture at a large scale. Economic mechanisms and the lack of political commitment have been the main obstacles to a profound reconstruction of the agricultural sector. The logic behind the Common Agricultural Policy and the German Agricultural Policy, which is based on post-war experience, is still applied. This reasoning holds that a state has to be self-sufficient in terms of food supply. Any deviation from the paradigm of independence was perceived to be a threat to state sovereignty. Organic farming was thus regarded as something which belonged to the realm of ecological fanaticism.

Today, organic farming is viewed with more good will, but the Federal Ministry of the Environment still considers it economically unfeasible. A coalition of the Ministry of Agriculture, the chemical industry, and the DBV, is able to reject any attempt to promote organic farming. Alliances of water suppliers and organic farming associations are not likely to alter this lobbying pattern. The counterpart of the agro-chemical lobby is the *Agrarbündnis*, an association of over twenty-two organizations concerned with organic farming. There are rivalries among the members, however, which even prevented them from arriving at a common objective. Some members, in particular the two large environmental organizations, NABU and BUND, seem to regard the *Agrarbündnis* more as a rival than an ally. At the same time, the ministry does not consider the *Agrarbündnis* to be a relevant actor, which is not surprising, since it only has one part-time employee who operates out of his home. Thus the *Agrarbündnis* has difficulty in obtaining access to the ministry.

Nevertheless, organic farming has proven successful at the local level. Wherever farmers are approached directly and are supplied with sufficient advice and financial compensation, the possibility for fruitful co-operation is increased. However, this is by no means an easy approach, and it requires a high degree of personal commitment, especially in the beginning. Initially, this approach is often confronted with skepticism and prejudice from local participants. So far, however, there have been excellent results, and levels of environmental pollution have been reduced to near zero. It remains true, however, that organic farmers have to sell their products in an unfavorable economic framework. While this framework will not change in the coming years, only time will tell if they can survive when, after six years, the compensation payments are terminated.

6.3.3 EFFECTS OF COMPENSATION PAYMENTS ON AGRICULTURAL PRACTICES

Compensation payments are not paid for organic farming only, but also to remunerate economic losses which occur as result of the restrictions in water supply protection areas. The Bavarian Ministry of Interior Affairs has compiled information about the success of negotiations for compensation payments in Bavaria. Negotiations are deemed successful if an agreement is reached between farmers and water suppliers on the amount of compensation payments, and if the authorities (regional government) agree to these payments. In cases where no success has been possible thus far, voluntary payments are often made by water suppliers anyway. The ministry did not collect data on the amount and form of payments; it noted that many water suppliers were reluctant to disclose this kind of information. No further investigations of the amount of payments are currently planned. According to information provided by the ministry, a total of 5,509 applications have been made for compensation payments in water catchment areas of which 3,743 (or 68%) have been agreed upon by the regional governments (*Regierungspräsidenten*). There are few instances of failed negotiations. Among seven regions, they include only four such instances in the region of Central Franconia (*Mittelfranken*), and two in the region of Upper Bavaria (*Oberbayern*)[18].

It would appear from the above that the measures taken to date have had a limited impact on agricultural practices, and indeed, anecdotal evidence supports this view. In essence, the legislation provides for substantial compensation payments to be made in return for limited modifications of agricultural practices (e.g., a 20% reduction in fertilizer use). The legislation does not provide for measures to enforce environmentally acceptable, sustainable forms of agriculture. The possibility of imposing a permanent grassland culture alone constitutes such a measure and is not sufficient on its own.

The pertinent question is, what constitutes 'good agricultural practice'. A standard definition is contained an official policy document of the government of Bavaria:

[18] Memo from the *Bayerische Staatsministerium des Innern* of October 27, 1992.

An agriculture following good practices aims to provide products that do not give rise to health concerns, are of high quality, and are produced at low cost, while respecting legally adopted ecological regulations. [19]

Two aspects of this require closer scrutiny: one is the term 'economical' (*kostengünstig*); the other is the focus on legally binding restrictions on agriculture. Together they create a regime in which environmental legislation establishes a framework for farmers to operate according to economic criteria. Shortcomings in the environmental outcome would thus be the responsibility of environmental policy and law, including its ineffective administration. This view ignores, however, the fact that the economic and regulatory framework for the agricultural sector is agreed upon at the EU level (with all its distortions and environmental externalities) and that local or regional water protection policy cannot provide an effective counterweight. It should therefore not seem surprising that the measures adopted so far have not had a high degree of environmental effectiveness (Kahlenborn and Kraemer, 1998: 20).

6.3.4 SUPPORT FROM AUTHORITIES

The federal Ministry of Environment has set several long term objectives regarding water resource management:

- to maintain or restore the ecological balance of waters;
- to ensure the drinking and industrial water supply in terms of quantity and quality; and
- to ensure that all other uses serving public welfare continue to be possible;

The water resources policy is based on following principles (BMU, 1994: 2):

- priority of prevention;
- co-operation of all parties concerned; and
- allocation of costs an the basis of the polluter-pays-principle and full covering of the costs while the tasks are fulfilled in a subsidiary and decentralized way.

The Ministry of the Environment further stresses that 'preventive environmental policy calls not only for fending off imminent dangers [...] but primarily for the protection and considerate use of the natural base (BMU, 1994: 2).' These principles are a striking contrast to such practices as compensation payments, which turn the polluter-pays-principle on its head.

The discussion of a national pesticide program has revealed that there is a chance for water suppliers to successfully demand implementation of these principles. In Germany the legal capacity which allows member states to deviate from European Law has not been exhausted. Recently the European Court of Justice approved the Danish national

---

[19] Article 5 of the Joint Declaration of the Bavarian State Ministry of the Interior and for Food, Agriculture and Forestry (*Gemeinsame Bekanntmachung der Bayerischen Staatsministerien des Innern und für Ernährung, Landwirtschaft und Forsten*) of June 6, 1988, No. IIB3-4532.5-0.36 and PI-4500-53 (AllMBl. (12), pp. 521-523). (English translation by the authors.)

taxes on pesticides, ruling that the law constitutes no infringement of the Treaty's obligations.

6.4 Conclusions

The main problem we have identified here is that water suppliers have to act within in a framework - i.e. the EU's Common Agricultural Policy - that does not allow them to tackle actual causes of the problem. Nevertheless, lasting solutions must be found. The most promising option seems to be co-operation in support of organic farming. This option requires considerable financial capacity on the part of water suppliers, because they are urged to compensate for economic losses brought about by the Common Agricultural Policy. On the other hand, it is so far the only measure which effectively reduces the pollution of our groundwater resources.

Environmental Effectiveness. The environmental effectiveness appears limited among the measures taken in the majority of cases so far (e.g., a 20% reduction in the use of nitrogen fertilizers). It is unclear whether reductions alone are insufficient to attain the environmental objective of clean groundwater, or whether the time lag between changes in agricultural practices and any measurable change in groundwater quality is too long for positive effects to be discerned. It must be remembered that changes in farming practices have been imposed for only a few years.

A related problem is monitoring for compliance. Short of very intrusive detective work, it is very difficult to ascertain whether a particular farmer complies with restrictions and obligations imposed. This is especially true of quantitative restrictions, where the actual quantities applied would have to be measured. The monitoring of total bans is regarded as a relatively simple matter; nevertheless, in some cases farmers are still said to be using substances that have been banned for some years already, including Atrazine.

Franz Otillinger, an official of the water supply company of Augsburg in Bavaria, conducted an investigation of the various options for restoring the quality of the city's groundwater reserves. The study focused on various land use options in order to assess their effectiveness in groundwater protection. Otillinger's results showed that restrictions which can be imposed within a water protection zone are not in themselves sufficient to restore the water quality in an aquifer. He recommends that water suppliers should try to buy the agricultural land in the immediate protection zone II in their catchment areas, and to promote an extensive agricultural regime through leases (Otillinger, 1992). In recognition of this, the *Land* water law of Lower Saxony explicitly provides for the revenue from its abstraction charge to subsidize the purchase of agricultural land by water suppliers.[20]

---

[20] *Achtes Gesetz zur Änderung des Niedersächsischen Wassergesetzes (Einsparen von Wasser und Förderung des Gewässerschutzes)* [Eight Law Amending the Water Law of Lower Saxony (Water Saving and Promotion of Water Resource Protection)], inserting the new Section 6, article 47 to 47h, into the Water Law of Lower Saxony.

Economic Efficiency. The economic efficiency of the measures described is difficult to judge and depends on the criteria employed. At present it is not possible to assess the total cost of the schemes in relation to their environmental effectiveness. Even if such an approach were possible, it could not take account of the fact that the measures are not designed to be economically efficient according to objective criteria, but are the outcome of a political bargaining process. Nevertheless, there has been intensive debate in Germany on the compatibility of the first model, Baden-Württemberg's 'water penny', with the polluter-pays-principle of environmental policy.

The debate, which was in fact a controversy, was started by Holger Bonus (1986), an economics professor who interpreted the water penny in light of the 'Coase theorem' (cf. Coase, 1960). The polluter-pays-principle has its basis in the cause and effect relationships of environmental pollution: those responsible for releasing harmful substances into the environment should pay for the consequences, in the form of either compensation, restitution, cost of prevention, taxes, fees or other charges. Bonus sought to dilute this traditional basis of the polluter-pays-principle by switching the emphasis away from the pollution caused by agricultural practices to the aspect of water resource consumption. In essence, he proposed to treat resource consumption (water abstraction) as equivalent to resource abuse (pollution by agro-chemicals). This allowed Bonus to declare that water users share the responsibility for the scarcity of clean water resources and enabled him to apply the Coase theorem to the case. According to this theorem, the initial allocation of rights to a resource does not affect the pattern of resource use or abuse - albeit under a wide and restrictive set of assumptions. Bonus claimed that it would not effectively make a difference whether farmers were permitted to pollute groundwater resources but would receive compensation for not doing so, or if water users had a basic right to clean groundwater and could either enforce restrictions on farmers or demand compensation for additional treatment costs. This proposition was fiercely attacked by other economists and environmental scientists. The ensuing debate was as much about economics as about the polluter-pays-principle itself. In the end, Baden-Württemberg's water penny was found to violate the polluter-pays-principle.

Resource consumption taxes such as water abstraction charges are nevertheless compatible with established principles of environmental policy, according to a reply by the EU Commission to a question from Rolf Linkohr, MEP.[21] The controversy would probably not have arisen if Baden-Württemberg had established a proper resource consumption tax that would avoid the distortions caused by a tax rate differentiating among the various uses of water, and if the revenue were made available to farmers engaging in sustainable farming practices.

General Assessment. In general terms it can be said that the German water industry has become increasingly politicized over the last twenty years. The adoption of favorable positions toward protective water resource management is not a new phenomenon. Recently, however, such positions have fallen more and more in line the stance of the

---

21 Reply by Clinton Davis for the Commission of 15 June 1988, OJ C49 of February 27, 1989, page 2.

environmental movement, at least where water resource protection and the control of toxic substances are concerned. Increasingly therefore, the conditions are good for the formation of (issue) alliances between the water industry and segments of the environmental movement. Given the widespread concern within the environmental movement about food quality, and its rejection of many aspects of current agro-industrial practices, such alliances may well focus on finding alternatives to the use of chemicals. Co-operation between water suppliers and organic farming organizations is already taking place, as in Osnabrück, a region with severe groundwater pollution levels.[22]

The growing co-operation between organic farming organizations and consumer protection organizations, and the increasing number of agricultural experts employed by water suppliers to advise farmers, point to greater agricultural competence within the water industry. In the near future, this might increase the quality and credibility of the water industry's contribution to the controversy over good farming practices. At present, good agricultural practices are determined by the established agricultural community, usually without regard to the environment. In years to come, this near-monopoly in norm setting might more effectively be challenged by the water policy community as a whole and by water suppliers in particular.

References

American Chamber of Commerce (1995) *EC Environment Guide*. Brussels: American Chamber of Commerce.

Bundesverband der Gas- und Wasserwirtschaft and Deutscher Verein des Gas- und Wasserfaches [BGW and DVGW] (1993) *Erfahrungen mit dem 12-Punkte-Programm aus der Sicht der Wasserwirtschaft in Nordrhein-Westfalen*. Bonn: Bundesverband der Gas- und Wasserwirtschaft and Deutscher Verein des Gas- und Wasserfaches.

Bundesministerium für Umwelt, Naturschutz und Reaktorsicherheit [BMU] (1994): *Wasserwirtschaft in Deutschland*. Bericht zum Tag des Wassers der Vereinten Nationen am 22. März 1994 zur Vorlage auf der Internationalen Wasserkonferenz in Noordwijk., Bonn: Bundesministerium für Umwelt, Naturschutz und Reaktorsicherheit. (Also available in English).

Bonus, Holger (1986) Don Quichote, Sancho Pansa und der Wasserpfennig. [Don Quichote, Sancho Pansa and the Water Penny.] *Wirtschaftsdienst* (XII): 625-629.

BUND (1994) *Zukunftsfrage Wasser*. Stuttgart, Eigenverlag.

Coase, Ronald H. (1960) The Problem of Social Cost. *Journal of Law and Economics*, 3 (October): 1-44.

Deutscher Bundesrat (1995) Beschluß. Verordnung über die Grundsätze der guten fachlichen Praxis beim Düngen (Düngeverordnung). *Drucksache 402/95* (15 December 1995).

Deutscher Bundestag (1995) Antrag der Fraktion BÜNDNIS 90/DIE GRÜNEN. Erforderliche Maßnahmen zur Umsetzung der EU-Nitratrichtlinie im Rahmen der Düngeverordnung. *Drucksache 13/3064* (21 November 1995).

22 *Zeitung für kommunale Wirtschaft*, 1992 (11): 4. *'Ökologischer Anbau bevorzugt'* [Ecological Farming Preferred.]

Haigh, Nigel, (1992): *Manual of Environmental Policy: The EC and Britain.* Harlow: Longman (loose leaf).

Höllein, Knud (1995): *Förderung des Ökologischen Landbaus im Einzugsbereich des Wassergewinnungsgebietes Mangfall der Wasserwerke München.* Fachtagung vom 21.2.1995 über ökologischen Landbau und Wasserschtutz. Hannover: Wilhelm Jungman, 27-29.

Jänicke, Martin, and Helmut Weidner (eds.) (1995) *Successful Environmental Policy - A Critical Evaluation of 24 Cases.* Berlin: Edition Sigma.

Kahlenborn, Walter and R. Andreas Kraemer (1998): *Nachhaltige Wasserwirtschaft in Deutschland.* UBA Forschungsvorhaben 102 04 110. Berlin: Ecologic.

Keohane, Robert O., (1984) *After Hegemony. Cooperation and Discord in the World Political Economy.* Princeton (N.J.): Princeton University Press.

Kraemer, R. Andreas and Frank Jäger (1997): Deutschland. In: Nunes Correia, Francisco and R. Andreas Kraemer (eds.) *Eurowater I: Institutionen der Wasserwirtschaft in Deutschland - Länderberichte,* pp: 15-187. Berlin: Springer.

Kraemer, R. Andreas and Walter Kahlenborn (1998, fortcoming): Regional Sustainability through Land and Water Management in Germany? In: Enid M. Barron and David Berson (eds.) *Agriculture and Sustainability in Europe.* Dordrecht: Kluwer International.

Otillinger, Franz (1992) Landwirtschaftliches Sanierungskonzept der Stadtwerke Augsburg. [Agricultural Restoration Concept of the Stadtwerke Augsburg.] *Wasserwirtschaft, 82* (3): 109-115.

Rohmann, Ulrich (1992) *Die SchALVO als Instrument zum Schutz örtlicher Grundwasservorkommen?* Karlsruhe: DVGW-Forschungsstelle am Engler-Bunte-Institut der Universität Karlsruhe.

Young, Oran R. (1994) *Protecting the Environment in a Stateless Society.* Cornell: Cornell University Pres.

Zullei-Seibert, N. (1996) Belastung von Gewässern mit Pflanzenschutzmittlen: Ergebnisse einer Bestandsaufnahme in Deutschland und der Europäischen Union. In: Institut für Grundwasserwirtschaft, *Conference: Grundwasserschutz, Konzepte '96, Dresden, 14/15 February 1996,* pp: 179-201. Dresden: Technische Universität Dresden, Eigenverlag.

7. POLICY FRAMEWORK - THE NETHERLANDS

STEFAN M.M. KUKS
Center for Clean Technology and Environmental Policy, University of Twente, Enschede, The Netherlands.

7.1 Introduction

7.1.1 WATER SUPPLY IN THE NETHERLANDS

Organizational structure. In the Netherlands the drinking water supply is primarily considered to be the government's responsibility, although it is organized as a public utility. While countries like France and the United Kingdom have opted for privatization of the water supply, the Netherlands still leaves open a supervisory role for public authorities. The general opinion of both the government and the water supply companies themselves is that they disapprove of privatization in the water supply sector. The Dutch government officially declared itself not to be a proponent of full privatization of water supply companies (VROM, 1993). This statement is motivated by the conflict the government sees between the interests of consumers and those of shareholders. The consumers' interests lie in having a guaranteed delivery of water at a reasonable price level; public health interests lie in having drinking water of good quality that meets the drinking water standards. Both of these require a drinking water price based on actual costs. The Dutch solution, which takes the form of commercial operating companies with public authorities as majority shareholders, is considered by both the government and the water supply sector to be an adequate organizational structure to meet these requirements.

A the time of writing (1997) the Netherlands has 28 water supply companies, only one of which is a private company. The others are public companies in which public authorities participate. About 15% of them are municipal companies, the remaining 85% being regional companies. In general the municipal companies in urban areas are the oldest, because the urban water supply started to become organized in the second half of the nineteenth century by linking houses to a central water main. Most of these companies are already 100 to 125 years old. Much younger are the water supply

Geerten J.I. Schrama (ed.), Drinking Water Supply and Agricultural Pollution, 149-190.

companies that also intend to serve households in the rural areas. Most of these companies, which in general have a much more extended service area than the municipal ones, were founded in the 1930s. These are also the larger companies in terms of organizational and financial capacity. Some of them serve almost about 90% of the households in a province, like the WMO in Overijssel and the WML in Limburg. Recent decades have seen many reorganizations, resulting in larger companies in the water supply sector, leading to a reduction in total number of water supply companies in the Netherlands, with a concomitant increase in the area serviced by the remaining companies. Because of these reorganizations the water supply sector has grown into organizational units of a size that is considered to be large enough to produce and provide water of good quality (Vewin, 1991a; Vewin, 1993; Langendijk, 1994).

A water supply company is in fact an administrative unit that organizes the abstraction of surface or groundwater, the production of drinking water from the extracted water, and its distribution or delivery. Abstraction can be from groundwater sources or from surface water. Groundwater is abstracted in well fields, where each well field has a pumping station in which the water extracted from the wells is collected before pumping it to a production unit where the groundwater will be treated, or purified if polluted, to meet drinking water standards, or other standards if the water is to be provided to another destination, such as an industrial application. Water supply companies in the Netherlands are not involved in water treatment in the sense of waste or sewage water treatment. This activity is organized by Water Boards in the Netherlands, which are responsible for the exploitation of waste water treatment works and the treatment of water that has to meet surface water standards. Despite this, there are water supply companies, mainly in the western part of the country, which use surface water sources for the production of drinking water, which they have to treat very intensively. As far as the distribution and delivery of drinking water is concerned, 99% of the households in the Netherlands are connected to a water main (Vewin, 1993; Van den Nieuwenhof, 1995).

In the 1970s and 1980s, the question of what might be the optimum structure of a water supply company started to gain attention. To have a viable water supply unit, the Council for the Drinking Water Supply judged that such a unit should have at least 100,000 connections, and also at least two well fields that are connected with each other. In the first place they were thinking of companies serving an area with the size of a whole province, since the provincial governments have responsibility for the water supply and are also the licensing agency with respect to the abstraction of groundwater, while the Inspectorate of Public Health and Environmental Quality is also organized into provincial divisions (Langendijk, 1994). It is the Water Supply Act that allocates to provinces the competence to supervise the water supply sector and to impose reorganization plans, if they think that is necessary for an efficient water supply and for public health reasons. Most provinces took advantage of this power in the1980s, since problems with the water quality and quantity were increasing at that time.

In 1993, the Dutch government - in its the Policy Document on the Supply of Drinking Water and Industrial Water - emphasized that the national policy should be lie in the direction of a strong industry, containing sufficient, large-scale companies whose service areas need not be restricted to the borders of the provincial territories. Although crossing these borders is not considered to be a problem, most water supply companies restricted their radius of action to service areas on a regional scale - mostly smaller than a provincial territory - because of the infrastructure costs that would be involved with transport over a longer distance (VROM, 1993).

Limits to the potential use of groundwater sources. Water supply companies in the Netherlands use about 70% groundwater sources and 30% surface water sources. Half of the surface water sources originate in the Rhine, the other half coming from the Meuse. The amount of surface water mentioned also includes surface water that is purified through infiltration in the soil, and dunewater. Of the water produced by water supply companies, 95% is delivered as drinking water and 5% as water for other purposes, such as industrial applications. Besides that, industries also pump up their own groundwater, which is not included in the percentages above. Neither is the amount of groundwater or surface water abstracted by farmers for irrigation. Only the amount of water produced and delivered by the water supply companies has been considered (Vewin, 1993; VROM, 1994).

In the Netherlands, the domestic use of water has increased by 50% over a period of 20 years (1970-1990), while the population has increased by 12%. The average domestic use per capita per day was 97 liters in 1970, and 125 liters in 1989. A further increase, up to about 135 liters in 2010. Is expected. In 2000, the drinking water use will have increased by about 14% compared to 1990, a figure which might double before 2060, if the volume of domestic use remains unchanged (in the absence of water conservation measures).

Between 1970 and 1980, the total volume of industrial water use has decreased in absolute terms since the introduction of the Surface Water Pollution Act in 1970, despite a large increase in industrial production during that period. This started to change from the beginning of the 1980s. Water conservation measures were then no longer able to compensate for the continuous growth in production. On this basis it is expected that the industrial use of water will increase in the future. At the moment, much high quality drinking water is still being used for purposes that don't require this quality. Using water of lesser quality, produced from surface water, could diminish the pressure on groundwater use, with the additional advantage to industry that it can save 40% on its water bills (Vewin, 1989).

At the beginning of the 1990s, the government started to intensify its water conservation policy and decided to increase the amount of surface water as a source for the water supply (VROM, 1990b). An important motive for this was the increasing environmental problem of dehydration in Dutch nature conservation areas, due to sinking groundwater levels. In 1991, the Dutch water supply companies - that cooperate in the national association of water supply companies Vewin - produced an environmental document, in which the goal was set of saving 10% domestic water use in 2000 compared to 1991 (Vewin, 1991b). Although the available quantity of surface water would not be a problem, the quality of this water is lower than that of groundwater sources, and much more susceptible to pollution from abroad. Because of incidents in the Rhine and the Meuse it frequently happens that water supply companies have to suspend their intake of river water for a while.

What about the vulnerability of groundwater sources that are used for the production of drinking water? There are about 220 well fields that can be classified in terms of their vulnerability. Most vulnerable are about 70 well fields with no natural protective layers of clay (sandy areas) which together account for 25% of the water supply in the Netherlands. In another 50 well fields the groundwater is partly protected by relatively poor covering layers of clay. In the other 100 well fields the groundwater is abstracted from relatively deep levels and is very well protected by quite good subterranean covering clay layers. Most groundwater well fields are situated in the northeast, the east, the middle and the south of the country. In the west, there is too much salt in the groundwater at quite shallow levels, which makes it difficult to produce drinking water from it (RIVM, 1991).

In July 1995, the Dutch environmental minister wrote in a letter to the Parliament, that she was concerned about the quality of the drinking water sources, both groundwater and surface water. Research had revealed that at least 15% of the well fields and half of the surface water abstraction sites were more or less polluted. Groundwater sources were becoming increasingly polluted with pesticides and nitrates. The minister expressed her concern about the increasing efforts that were needed to deliver a drinking water product of good quality. Water suppliers were being forced to increasingly greater expense in this regard. In the same letter the minister also concluded that there appeared to be no problem with the quality of the drinking water in the Netherlands. The most recent data (1993) show that only incidental violations of the quality standards have been recorded. In no case was there any danger to the public health and measures had been taken to ensure that these incidents would not recur (VROM, 1993-1997; VROM, 1995b; VROM, 1996). In September 1995 the Dutch government issued a new Policy Document on Drinking Water Supply and Industrial Water Supply, in which it announced that more efforts would be put into the protection of groundwater and surface water sources over the coming years (VROM, 1995a).

In the Netherlands the water supply companies are regarded as one of the target groups of Dutch environmental policy. On the one hand they contribute to some environmental problems, such as dehydration, the emission of heavy metals from obsolete water pipes, and the disposal of polluted treatment sludge. On the other hand they are considered to be 'environmental producers', like the waste disposal companies and sewage water treatment companies. They are environmental producers because they are upgrading the quality of natural water. They also play a role in tracing groundwater pollution, and they are expected to communicate reported pollution to government agencies and inspectorates to enable them to frame adequate regulations and enforcement measures. For water supply companies, which are traditionally used to concentrating on geohydrological mapping and water treatment, it is a new field to build up knowledge on agricultural practices within the areas where their well fields are located (Van der Schot, 1995).

Regulation of the water supply. Water supply companies primarily have to deliver drinking water that meets the quality standards as provided by the Water Supply Act. These standards are national requirements, which means that no regional differentiation of standards is permitted. Water supply companies have no responsibility to protect groundwater from being polluted. It is left to the twelve Dutch provinces to create conditions for good quality groundwater and surface water. The national government provides national planning for water management and requires the provinces to implement this plan in terms of their own provincial water management plans. These are plans in which surface water management and groundwater management are integrated. Provinces have delegated the operational management of surface water to water boards, who regulate all discharges to surface water in collaboration with the municipalities, who manage the sewage systems.

The provinces cannot delegate the operational groundwater management, since there is no regional agency like a water board to take over the implementation work and until now it there has been no serious consideration of delegating groundwater management to water boards. As far as the groundwater quantity is managed, provinces are enabled by the Groundwater Act to issue permits for groundwater abstraction. They also have to impose a levy on each 1,000 liters of groundwater abstracted. The permit and the levy system are especially used to prevent exhaustion of groundwater resources and the associated dehydration problems.

The quality management of groundwater is regulated under the Soil Protection Act, which gives to provinces the responsibility to designate groundwater protection areas and to regulate activities in those areas that might harm the groundwater quality. Groundwater protection areas are designated to protect groundwater that has to be used for the production of drinking water. In most areas regulations are in force for the agricultural use of fertilizers and pesticides. The Soil Protection Act also provides that water suppliers have to financially compensate farmers in these areas in so far as they are more stringently regulated than farmers outside the protection areas.

Provinces also have a role in supervising the efficiency of the water supply sector, a competence that has been settled on them by the Water Supply Act. Groundwater is considered to be a collective property, which should justify public interventions in the way water suppliers operate. The supervisory role of the provinces also means the service area of the water supply companies mostly do not cross provincial borders, no matter how large or small they are.

In other words, the provinces have a primary responsibility to protect groundwater that is needed for the production of drinking water, and they have a supervisory authority with respect to the efficient operation of water supply companies. However, provinces do not have the authority to impose drinking water quality standards. Only national and European requirements have such authority, and these are addressed directly to the water suppliers by the national government (Glasbergen, Groenenberg and Roorda, 1989).

In this study of the Netherlands, we see 'water supply authorities' as water supply companies that aim to supply the public with drinking water that has to meet the water quality standards provided by the national government and the European Union.

Although water supply companies are regulated by drinking water standards, they do not themselves have any regulatory competence to implement these standards by regulating farmers in order to prevent agricultural pollution of drinking water sources. Nevertheless, we are interested in the control capacity of water supply companies in bridging this 'regulatory gap'. Related to this: are water supply companies interested in encouraging preventive action and controlling agricultural operations in their water catchment areas, or do they rely more on engineering the water quality by purification techniques? The factors driving the water supply companies in the direction of encouraging prevention and their capacity to control farmers will be discussed.

7.1.2 DUTCH CASE STUDY

For this case study of water supply authorities in the Netherlands and the way they try to prevent agricultural water pollution, we selected five water supply companies in five different provinces. First, we selected five of the twelve provinces in which intensive agricultural operations take place in areas where extremely vulnerable groundwater systems are used for drinking water supply. These are the provinces of Drenthe, Overijssel, Gelderland, Noord-Brabant and Limburg, which together include, generally speaking, the east and the south of the Netherlands. Secondly, in each of these provinces we looked at the activities of the water supply company that is most involved in problems with the agricultural pollution of drinking water sources. In all provinces that were studied, this was the largest or one of the largest companies. We have interviewed one or more representatives of these water supply companies to acquire data for this case study. We also asked them to identify the most closely involved representatives of the agricultural community and of the provincial authorities in each of the provinces. In all provinces, the agricultural community has been represented by a regional division of the national agricultural board. We also interviewed these representatives and the provincial officials involved.

While writing this case study, the interview data have been used in two ways. On the one hand we tried to describe what was going on in each of the provinces as far as possible, according to the common sense views of the respondents. On the other hand we tried to aggregate the interview data from the five provinces to give a description of three perspectives on the policy problem, one perspective for each of the three parties involved (water suppliers, agricultural communities and provincial authorities).

We have focused the case study on a ten year period from 1985 until 1995, knowing that 1987, 1989, 1991 and 1993 were important years for the protection of groundwater. In 1987, the national government started to regulate agricultural practices with standards to restrict the use and spreading of animal fertilizers on agricultural fields. In 1989, many provinces in the Netherlands came up with an ordinance for more stringent regulation of fertilizer use in groundwater protection areas. In 1991, the Dutch government succeeded in finalizing a negotiated agreement with the agricultural community on a reduction of the use of pesticides. And in 1993, the implementation of the European Groundwater Directive resulted in the designation of the Netherlands as a whole as a vulnerable area in which the upper groundwater throughout the country should be protected by meeting a nitrates standard of 50 mg/l. Most interviews for the case study took place at the end of 1994 and in the first half of 1995.

Figure 7.1: Map of the Netherlands with the five selected provinces.

The results of the Dutch case study are reported in chapters 7 and 8 of this book. Our leading question is: to what extent are water supply authorities in the Netherlands trying to prevent agricultural water pollution, and to what extent do they succeed in that? Before answering this question in chapter 8, we will use chapter 7 to describe three contexts within which water supply companies choose their strategy. The first is the 'problem context', which should give an idea about the magnitude of problems with agricultural water pollution in the Netherlands and the need for action as observed or advocated by several stakeholders, such as authorities, water suppliers, environmentalists, farmer groups. An important input for the definition of problems are the European standards for the maximum concentrations of nitrates and pesticides in drinking water and groundwater. We expect that the problem context, as perceived by water suppliers, will affect their mission and orientation, and through that the strategy they choose.

A second context is the 'regulatory context', by which we try to identify regulations that give direction to the actions of water supply companies, not only through their mission and orientation, but also by setting their competence and discretionary power as an organizational resource and part of their ability to control. Although water supply companies are completely responsible for the drinking water quality, they have no responsibility for the quality of groundwater. So, if they want to control agricultural pollution of groundwater, they will meet other agencies with formal authorities in this field. In chapter 7 we describe how the Dutch national government is trying to regulate farmers and the amounts of fertilizers and pesticides they use. The national legislation on fertilizers and pesticides sets a regulatory framework within which water supply companies can develop their own strategy.

A third context that is assumed to be of influence on the strategy of water supply companies, is the 'network context' by which we seek to describe how water suppliers in general are dealing with agricultural issues and how the agricultural community is dealing with environmental issues. We expect that the way both communities are interacting (structural dimension), share cognizance (cognitive dimension), and sympathize with each other (affective dimension) will affect whether and how water suppliers are going to encourage farmers to take preventive action. Here, we use some theoretical concepts of the policy network approach as described in chapter 2, such as the concept of 'interrelatedness' for describing the structural dimension of networks, and the 'commitment' concept for describing the cognitive and affective dimensions of networks. The extent of interrelatedness in a policy network can be described as the extent and intensity of the interactions between individuals, groups, and organizations within the network. The extent of commitment within a network can be described as the extent to which individuals, groups, and organizations within the network sympathize with each other's objectives, as far as these objectives are relevant to the policy area (Bressers and Kuks, 1992).

In chapter 7, we focus on the actions of water supply companies in the five selected provinces. First, we give a general overview of the way in which provincial authorities and water supply companies were involved in policies for groundwater protection areas during the ten year period of our study. We learned during the case study that two different periods could be distinguished in all five provinces. In a first period, from 1988 to 1991, provinces delineated groundwater protection areas, ordained regulations for agricultural operations in these areas, while most larger water supply companies negotiated with representatives of the regional agricultural communities on agreements for financial compensations, trying to make the regulations for the protected areas work. In a second period, from 1991 to 1995, expectations about the effectiveness of the ordinances with compensation agreements declined, while expectations about the effectiveness of another kind of policy strategy ('stimulation policy') grew. Although it is possible to discern a general picture in all provinces, we also found differences among the five selected provinces, which at least are related to different ways in which the problem context is perceived. Therefore, we also describe the problem context and the policy approach followed for each of the provinces separately.

Secondly, we discuss the extent to which water supply companies are motivated to encourage the farmers to take preventive action. The motives of the water suppliers - or the factors driving them - are closely related to what has been called in chapter 2 their 'mission and orientation', which is an element of their control capacity. They can be understood in relation to the three contexts we describe in chapter 7. The seriousness and size of water problems (problem context), the presence of specific regulatory provisions or gaps (rule context), and the way water suppliers and farmers get along with each other (network context), all have some influence on the mission and orientation of water supply companies.

Thirdly, we take a look at the organizational resources which are available to water supply companies, and the way each of these resources might help them to gain control over water pollution. We are interested not only in the resources that are available to water suppliers, but also in resources to which farmers are susceptible and which might change the farmer's behavior, if they were to be transferred to them. The resources we distinguish are organizational capacity, financial capacity or money, information, confidence, authority, and time (cf. Klok, 1995).

In drawing our conclusions on our case study, we suppose that the motives and resources of water supply companies determine their strategies on how to control the agricultural pollution of drinking water sources. This strategy might be a remedial, a preventive one, or both. As far as a strategy is prevention oriented, part of the strategy are the instruments they use to influence farmer's behavior. The application of instruments will be considered as the transfer of specific resources from water suppliers to farmers. The susceptibility of farmers to resources, and the extent to which this results in their adoption of preventive measures, will be our measure of the presence of control capacity on the water supplier's side towards prevention.

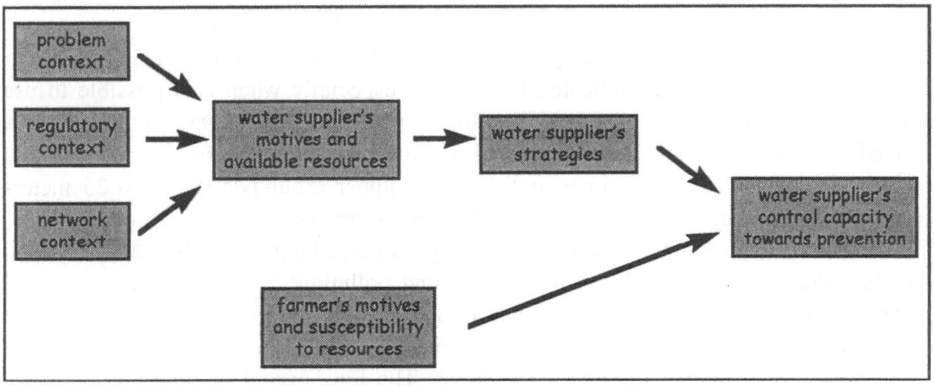

Figure 7.2: Research model for the case study on the Netherlands.

7.2 Regulating nitrates as a source of groundwater pollution

7.2.1 NITRATES AND THE EU DRINKING WATER STANDARDS

Nitrates are a very serious problem in the Netherlands, although the facilities for the purification of groundwater from nitrates are better than those for pesticides. The maximum level of acceptance for nitrates in drinking water is 50 mg/l. This is required by the EC Drinking Water Directive, that has been implemented in Dutch legislation in 1984 ('Drinkwaterbesluit'). The revised EU Drinking Water Directive of 1994 has been implemented in Dutch legislation in 1995. Of the 220 groundwater well fields in the Netherlands, about 70 are very vulnerable to agricultural pollution (nitrogen and pesticides). In the period 1970-1986, an evident increase in the nitrate level in the untreated groundwater appeared in more than 25% of the vulnerable well fields in the Netherlands. In the mid-1980s at two well fields (Montferland in Gelderland and Reuver in Limburg) radical interventions were necessary to decrease violations of the drinking water standards in the untreated groundwater. In some cases the siting of the well field had to be changed, either by moving to sites with less intensive agriculture or by abstracting water from a greater depth (RIVM, 1991).

In the mid-1980s it was calculated that - in the absence of any manure policy - 25% of the well fields would show nitrate concentrations of more than 50 mg/l in the short term. In 1989, an evident increase of nitrate concentrations was observed at well fields, especially in the east of Overijssel, the east of Gelderland, the east of Noord-Brabant, and the south of Limburg. It was expected that within ten years a large number of groundwater sources would require nitrate purification. In 1989, the 'Advisory Committee on Perspectives for the Agricultural Sector in the Netherlands' reported that, as soon as the year 2000, 50 million m^3 of groundwater would need nitrate purification. The groundwater abstracted for drinking water purposes amounted in 1990 to about 800 million m^3. In 1989, it was expected that in 2050 the standard for nitrates in drinking water would be exceeded at about 35 pumping stations (V&W, 1989).

In general, the groundwater quality is still sufficient to prepare drinking water without any excessively complicated technology, especially when it is possible to use water from deep aquifers, covered by protective layers. However, in the shallow groundwater, even around the water well fields, a lot of pollution has been found already. In 1989, the nitrate concentration in the upper groundwater (10 to 25 meters below the surface) appeared to exceed the drinking water standard (50 mg/l of nitrates or 11.3 mg/l of nitrogen) in about 70% of the agricultural land in sandy areas. The soil structure determines the mobility of underground pollution. The age of the abstracted water might vary from hundreds of years in areas where the groundwater is very well protected underground, to less than 25 years in areas where hardly no protection for the groundwater is available, like the sandy areas. The long period of time during which water travels underground explains why the increasing agricultural use of nitrogen, which started about 40 years ago, first came out in the 1980s. A lot of nitrogen from the past is still on its way, and that process is irreversible. It is expected that, despite policy measures, the same nitrate concentrations will show up in the groundwater, at least until 2010. Therefore it is also expected that purification techniques will be unavoidable for the next decade (RIVM, 1991; Vinkers, 1991; Van den Nieuwenhof, 1995).

7.2.2 REGULATORY FRAMEWORK FOR NITRATES AT THE NATIONAL LEVEL

On the national level in the Netherlands, there are two related legal frameworks that create a basis for a national manure policy: the Soil Protection Act and the Fertilizers Act. Both items of legislation came into force in 1987. However, while both acts were signed by the ministers for agriculture and the environment, the Soil Protection Act was formulated mainly by the Ministry of Environment. This act focuses on the environmental effects of excessive manuring, and seeks to regulate the process in which manure is applied. These standards do not directly affect the livestock holding process itself, and thus the production of manure. Instead, by setting standards to the application of manure, the concept of manure surpluses is defined, with the attendant problems.

The Fertilizers Act was formulated mainly by the Ministry of Agriculture. This act focuses on the problems with surpluses and on the production of manure itself (the livestock holding process). The policy instruments created by this act can be typified as financial and communicative incentives. The preference for these types of instruments may be explained by the existence of a strong 'interrelatedness' and a strong 'commitment' within the traditional agricultural network, as it existed at least until the end of the 1980s (cf. Bressers and Kuks, 1992). As a result of this, the objective of economic growth for the target group dominated the traditional Dutch agricultural policy. Setbacks in growth were compensated by incentives and instructions to encourage technological innovations. The actors involved had a great confidence in technological solutions and in the ability of the target group to govern itself (Kuks, 1988; Dietz and Termeer, 1991; Baldock and Bennett, 1991).

7.2.3 MANURE POLICY (FIRST AND SECOND PHASE): THE SOIL PROTECTION ACT AND THE FERTILIZERS ACT

The Soil Protection Act. The introduction of manure standards in the Netherlands can be divided into three or four phases. The standards in the first phase (1987-1990) were chosen so as to avoid the production of surplus manure in the country as a whole (to avoid export of manure). This phase only achieved a nation-wide spreading of manure and a consequent reduction of environmental damage in concentration areas. While setting these standards, the government took account of 'the tension between what is desirable from environmental and hygienic viewpoints, the possibilities of finding practical solutions to the surplus manure problem and the financial and economic consequences for the agricultural sector' (VROM, 1986).

TABLE 7.1: National standards for the use of phosphate in animal fertilizers

Phase	Pasture	Cornfield	Arable Land
1987-1990	250[1]	350	125
1990-1995	200	250 (in 1991) 200 (in 1992/93) 150 (in 1994)	125
1995-2000	175	125	125
from 2000[2]	110	75	75

Source: VROM (1986).

1 Figures are phosphate in kg/ha/y.

2 Final standards based on 'balanced manuring'.

The Soil Protection Act conferred on the provinces the competence to mark out groundwater protection areas, for which a provincial groundwater regulation must have been passed by 1989. This competence enabled provinces to establish a supplementary, specific level of protection for vulnerable areas in which groundwater is used for the production of drinking water. For the provincial regulation of pesticides, the previous competence, based on the Pesticides Act, was maintained (Meijden, 1988; Brauw and Naaijkens, 1989).

In 1989, most provinces introduced standards for ground water protection areas that are close to the national standards for the third phase. So, for the period from 1989 until at least 1995, a difference was created between the regulation for farmers inside and outside the protected areas (Geleuken, 1991).

The Soil Protection Act states that farmers within the protected area should be financially compensated for the losses (costs of disposing of the surplus manure) they suffer as compared to farmers outside the protected area. Requests by farmers for compensation will be dealt with by the provinces, which can collect funds for this by means of a charge paid by those who abstract groundwater (mainly the water supply companies). The levy is related to the amount of water abstracted. Water supply companies can charge the consumers of drinking water for this levy. In fact, the consumers of drinking water pay for the production of a collective good (according to the profit principle).

The reason for the compensation provision was to prevent protests by farmers in groundwater protection areas who would be faced with more restrictions than farmers outside those areas. Legislators feared that the more restrictive standards in groundwater protection areas could not be enforced without compensation. Left wing political parties opened the discussion on a motion to reject the provision because it would implicitly acknowledge a 'right to pollute'. The provision was felt to contradict the 'polluter-pays-principle'. In the event, the 'equality principle' supervened. The consumers of drinking water seem to have had no voice in this political debate. They pick up the costs of pollution caused by farmers, as they do in other countries, such as England and Germany (Kuks and Neelen, 1991).

The Fertilizers Act. By setting manuring standards, the Soil Protection Act creates a disposal problem. The greater the restrictions on the spreading of manure, the more difficult it is to dispose of the surplus manure in a responsible manner. The Fertilizers Act provides, again through a series of implementation orders, a supplementary policy which is intended to offer a solution to the surplus problem. The most important objectives of regulation by the Fertilizers Act are:

1. the determination and registration (*manure accounting*) of the surplus manure produced by the farming community;
2. the disposal of manure surpluses (*manure banks*);
3. expansion of manure production is forbidden (*reference quantity*); and
4. division of the costs of combating the problem between the farms with manure surpluses (*surplus levy*).

Keeping account of the quantities of manure is compulsory for those producing manure (livestock farmers), for managers of manure storage plants and manure processing installations, and for manure traders. The manure accounts are kept unofficial forms. This bookkeeping should show whether a farm has a surplus of manure and, if so, how large this is. If there is a surplus, the accounts, the consignment notes accompanying loads of manure sent for disposal, and the actual amount of manure kept in storage, should proof that the surplus has not been spread on the farmer's own land. The General Inspection Service (AID), an agency of the Ministry of Agriculture, checks whether the accounts have been correctly administered.

The Manure Act also provides for the introduction of manure banks, which serve to promote the efficient disposal of manure. Manure banks mediate in the transaction, processing or destruction of manure. They are obliged to take over a surplus on request by a member of the farming community, and they will try to dispose of the surplus to arable farmers or market gardeners. If they cannot get rid of the surplus, then it must be processed or destroyed. The costs of storage, marketing, and processing are charged to the owner of the surplus manure.

As long as there are insufficient solutions to the manure problem in the Netherlands, it is imperative that a continued increase in the production of manure is avoided. An expansion ban has been introduced to combat this problem, which entails that the farmers producing over 125 kg/ha/y of phosphate may not increase their production. To determine whether a farm has increased its production of manure, the reference situation of each manure-producing farm was set at December 31, 1986 (reference quantity). Farming land is therefore subject to not only a milk quota (in accordance with the EC 'superlevy'), but also to a manure quota. Both quotas are tradable in conjunction with the land.

The livestock holding farmer who produces more than 125 kg of phosphate will be subject to a surplus levy. The proceeds from this levy will be used to finance measures, which contribute to solving the manure surplus problem. This applies in particular to long-term solutions, such as the creation of provisions for efficient disposal, transportation and processing of manure surpluses, and financing of the manure banks. The surplus levy is progressive: no levy is charged on a production of less than 125 kg phosphate; however, production of more than 200 kg phosphate will result in double the normal levy charged. Furthermore, discounts may be granted on certain conditions, for instance, if a low-phosphate fodder is used, or if the manure producer him/her self has found a more or less permanent way of disposing of the manure. The discounts have been introduced as an incentive for farmers to find their own solutions for the surplus manure.

7.2.4 MANURE POLICY (THIRD PHASE) AND THE EU GROUNDWATER DIRECTIVE

In 1994, the national government was hesitant to pass legislation for the implementation of the third phase. It argued that a tightening of the manure standards would not make the targets for the environmental quality more feasible. They expected a delay in the policy's effectiveness of about 15 years (2015 instead of 2000), because of the problems with handling manure surpluses in the Netherlands. They also argued, that the manure standards, as planned for the third phase, only regulate the amount of phosphates in animal fertilizers. Emissions to the groundwater due to the use of artificial fertilizers containing nitrogen, are not affected by these standards. And the introduction on the market of cattle fodder containing less phosphate created an imbalance in the ratio between phosphates and nitrogen in animal fertilizers.

The government has greater expectations of a 'mineral policy' instead of a 'manure policy', which means that they prefer standards that regulate the loss of minerals (including phosphates, nitrogen and other minerals) at a farm. Such a policy should affect the total use of fertilizers (animal as well as artificial). A crucial part of this policy is that farmers have to work with an accounting system, called a 'mineral balance', in which they record the input and the output (including losses) of minerals at their farm. Encouraging the prevention of and search for mineral losses can make farmers more aware of the seriousness of the manure problem than a new tightening of the standards. Such a mineral policy should include a policy settlement in which the agricultural sector commits itself to reduction targets for the coming years. The implementation of these reduction intentions should be left to the sector itself, which could be effective (looking to the effectiveness of the settlement which aims to reduce the use of pesticides). The target group will be moved by this policy in the direction of more self-regulation (VROM, 1993-1997).

Meanwhile, in June 1991, the EU Council of Environment Ministers passed a directive for nitrates in groundwater. The purpose of the directive is to reduce the agricultural emission of nitrates to ground- and surface water. The directive prescribes, that nitrate concentrations in the upper groundwater should not exceed a level of 50 mg/l, which corresponds to the drinking water standard. To guarantee this, the EU wants to impose a manure standard of at most 170 kg nitrogen from artificial or animal fertilizers per ha, to start in 1999. In the period 1995-1999 a maximum of 210 kg/ha should be realized, unless it can be proved that a higher concentration would not harm the quality of the environment. A member state that wishes to deviate form the standard, has to propose its arguments to the European Committee. Enforcement is not based on the actual loss of nitrogen as found in the groundwater, but on the loss as recorded in the farmers' administration.

As a consequence of this directive, the Netherlands as a whole have been considered to be a vulnerable area since 18 December 1993. This means that not more than 50 mg/l of nitrate is allowed in the upper groundwater throughout the country. The EU directive should be implemented in the Dutch manure and mineral policy for the third phase by means of a tightening of the standards for the use of animal and artificial fertilizers. The implementation should have been in place before 18 December 1995. The Dutch government has given an explanation to the European Commission as to why it expects problems with the implementation of stricter standards. It also has explained that the Dutch government wishes to introduce a mineral policy, although they believe that it cannot be passed as legislation before 1997. The Dutch government stated that, in its view, there were three implementation problems: firstly, the necessary policy instruments would not be available before 1997; secondly, the capacity for handling manure surpluses is still too small; and thirdly, the gap between what is needed for the environment and what is feasible from an agricultural point of view can hardly be bridged.

In Brussels, the Dutch government stated that it preferred a system of mineral accounting with an compulsory declaration to achieve a reduction in nitrates losses. This system uses standards for the spreading of manure are used, but standards for the loss of minerals. Initially these standards would be more relaxed than the standards required by the EU. However, after a period of time the standards proposed by the Dutch might be more successful (effective in reaching reductions). The EU still doubts the effectiveness of the proposed Dutch instruments. In the Netherlands both the agricultural sector and the national and provincial governments feared that the EU would obstruct the Dutch proposal.

The Dutch policy makers realized that balanced manuring - with the implication that farmers would apply more minerals to the soil than necessary for crop growth - is not feasible. In the Netherlands, with its unstable weather patterns, specific soil structure and intensive agricultural production, this would lead to a serious production loss. Yields of twelve tons of raw materials per ha per year cannot be achieved without an excessive supply of minerals. The idea of balanced manuring was amended to allow a certain loss of minerals (LNV, 1993; VROM, 1993-1997).

In the end, in October 1995, the Dutch government presented its proposal for the third phase of the Dutch manure policy. The policy in this proposal is aimed at farms who pose the greatest risks to the environment (farms with high stocking densities). The government wants these farms to start with minerals accounting. Minerals accounting is an input-output bookkeeping system, that relates total applications of fertilizers to production. If the losses, which must be reported yearly, exceed the standards for phosphate and nitrogen set for that particular year, a fine is levied on the surplus. Legislation aimed at the introduction of minerals accounting, is being prepared. It is to enter into force on 1 January 1998. From then on, farms with stocking rates of more than 2.5 LU (Livestock Units) must report their mineral losses. In 2002, the threshold will be lowered to 2 LU.

When minerals accounting is introduced in 1998, the loss standard for nitrogen will be 300 kg/ha. In 2005 it will be set at 200 kg/ha, and in 2008/2010 at 180 kg/ha. In other words, the government will not tighten the standards before 1998. The European Union standard will not be met before 2010 (LNV, 1995).

The Dutch government's proposals provoked a storm of protests, not only among the farmers, a large group of which were opposed to the rigor of the proposed regulations, but also among the environmental movement and the water supply companies. The water supply companies, unified in Vewin, think that these proposals are quite insufficient to meet the European directive for nitrates in groundwater and, in fact, allow groundwater pollution for an extended period. The Vewin fears that water supply companies in the near future will have to invest heavily in nitrate purification systems, which might cost at least 200 million guilders per year. If that is the case, farmers should partly compensate the costs, in Vewin's view, for example by paying a levy on nitrogen losses which could be destined to reward farmers, if they sanitize the loss of nitrogen. Water supply companies believe the drinking water consumer cannot be the one to pay for the purification costs as a consequence of the excessively weak manure policy. Otherwise the water price in the provinces with intensive agriculture, like Limburg, Noord-Brabant,

Gelderland and Overijssel, would rise to two or three guilders per 1,000 liters. For an average household the extra costs due to the manure policy, could be 150 Dutch guilders a year, according to calculations done by the water supply sector. The water supply companies advocate advancing the implementation of the European directive and no longer waiting until 2005 to meet the standard for nitrogen use.

The environmental movement expects that the European Commission will not approve the Dutch proposals for a manure policy, since they do not meet the European nitrate directive on 60 to 70% of the sandy areas, which is 20% of the total agricultural land in the Netherlands. More than half of the groundwater well fields are in this area. If the Dutch government will not comply with the European directive, Brussels can impose a penalty, and every other interested party (environmental organizations or water supply companies) can even go to court to appeal for compliance (VROM, 1993-1997).

Almost 600,000 farms, or 13% of the total number in the European Union, do not meet the requirements of the European directive for nitrates in groundwater, and have to modify the way they farm. In the Netherlands, 63% of the farms do not meet the standard. These conclusions are contained in a report by the Dutch Agriculture Economic Institute (*Landbouw Economisch Instituut*, LEI) to the European Commission (November 1995). Only the former twelve member states were included in this study, not Austria, Sweden and Finland. On average, the 600,000 farms that do not meet the standard are producing 350 kg/ha of nitrogen. The average production for all farms in the EU amounts to 70 kg/ha. The differences among farms, as well as among member states are huge, and member states with the highest nitrogen surpluses are Belgium, Denmark, Germany, France (Brittany), Luxembourg and the Netherlands. With 321 kg/ha the Netherlands head the ranking of member states, with Belgium at 170 kg/ha following, while lowest in this ranking is Portugal with 6 kg/ha. The number of farms in the Netherlands that do not meet the Nitrate Directive is about 60,000, or 63% of the total number, while in Belgium this number amounts about 50%. Together with France (Brittany), Spain (Galicia) and Italy (Lombardy), these member states also have the highest livestock density. In contrast to the Netherlands, Belgium, Denmark and a large part of Germany, high livestock densities and high nitrogen losses in the other member states are concentrated in rather small regions (VROM, 1993-1997).

7.3 Regulating pesticides as a source of groundwater pollution

7.3.1 PESTICIDES AND THE EU DRINKING WATER STANDARDS

Two types of norms are available for the quality of drinking water. On the one hand substances like nitrates or arsenic are not allowed to exceed certain concentrations in the water, while on the other hand just absolute standards are used for substances like pesticides, referring to the daily intake, regardless of their concentrations in drinking water or food. For a lot of known substances, toxicologists have assessed the amounts in which they will be toxic to the human body. In case of toxic substances, maximum concentrations are set at the level of one hundredth or one thousandth of the safe amounts for a daily intake without harm to one's health. For all pesticides that are applied to consumer products, these limits have been very accurately assessed. A pesticide only will be approved, if the residue on the product stays below the safety limit, which is why it is possible for us to eat an apple from a tree, although it might have pesticide residues on its skin. Besides the toxicology of consumer products, like fruits and vegetables, pesticides can also be toxic to the environment. To review a pesticide on this property, one has to know how long the pesticide might be active or resistant underground, how harmful it might be to underground organisms, and how much of it is allowed to leach to the groundwater.

In the 1970s, a standard for pesticide residues was introduced in the Netherlands at a maximum of 0.1 μg/l drinking water. This standard was chosen after some utilities had found pesticides in the groundwater used for drinking water production, which made consumers and water suppliers very nervous. Their immediate conclusion was, that pesticides should not be allowed in drinking water at all, which seemed to be a logical conclusion in those circumstances. The standard has been fixed on the basis of the detection limit, which is the smallest amount that can be measured. Lower values than 0.1 μg/l were not measurable, given the state of the art at that time, which in fact made it a zero-base standard. This is called the 'precautionary principle'.

After the concern expressed by the water supply companies, this new standard brought action from the producers of pesticides and the Dutch trade association Nefyto. In their opinion, the setting of this standard strongly reduced the variety of pesticides that could be brought onto the market, since the standard indicates that many of the pesticides on the market are leaching in greater or lesser quantities to the groundwater. Nefyto reasoned that, even if some pesticides were no longer to be used, they would still show up in the groundwater since they are already underground. Nefyto then concluded, that it would be better to develop standards for each pesticide separately, since it is not the presence in water, but the concentration that sets the toxicology.

In 1980, the European Commission fixed the standard at 0.1 μg/l in a European directive for drinking water. Since then the chemical industry has been permanently exerting pressure on the European Commission to change the directive. The industry complains, that in fact they are paying for the production costs of drinking water, causing them to lose billions of money, while the standards are quite arbitrarily chosen. An alternative would be to set differentiated standards according to the toxicology of each type of pesticide, based on the WHO Guidelines. Such toxicity based standards would mean a considerable relief of the restrictions for many pesticides.

By contrast, environmental organizations and water supply companies oppose to the idea of having differentiated standards for 700 different types of pesticides, which would be too complicated to enforce, in their view. Dutch water supply companies prefer to adhere to the precautionary principle, as do consumers' organizations. Some people even suggest being more strict than cautious, since technical progress has provided more precise measurement devices nowadays. In 1993, an audit by the Dutch Consumers' League found residues of the pesticide atrazine in the drinking water on three locations in the Netherlands. Although the residues were below 0.1 μg/l, the Consumers' League reasoned that substances like atrazine should not be present in the drinking water at all, if one adheres to a precautionary principle (VROM, 1993-1997).

In the meantime, both Dutch and European authorities have been very pragmatic in enforcing the 0.1 μg/l standard. Violations of the standard do not always result in legal action, as they are reviewed first for their toxicological relevance, and the adoption of enforcement measures is considered only thereafter. In 1993, research by the National Institute of Public Health and the Environment (RIVM) showed, that 65% of the agricultural land in the European Union is polluted with pesticides.

In the Netherlands, groundwater is investigated for pesticides only incidentally. There are not enough data available for a national survey. In recent years excessive high concentrations of about 19 different pesticides have been found in the shallow groundwater, while excess concentrations of six pesticides have been found in drinking water (RIVM, 1991).

To work with a 0.1 μg/l standard for pesticides, is also troublesome for the water supply companies. They are responsible for purifying water to the point where pesticide concentrations are reduced to the standard level. They have to incur great purification costs to do that, which is reflected in the price of drinking water. The increasing possibilities for the detection also has the effect, that even more substances are being discovered, and that, as a consequence, increasing investments in purification techniques are required. The European Commission estimates that investments in the drinking water supply in Europe will increase by 9% per year (8% in the Netherlands). By 2000, it is expected that investments in the drinking water supply sector will be twice the amount spent in 1993, according to a speech by EU Commissioner Paleokrassas at the EU Conference in September 1993 on the revision of the Drinking Water Directive. 'We have to investigate the relation between public health and the costs of drinking water, although public health will be in the first place. It is more a matter of who is going to pay for the costs,' the Greek Commissioner explained. He also emphasized, that there is no doubt that consumers in the future will have to pay more for their drinking water. To show how inexpensive drinking water is, he emphasized that the European consumer still gets 1,000 liters of water for the price of one liter of beer.

In June 1994, the European Union passed a revised directive for pesticides in drinking water, in which it was opted to apply in most cases the enlarged toxicological standard for the leaching of pesticides into groundwater. The standard of 0.1 μg/l was to apply only in areas where water is used for the production of drinking water. As a transitional arrangement, the member states are allowed to apply the enlarged toxicological standard until 1999. In special cases, member states are allowed to designate areas in which the drinking water standard has to be applied, although groundwater in those areas is not directly used for the drinking water production. This is an arrangement that fits quite well with the principles of Dutch environmental policy, and the Dutch government made use of it. It was a great surprise to the organization of pesticides producers, Nefyto, to see that within a few hours after the European ministers in Brussels agreed on the directive in Brussels, the Dutch government in The Hague came out with a legislative proposal to designate the entire Netherlands, because of the enormous amount of water, as an area where groundwater should be of appropriate quality for the supply of drinking water. In July 1995 the Dutch government incorporated the EU directive for drinking water into the Drinking Water Decree of 1984.

Meanwhile, the Nefyto fears that the Dutch government has let itself in for a lot of enforcement problems. As Nefyto put it, how can the Dutch government be that stringent, while there is free trade within Europe? While the Dutch drinking water standard bans specific pesticides, they might be allowed for use in other member states. The European directive implies, that a producer of a pesticide that is forbidden in the Netherlands, but has been approved in Greece, should also be allowed to use such a compound in the Netherlands. However, a spokesman of the Dutch Ministry of Agriculture, which is responsible for the approval of pesticides, stated that this would not mean that such a fact would oblige the Dutch government to approve that pesticide. Member states can have their own approval policies, related to specific circumstances. The Dutch Ministry of Environment thinks, that about 20% of the already registered pesticides have to be reviewed and might be banned. In the context of the approval policy, they will consider which substances will be indispensable for agricultural practice, and which might be remained for that reason (VROM, 1993-1997).

About 700 different sorts of problematic pesticides exist that need regulation. The discussion is about their admissibility. In contrast to nitrates, pesticides are not allowed at all in the groundwater, not even in a certain amount. Therefore, regulation of pesticides needs to provide precautionary measures.

There is a large variety of pesticides. Many of them are still not detectable. So water suppliers are anxious, that they may unaware of their presence in the groundwater. Not even all the effects of the detectable pesticides are known, which is why water suppliers adhere to the principle that pesticides are not allowed to be in the groundwater (VROM, 1993-1997).

7.3.2 REGULATORY FRAMEWORK FOR PESTICIDES AT THE NATIONAL LEVEL

Groundwater protection first gained attention in the Netherlands at the beginning of the1970s. The Provincial Administration Act conferred on the provinces the competence to take care of the protection of groundwater. Since then provinces have to mark down 'water extraction areas', called 'ground water protection areas' since 1987. The regulatory consequences of this provincial competence concentrated on point source pollution, like leaking oil tanks and crashed tank trucks. No one was thinking at that time of regulating agricultural pollution, except for pesticides. From the early 1970s on, the 1962 Pesticides Act already banned the use of certain pesticides in water extraction areas. This had to be enforced by the provinces.

The Pesticides Act prescribes the approval and use of pesticides. A pesticide may only be stored, used or placed on the market after it has been evaluated for damaging side-effects to humans, animals, plants and the environment. A series of regulations issued under the Act lay down various specific requirements, such as the permissible level of residues, safety instructions pertaining to sale, storage and use, and the procedure to be followed in approving a pesticide. The competent authority for approving pesticides is the minister of agriculture (or another minister if the preparation is for use in another sector). Technical advice on pesticides submitted for approval is provided by the Pesticides Approval Commission, a body consisting of representatives from the Ministries of Agriculture, Welfare, Public Health, Environment, and Social Affairs. Approval is for a maximum of ten years, though in most cases a shorter period is specified - typically between three and five years - and the uses to which a pesticide may be put are specified. The approval can be withdrawn instantly, if it transpires that the pesticide does indeed have damaging side-effects. This has happened on several occasions.

For a long period of time, the approval or registration policy based on the Pesticides Act functioned as the crop protection policy. As a result of the increasing criticism of the negative effects of pesticide use (on the environment and the drinking water supply), the government developed a crop protection plan in which the negative effects of pesticide use are discussed. The development of this 'Long term crop protection plan' (*MeerJarenPlan Gewasbescherming*) took a long time, due to the many conflicts that occurred between the departments involved. The government commenced drafting the plan in 1987. Finally, in 1990 a draft was issued and was sent to all the parties involved. The final draft was submitted to Parliament in June 1991. The Committee on Agriculture and Environment discussed the plan in April 1992. As a result of a hearing, the committee asked the government to come to an agreement with the agricultural interest organizations and the industry. In July 1993, the Dutch government signed the negotiated agreement concerning crop protection with the agricultural sector to regulate the use of pesticides. The settlement aims at a decrease in the structural dependence of crop protection on pesticides, and at a general reduction in the use of pesticides. Besides these generic goals, more specific goals were set for the year 2015. Firstly, the use of pesticides should generally be reduced by 50% by 2000 (compared to the amount used in 1987). Secondly, the use of harmful pesticides should be sanitized. Indicators for the predicate 'harmful' are the 'underground mobility', the 'persistence' (degradability), and the toxicity (how harmful they are in a water system). In fact, pesticides that have the potential to leach to the ground water should be due for review and re-registration. It has been left to the sector itself to determine how to reach these goals (LNV, 1991).

The main goals of the pesticides policy can be identified as follows:

1. acquiring more knowledge of the environmental and health risks of pesticides use;
2. improving the efficiency of crop protection, in order to achieve reductions in use;
3. creating better alternatives for the pesticide products that are already available; and
4. prohibiting specific pesticide products.

The pesticides policy mainly targeted at the production of and trade of pesticides and their use or application. Looking at the use of pesticides, there can be effects in two ways: on the residues of pesticides in agricultural products for human consumption; and on the residues of pesticides in the groundwater, which are harmful to the environment in general and to drinking water production in particular.

Pesticides are used by farmers in order to benefit the growing crops. Farmers' participation in the policy making process is driven by the objective of preventing any restrictions on crop protection that may affect the yield. On the other hand, it is in the interest of environmental actors (authorities as well as NGOs) to restrict or prevent the environmental and health effects of pesticides use. A third party in the policy process is the chemical industry. Their main interest is to safeguard the trade in pesticides. This does not imply that they are opposed to the participation of environmental actors. In general the industry is sensitive to matters of reliability, and motivated to work on the prevention of environmental and health effects for which it could be held responsible in the future.

To farmers the use of pesticides is part of their economic activity. Pesticides are regarded as being essential means to reach a certain degree of productivity. Therefore, the pesticides policy is not based on prohibiting the use of pesticides, but on their careful handling. Inasmuch as pesticides may be prohibited, this is done through restrictions on the approval of pesticides which affects the production and trade in pesticides in the first place, with the industry as a target group rather than the farmers. Policy makers consider it to be the producer's responsibility to instruct farmers as how to handle pesticides in order to prevent environmental and health effects as far as possible.

To farmers it is important also to get information about the effects on crop revenues of different ways of pesticides use. In the negotiated agreement on crop protection it is agreed that farmers will reduce the use of pesticides in general on the condition that this will not harm the protection of crops. Central instruments in this policy strategy are education and training activities. It is assumed that farmers will only cooperate, if they can be sure that alternative practices of pesticides handling (for example crop rotation, using more resistant crop types, using pesticides with lower risks) will not harm crop benefits.

The traditional pesticides handling practice shows, that farmers in general are not very knowledgeable about the effects of pesticides, and that they commonly apply more pesticides than necessary to be sure of the crop protecting effect.

Looking at the production and trade of pesticides, the registration procedure sets standards for the approval of pesticides. It therefore has the character of a policy instrument based on prohibitive steering. Through registration the policy makers try to regulate the availability of alternatives for crop protection. The aim of this instrument is not only the prohibition of harmful pesticides, but also the encouragement of the production of new and better alternatives. In that way, the registration procedure also works as a communicative incentive, as well as a financial incentive towards the pesticides producing industry (the target group). The procedure requires that producers investigate all the effects a pesticide may have, before they are allowed to introduce it on the market. In this way it works as a communicative incentive: the procedure generates more knowledge on the effects of pesticides. Related to this, producers are motivated to invest in innovations in crop protection. In that way, the procedure works as a financial incentive. As an effect of the pesticides policy, a new generation of pesticides is beginning to make its way on to the market. A number of companies are investing heavily in the development of low-risk pesticides.

The pesticides policy recognizes the reliance of farmers (and market gardeners) on the availability of technical means (the state of the art). This explains why a policy leaning on optimization of pesticides use can only be partly successful. Pesticide management by farmers themselves assumes, that farmers are able to completely control the pollution of groundwater by pesticides. However, to a great extent the industry producing the pesticides can be made responsible for the effects of pesticides application. They are expected to have more expertise than farmers concerning the environmental and health effects of pesticides. Therefore, the registration procedure is still a crucial element of the policy. It motivates the producers as a target group to invest in research and development for the production and marketing of better products.

Farmers' organizations are aware of the importance of the producers as a target group. It is in the interest of the farmers to shift the discussion about pesticides effects to a discussion about the production of pesticides, instead of a discussion about their use. As long as this discussion does not harm their incomes, farmers' organizations are very cooperative in the policy making process. They accept agreements on self-regulation - as made in the 'Long term crop protection plan' - to avoid policy instruments that will intervene more stringently in the application process. In other words, it is in the interest of the farmers that the policy makers' attention will remain focused on the industry (Smeets, 1990; Reus, 1991; Brand, 1994).

7.3.3 PESTICIDES USE AND THE 'LONG TERM CROP PROTECTION PLAN'

As described before, the Dutch 'Long term crop protection plan' aims at a decrease in the structural dependence of crop protection on pesticides, and at a general reduction in the use of pesticides. Besides these generic goals, more specific goals were set for the year 2015. Firstly, the use of pesticides should generally be reduced by 50% by 2000 (compared to the amount used in 1987). Secondly, the use of harmful pesticides should be sanitized. In fact, pesticides that have potential to leach to the groundwater should be due for review and re-registration. It has been left to the sector itself to decide how to reach these goals.

As far as the first goal is concerned, the plan has been quite successful, in that a reduction in use of the amounts of pesticides (in kilograms) has been achieved. In 1995, the annual emission of pesticides to the soil and the groundwater had been reduced by 80% compared to 1986. The target was a reduction of 40 to 45%. The effect of the remaining pesticides on the environmental quality has not been reviewed (LNV, 1996). Environmental organizations think this achievement can be explained in terms of the enormous reduction in the use of soil disinfectants, while the use of other harmful pesticides has not actually been reduced. Meanwhile, new more concentrated and more harmful pesticides have also entered the market, making the result in terms of reductions in kilograms less laudable (Vereniging Milieudefensie, 1996). In 1995, the Centre for Agriculture and Environment (CLM) and the Agriculture-Economic Institute (LEI), in a preliminary evaluation, recognized the achievement of reductions, although they warned that fewer kilograms used does not automatically imply improvements to the environment. The government would give a misleading signal to the agricultural sector, if it were only to count kilograms and ignore the results in terms of environmental quality (VROM, 1993-1997).

As far as the second goal of the 'Long term crop protection plan' is concerned (sanitation of harmful pesticides), environmental organizations do not believe that a reduction has been achieved; instead there has been an increase of 2% (Vereniging Milieudefensie, 1996). This has been explained in terms of delays in the European harmonization of pesticides policies, and the fact that producers of pesticides went to court to appeal against the sanitation of their products. The illegal use of pesticides stayed outside the statistics. Assessments of the amount of illegal pesticides in the Dutch market vary from 2 to 10%.

For the sanitation of harmful pesticides the approval policy is important. The European Union is working on the revision of a substances list, as a base of allowance of pesticides in the EU in future. Probably only 89 of the 700 pesticides in use will meet the new requirements. Pesticides that are no longer necessary, or pesticides that are evidently harmful to humans or the environment will be banded or restricted in their use on the basis of Directive 91/414/EU on the marketing of phytosanitary products. According to the European environmental movement, consumers' organizations, and labor unions in the agricultural sector, this will not necessarily enhance environmental protection, as there is no guarantee that the 89 remaining pesticides are less harmful. They want the EU to consider more closely the environmental effects of pesticides when revising its list of pesticides (Verdonk and Velders, 1996; VROM, 1993-1997).

7.3.4 REGISTRATION OF PESTICIDES AND THE 'PESTICIDES APPROVAL COMMISSION'

Until 1993, the approval of pesticides was in the hands of the four responsible ministries that participated in the Pesticides Approval Commission. Because of ongoing interdepartmental conflicts about what environmental and public health criteria should be applied, and the consequence that pesticides were neither approved nor banned, the Parliament intervened, and in 1993 it established a new, independent Pesticides Approval Commission. At the same time, Parliament passed the 'Long term crop protection plan', while also approving the requirement that 90 environment critical pesticides should be sanitized before 1995. Three years later (in 1996), the result is that those 90 pesticides have not been sanitized and that even 70 new pesticides have been approved without reviewing them on environmental criteria. Some of the newly approved pesticides are very persistent and toxic substances. This could happen because the ministries involved in the previous commission did not establish any criteria for the newly established commission and they also neglected to supervise the work of the new commission. An evaluation report on the approval policy since 1993 by a task force of departmental officials, makes it clear that obvious standards and criteria were missing in some cases where the commission was forced to make a balanced consideration of the interests involved. The task force writes, that the consideration of interests should not be a matter for the commission, and should be left to the political arena. The responsible minister for agriculture concluded, after reading the report, that the approval policy had failed with respect to the involvement of politics in the implementation work of the Pesticides Approval Commission. The discretion of the commission needs to be restricted as much as possible, he judged. By contrast, the Agricultural Board and the producers of pesticides think, that the Pesticides Approval Commission has insufficient discretionary competence (Brand, 1994; VROM, 1993-1997).

7.4 A policy network point of view: the dynamics of agricultural issues in the water supply sector

7.4.1 WATER SUPPLY SECTOR GROWING INTO A POLICY COMMUNITY

Environmental interests have only recently gained attention within Dutch groundwater management. Initially, groundwater was managed only to supply drinking water and for the related health care aspects. That was why the ministry for Public Health introduced the Water Supply Act in 1957, not only to make demands on the quality of drinking water, but also to institutionalize the organization of the drinking water sector. This act formalized an already existing practice by which provinces could permit or forbid the establishment of new or the expansion of existing water supply companies.

Until the turn of the century, most of the water supply companies were local and private initiatives. The beginning of the twentieth century saw the establishment of companies covering a wider area, often with the participation of municipalities. Municipalities and private persons were not willing to cooperate in all cases, however. To guarantee the efficiency of water works all over the country, several provinces developed regulations governing the water supply on their own initiative, a practice that later found support in the Water Supply Act.

Since 1975, most of the provinces have made plans for a further concentration of water supply companies. The number of companies has been reduced: from 102 in 1980 to 73 in 1988, to 49 in 1990, to 41 in 1993, and to 28 in 1997. But there are plans for a further reduction still, to about 30. The provinces and the drinking water sector itself, as represented by the Vewin (National Association of Water Supply Companies), agreed that the structure of the drinking water sector in its present form does not fit the demands for securing clean water supplies in the future. They believe that the water supply companies can only fulfill their tasks, if they have a strong organization, which implies adequate technological know-how and financial capacity (Vewin, 1991a; Vewin, 1993; VROM, 1993-1997; VROM 1995b).

Those supply companies that use surface water for the production of drinking water, are faced with greater problems in guaranteeing good quality than those who use groundwater. The latter consider themselves to be relatively invulnerable, which is why they offer a lot of resistance to reorganization plans. They argue that they have never had problems with the supply of water, and that they have always had a good quality, that their charges are reasonable, and that their customers are always satisfied (Van der Knaap, 1987). The smaller companies, in particular - most of which do not have the provincial authorities participating as a shareholder - try to maintain their autonomy. But the burden of proof lies with them to demonstrate, that they can still operate in an efficient way. The Vewin is very cautious in taking a stand (Vewin, 1989). Actually, they support the idea of developing more professionalism and efficiency in the drinking water sector. But the Vewin tries to avoid a confrontation with the smaller companies, since it wants to be an organization that represents the entire drinking water sector (Achttienribbe, 1993; Langendijk, 1994).

We may conclude that, until the 1950s, the water supply sector can be qualified as a fairly fragmented network. Though the companies shared a common purpose, they acted individually. Water supply came into being due to local initiatives, and for a long time they sought to maintain their autonomy. After 1957, the sector became more integrated due to two developments. First, reorganizations and mergers between companies led to scale enlargements in the sector and an increasing interrelatedness. Second, the interrelatedness and commitment within the sector increased as a result of the rise of national organizations which were intended to promote the interests of the entire sector, such as the Vewin and the KIWA (a research institute for technological innovations in the water supply sector). The need to increase efficiency in the sector encouraged a strong policy community with a strong technocratic approach to the problems the sector was faced with. Thus, these institutional changes were generated from inside the policy community, based upon a common perception of the way forward.

7.4.2 WATER SUPPLY SECTOR ENCOUNTERING THE AGRICULTURAL SECTOR IN AN ISSUE NETWORK

The presence of nitrates in groundwater poses a serious threat to the drinking water supply. More than 50% of Dutch water extraction locations (especially those in the east and the south) are likely to become unsuitable as drinking water sources in the near future. They are situated in areas with a sandy soil and many intensive cattle farms . Although there were already indications that excess manure was being produced in some regions in the Netherlands as early as the mid-1960s, it took until the 1980s before the government began to deal seriously with the manure problem. The delay was due to a demarcation dispute between the Ministries of Environment and Agriculture. The Soil Protection Act, which came into effect in 1987, is the first Dutch legislation that aims at integrated protection of the soil and groundwater. It is primarily aimed at preventing excessive manuring, by intervening in the spreading of manure on agricultural land. The Act contains a number of standards to fix the amount of manure which is allowed to be spread on the land. These standards apply nationally.

As described in section 7.2.1.1., the Soil Protection Act provides additional protection for areas in which groundwater needs to be withdrawn for the supply of drinking water.

Provinces are authorized to enforce more restrictive manuring standards in these areas. Farmers within the protected areas should be financially compensated for the losses by means of a charge paid by those who abstract ground water (mainly the water supply companies). The reason for the compensation provision was to prevent protests from the farmers in groundwater protection areas, who are faced with greater restrictions than farmers outside those areas.

The drinking water sector was thus confronted by a very strongly organized opponent from outside the sector - the farmers. Although the agricultural sector constitutes only five percent of the active labor force in the Netherlands, it has had considerable influence in Dutch politics. Since 1954, the agricultural sector has been nationally organized in the Agricultural Board. The agricultural lobby was very effective, which is, for example, reflected by the fact that it usually is consulted at a very early stage of policy making. The Agricultural Board also has regional boards in each province, that consult with the regional or local authorities, particularly about planning and land use. The regional boards played an important role in determining the manure policy for groundwater protection areas. They negotiated with the provincial authorities and the water supply companies about the manure restrictions that are to be applied in these areas, the disposal of the resulting manure surpluses, and the compensation payments for farmers (Dekker, 1988).

While the agricultural sector is strongly developed at all levels - national, regional, and local - the drinking water sector was for a long time not so well organized. Traditionally, water supply companies are proud of their autonomy. The Vewin always played a modest political role, being only active in emergency situations, such as the 'bentazon affair' in 1988, while fulfilling its main task of providing service to the drinking water branch. At the end of the 1980s, the Vewin became more politically involved and responsive, although the political discussion on the Soil Protection Act had already been concluded. When the presidency came into the hands of the governor of the province of Zuid-Holland, the Vewin succeeded in moving the negotiations concerning the compensation in groundwater protection areas from the regional level (with one negotiator from the Agricultural Board representing the whole country negotiating with each water supply company separately) to the national level. Since then, the framework for regional negotiations is the state level.

We conclude that external interests and pressure on the water supply sector in some sense strengthened the sector as a policy community. It became a more tightly organized policy community, especially in terms of an increasing commitment between the members of the community. In the meantime, however, the issue of agricultural pollution of groundwater confronted the water supply community with another strong policy community, the agricultural sector. This sector is very experienced and has a long tradition in lobbying and negotiating strategies. The agricultural sector as a network is not only characterized by a strong commitment, but also by a strong interrelatedness. This strong interrelatedness forced the water supply sector to participate in negotiations on a more aggregated (regional and national) level. In fact, it strengthened the interrelatedness within the water supply sector (Kuks, 1988, Den Breejen, 1991).

7.4.3 CONSULTATIONS BETWEEN WATER SUPPLY COMPANIES AND FARMER ORGANIZATIONS

The discussion of the compensation payments shows, that the regulatory strength ('control capacity') of the authorities is very limited in the case of groundwater quality management. Although they try to regulate by means of ordinances and prohibitions, the enforcement of these rules is difficult. Therefore, a system has been chosen in which private organizations (water supply companies) have to participate in the enforcement of the rules.

The drinking water sector is beginning to define its role, however. In 1990, the director of Vewin stated:

> *This is a considerable change for organizations which traditionally are engaged in pumping and presenting the bill. However, if you want to create groundwater protection areas, then others may have fewer opportunities to use the soil in these areas. If that is the case, groundwater protection can only be realized by offering compensation to them. This is not a new phenomenon: the drinking water sector already contributes one third (about 20 million Dutch guilders) of the costs of the Rhine Salt Treaty. In the Third National Policy Document on Water Management this is called 'paying for quality'.* (Jehae and Van Soest, 1990 [translation by the author])[1]

The compensation provision formally implies that farmers can claim their losses against the province, which in turn may charge the water supply companies that withdraw groundwater in that area. Water supply companies prefer to settle the matter in a friendly atmosphere. They also want to control the disposal costs of manure surpluses and to avoid unnecessarily high bills. For both reasons, several water supply companies have started to act in the matter of disposal of manure surpluses, which in itself is a very unusual task for a water supply company. An increasing number of companies are also attempting to buy out farmers in their most vulnerable areas.

Although methods exist for the purification of groundwater that is polluted with nitrates, the drinking water sector strongly opposes this option as long-term solution. The director of Vewin explains:

> *If the water supply companies started with complete purification tomorrow, the polluters could no longer be forced to change their behavior. We want to use the drinking water, and with that the consumer, as a crowbar for improving the environment.* (Velema, Boer and Verheul, 1989 [translation by the author])

The strategy of consultation, chosen by the water supply companies to deal with agricultural pollution of groundwater, had already been tested by the companies which use surface water from the large rivers for their drinking water production. Those companies are mainly located in the western part of the Netherlands. They have huge problems in maintaining a good water quality. A great part of the pollution in these rivers stems from foreign industries, which means that they are dealing with extra-territorial actors. The director of Vewin indicated that Vewin is reluctant to develop new

[1] See also V&W (1989) and Van der Schot (1995).

techniques for analyzing water for the purpose of tracing polluters. However, the data can often be used to exert pressure on polluting industries. For example, the City of Rotterdam has chosen to consult with polluting industries, even when they are located abroad, rather than taking judicial action. In 1990 a spokesman of the City of Rotterdam stated:

> *We try to handle the collected data very carefully, just because we want to get in conference with the discharging industries. Negative publicity will be applied only when the polluter is really unwilling. However, threatening publicity has proved to be a very strong instrument.* (Jehae and Van Soest, 1990 [translation by the author])

This demonstrates that the drinking water sector realizes that it has its own role to play in water management, and that it can often reach further than any other authority. This seems to be true for the case of point source pollution. But is this also true for non-point source pollution? Consultation with polluters was regarded to be the only solution to the problem, since the national and provincial authorities were unable to guarantee strict enforcement of regulations. However, the case of groundwater protection areas also demonstrates that the possibility of using a consultation strategy depends heavily on the extent to which non-point source polluters are organized. The agricultural industry is very well organized, and that may be one of the reasons why they became the first target group to be addressed by the Soil Protection Act in 1987.

In practice, the approved approach seems to have been based largely on self-regulation. Thus, the secretary general of the Ministry of Agriculture, when he signed the negotiated agreement on the use of pesticides, argued that the target group was also consulted on the reduction of the use of pesticides by farmers.

> *One can prohibit the use of pesticides, but the need for them won't disappear. Such a measure has hardly any effect, because the farmers would use other or illegal means which could be worse on a large scale. Actually, the reason for using excess pesticides nowadays is that farmers have too little knowledge about how to use them and they won't take risks. A policy strategy should be more directed to an increase in expertise by means of the branch itself.* (*NRC Handelsblad*, December 23, 1988 [translation by the author])

However, it remains very difficult for authorities and water supply companies to address the problem of non-point source polluters in less organized branches. One of difficulties of tracing polluters is that pollution reaches the groundwater only after a long period. Even if farmers entirely ceased manure spreading, it would still take fifteen to thirty years before all the current nitrate pollution has come out with the groundwater that is used for the production of drinking water.

Finally, we may formulate some conclusions on the way in which the water supply community dealt with external threats. On the one hand, water supply companies reacted in a very technocratic way by searching for technological innovations to satisfy the demand for drinking water of an acceptable quality. On the other hand, they tried to react against the Soil Protection Act which ignored the polluter-pays-principle. They did not succeed in their opposition, although they were strongly supported by environmental groups. The water supply sector and the environmental groups are united in their support for the polluter-pays-principle. In this sense, the strong relation between the water supply sector and environmental groups can be conceived as the existence of a broader policy community. However, the interrelatedness in this community is weak: no strong or intensive interactions exist between both sets of actors. In the end, the polluter-pays-principle was not applied due to a successful lobby by the agricultural sector and due to the lack of political organization and influence of the drinking water consumers. It appeared that the best feasible political outcome was to saddle the consumers with the costs of pollution prevention.

Another indication for the existence of common interests of the water supply sector and environmental groups, is that they both stress the importance of strict enforcement of the rules. However, water supply companies realize that it is very difficult to control the spreading of manure. The control capacity of the regulatory agencies (provinces) is limited in this respect. That is why the water supply sector expects better results through direct negotiations with farmers' organizations. After the settlement of the compensation provision in the Soil Protection Act, 'consultation with target groups' was left as the most promising strategy in the issue network in which the water supply sector and the agricultural sector both participate (Kuks, 1988; Bressers, Huitema and Kuks, 1995; Van der Schot, 1995).

7.5 Another policy network point of view: the dynamics of environmental issues in the agricultural sector

7.5.1 TRADITIONAL AGRICULTURAL NETWORK

This section analyses the policy network which is traditionally involved in the formulation of agricultural policy. We call this the traditional agricultural network. In fact, we describe the network as it existed in the Netherlands until the 1980s, before the development of the manure policy. In our analysis of the relationships in the agricultural network we make use of the 'interrelatedness' and 'commitment' concepts (cf. Bressers and Kuks, 1992). In the following section we examine what changes the network has undergone as a direct result of the manure problem, and the consequences this has had, and may have for the choice of policy instruments. In that section we actually handle the agricultural network and the related environmental policy in the 1980s and 1990s.

Three aspects of interrelatedness may be distinguished in the network relationships. In the first place it should be noted, that the discussion between policy makers and the agricultural target group is highly institutionalized. The Agricultural Board consults frequently, and on a more or less regular basis, with the Ministry of Agriculture and the agricultural specialists in Parliament. Each month, formal talks take place between the Agricultural Board and the minister. This way, the Agricultural Board is able to formulate timely responses to policy initiatives, or to initiate policy itself. The permanent committees for the agricultural sector are in the habit of first consulting with the Agricultural Board before any parliamentary debate on agricultural policy begins. This practice is described by a member of Parliament: 'When the minister returns from

Brussels he spends the entire day in discussion with the Agricultural Board. The permanent committee is sometimes only informed a week later.' (Dekker, 1988)

In the second place, it is notable that a considerable number of personal ties exists in the agricultural sector. Many administrators in the agricultural sector simultaneously occupy different positions in other organizations, and therefore represent the interests of the sector on a number of different fronts. Personal ties exist not only within the agricultural interest group (particularly between the Agricultural Board and its five component organizations), but also between the agricultural interest group and other organizations in the agricultural sector (such as trade cooperatives, industrial cooperatives, the banks and insurance companies with a strong tradition in the agricultural sector). Many agricultural administrators also occupy positions in political organizations at different government levels, or in Government management boards and semi-state institutions.

Thirdly, we have observed that the agricultural sector has an extensive infrastructure at its disposal which serves to support the farming industry. The sector has its own educational institutions, an impressive information network, a considerable number of renowned quality control institutions, an extensive research and development department, and its own daily and weekly newspapers with a wide readership in the agricultural sector, which are considered to be an important source of information. However, the question we must ask ourselves is, whether this can be considered as indicative of interrelatedness or of commitment. On the one hand the infrastructure increases the intensity of interaction, for instance, from the point of view of communication structures. On the other hand, this infrastructure deepens the understanding between policy makers and target group as an example of commitment at network level.

One reason for the strong interrelatedness in the agricultural network is, that the target group consists of a large number of small, self-employed entrepreneurs, who form a relatively homogeneous group with respect to the way they carry on their businesses. As this concerns small enterprises (family businesses), the need for interest groups is considerable. The agricultural market has to be protected and the industry needs support in order to increase its productivity, reduce its costs, and be in a position to exploit export opportunities. The homogeneity of this industrial sector means, that it can be successfully represented by one umbrella organization, the Agricultural Board. However, mutual dependence is an essential factor. The Ministry of Agriculture is expected to represent the interests of the agricultural sector in the Dutch council of ministers and in the European council of agricultural ministers. Therefore, the ministry is dependent on the target group to obtain information on administration in this sector, and on the interest group, as far as disciplining its members is concerned.

The strong interrelatedness in the agricultural network has led to what is known as corporatism. Policy formulation in the agricultural sector is in the hands of three actors: the Ministry of Agriculture, the agricultural commissions in Parliament and the Agricultural Board, which are also referred to as the iron triangle, who form a united, consensual front before taking a standpoint. The representation of agricultural interests is institutionalized in the Agricultural Board, whose authority is recognized by both target group and policy makers, and that has been granted civil authority to implement policy, and discipline its policy field autonomously. It is very difficult for outsiders (such as the Ministry of Environment) to break through this corporatism (Bressers and Kuks, 1992; Frouws, 1993; Termeer 1993).

We can also recognize forms of commitment in the agricultural network. The policy makers have a strong positive attitude to the target group of the policy. They clearly consider the representation of the interests of the target group to be the reason for their existence.

This strong commitment has two main sources. In the first place most of the actors in the agricultural network have remarkably similar backgrounds. Many actors (and officials of the Ministry of Agriculture) grew up in an agrarian family, have either had an agrarian education or have studied at the only agricultural university in the Netherlands, which has always maintained strong ties with the agricultural sector. Anyone who has found access to the agricultural network, transfers easily from one function to another. Administrators are generally recruited from the sector's own ranks. Top administrators in the agricultural sector generally have occupied other positions in the agricultural network. Until recently, this also applied to the officials who have successively filled the post of minister of agriculture.

Another reason for the presence of this strong commitment, is that policy formulation is very dependent on information generated by the agricultural sector itself. The Agricultural Boards, as well as the various research institutions (which form part of the infrastructure) ensure that the policy makers are well informed on the concerns of the target group of the policy.

The strong commitment in the agricultural network means, that considerable resistance is offered by policy makers to product restrictions in the agricultural sector. Maintaining a strong agricultural sector has become their own interest. Necessary changes are generally sought in technological innovations in this industry. Policy makers have absolute confidence in the self-governing ability of the agricultural sector (Bressers and Kuks, 1992; Frouws, 1993, Termeer, 1993).

7.5.2 CHANGES IN THE TRADITIONAL NETWORK AND CONSEQUENCES FOR THE CHOICE OF INSTRUMENTS

From about 1980 on, the agricultural network has been under pressure because of the manure problem. In this section we describe which new actors have been trying to gain access to the agricultural network, the response of the existing network to these 'newcomers', and the consequences this has had for the way in which policy has been pursued in respect of the manure problem. At least three phases can be distinguished in the history of dealing with the manure problem (Bressers and Kuks, 1992).

First phase (1972-1984): the network tries to deny the existence of the policy problem. As early as the end of the 1960s and the beginning of the 1970s it became clear that a manure problem existed, and that it would be necessary to tackle this problem nationally. However, publications with facts concerning the size of the problem, collected by the national statistical bureau (CBS), were kept back by representatives of the Ministry of Agriculture. Members of Parliament received insufficient information about the seriousness of the problem. They gained the impression, that the growing pace of the intensive cattle breeding would diminish, and that the main problem consisted of malodors, distribution and fodder composition. It would be possible to develop technical solutions to these problems (Frouws, 1990: 38). Apart from the Ministry of Agriculture, the Agricultural Board, as well as the target group itself played down the manure problem during this phase. Even at the end of the 1980s 35% of the farmers interviewed denied the existence of a manure problem; 57 % denied that manure could threaten the fertility of the soil (Vermeulen, 1992).

Using delaying tactics, the Ministry of Agriculture tried to block the introduction of manure policy measures. An interdepartmental struggle was going on. Although the Ministry of Environment started to formulate the Soil Protection Act in 1980, the Ministry of Agriculture initially disputed the range of this act. As they did not succeed, they suddenly introduced a draft for a new Manure Act in 1981. The Ministry of Agriculture, as well as the Agricultural Board feared losing the traditional role of initiator, which they used to have within the agricultural network. Moreover, they used the delaying tactic of establishing research committees, which were assigned the mission to study the seriousness and size of the manure problem, as well as the possibilities for its solution.

It is typical of this phase, that both ministries developed distinct regulations with respect to the manure problem in the Netherlands. The regulation prepared by the Ministry of Environment was characteristic of a situation of weak commitment and weak interrelatedness. The Soil Protection Act initially contained very ambitious objectives, which could only be realized by means of directives. By contrast, the Manure Act of the Ministry of Agriculture showed the characteristics of a situation in which a strong commitment and a strong interrelatedness exists. A preference can be seen for policy instruments stimulating self-regulation. While the Ministry of Agriculture has its own target group, which can be identified very clearly as an homogeneous group, the Ministry of Environment lacked such clear identifications with any specific target group until at least the middle of the 1980s. Because of a weak interrelatedness and a weak commitment with target groups, the Dutch environmental policies in the 1970s and the beginning of the 1980s are characterized by: a sectored approach to each environmental sector; a restricted role for target groups in the policy making process; optimistic and therefore ambitious, but mostly unrealistic objectives; and a preference for regulation by standards and permits as the dominant policy instruments.

Second phase (1984-1989): the network tries to delay the policy formulation process. In 1984 the then ministers of agriculture and environment reached the compromise, that the formulation of both acts would be attuned to each other, and that they would be signed by both ministers. At least this 'condemned' both ministries to cooperation, although it did not imply the end of the existing interdepartmental struggle. This was an initiative of the former minister of environment Winsemius, who wanted to 'unfreeze' the cold relationship that had existed for years. He also conducted a sort of 'tour', explaining his objectives to agricultural organizations and farmers all over the country. The double responsibility arrangement changed the struggle from a demarcation dispute into a debate about the contents of both regulations. The use of delaying tactics continued. Because both regulations were arranged as a framework regulation, it was possible to delay the decision making concerning the filling in of the framework. In this way the agricultural network gained time to look for technical solutions to the manure problem.

Because of the 'condemnation to cooperate' issued at the ministerial level, the traditional agricultural network became confronted in the second phase with new actors. Those new actors were trying to enter the traditional network, or to 'open the iron triangle'. First of all the Ministry of Environment announced its entry as a new policy maker in the agricultural sector. It is typical of this ministry, that its access to important target groups in environmental areas (agriculture, industry and transport) depends on other ministries, that are traditionally responsible for policy concerning these target groups. External integration is pursued by the Ministry of Environment as a means of compensating for its lack of power. Attending to environmental issues should become an integral part of every ministry dealing with environmental matters. Ministries and interest organizations of non-environmental policy sectors have reacted to this initiative by consulting their own environmental experts (Frouws, 1988: 47-48). Since the beginning of the 1980s, the Ministries of Agriculture and Environment have been engaged in interdepartmental talks on a collective approach to the manure problem. In addition to the ministry, new interest groups also announced their entry in the traditional agriculture network. From the beginning, the environmental interest groups proposed a reduction of the livestock numbers in the Netherlands, as a permanent solution to the manure problem. They wanted to urge the policy makers not to underestimate the manure problem.

The entry of new actors to the policy formulation network has sparked off two distinct reactions. The first was, that the traditional actors tried to keep their 'iron triangle' closed. Although the Ministry of Environment tried to draw the initiative for the manure policy to itself, the agricultural sector has been successful in ensuring that the Ministry of Agriculture functioned as co-policy maker. By means of the iron triangle, the Agricultural Board has been able to exercise a powerful influence on the manure policy. The Ministry of Agriculture has continually sought to avoid direct contact between the Ministry of Environment and the agricultural target group. It wished to be solely responsible for determining which information is received by the target group, and when. In this context, it is noteworthy how an official has described the reaction of the Ministry of Agriculture to the plans of the Ministry of Environment to instigate an investigation into the consequences of a possible reduction of the numbers of livestock:

> *Agriculture had serious doubts (...). They feared, and not without grounds, that the outside world would blame the Government for planning to reduce the numbers of livestock. The only purpose of this investigation, however, was to establish the consequences of a reduction. To pursue a good policy you have to know what the consequences of reduction might be.* (Logemann, 1990 [translation by the author])

A second reaction to the entry of new actors to the policy formulation network was, that a gradual division occurred in the traditional agricultural network. On the one hand there were signs of disagreement within the iron triangle. The manure legislation had already led to an open conflict between the minister of agriculture and the Agricultural Board. Members of Parliament no longer relied solely on advice from the Agricultural Board, but also included advice given by environmental groups in their decision making (Termeer, 1990: 99). The policy formulation process in the agricultural sector no longer occurred in isolation, and consequently it became politicized. Meanwhile, farmers' interest groups were dealing with a problem of legitimization (Frouws, 1990). Since policy makers were asking their advice on burdensome and restrictive measures, they feared that their legitimacy would decrease in the eyes of their members. This blocked the cooperation with the policy makers within the iron triangle. Completely against the corporatist tradition, the Agricultural Board came out with discordant advice, especially in reference to the allocation of production rights (as was the case with the advice on the so-called *Verplaatsingsbesluit*).

On the other hand, there were indications of a divergence between the iron triangle and the agricultural target group, particularly in respect of relaying information to the target group on the seriousness of the manure problem and the need for change. Different actors in the agricultural sector feared, that both the Ministry of Agriculture and the Agricultural Board had continually delayed providing information, with the result that adjustments achieved in the sector would probably be inadequate and would make a reduction in the numbers of livestock inevitable in the end. These actors believed that the policy makers should be clearer about the investments to be made by the target group, in order to maintain production levels in the future (Klep, 1989; 1990; Termeer, 1990). Numerous stock breeders were keeping more animals than they themselves considered necessary, reasoning that if the livestock percentages have to be reduced in the future, it would be better to have as many animals as possible on the census date (Klep, 1989). The formulation of a manure policy made it difficult for the iron triangle to keep its ranks closed. However, it did not mean a total break with the rules of corporatist decision making. The agricultural network still was in existence, and it unmistakably placed its mark on the manure policy.

The formulation of policy for the manure problem has ultimately led to a division into three major solutions: restriction on the spreading of manure, relocation of manure surpluses, and the encouragement of innovation. The solution that requires restrictions on the spreading of manure is the result of an initiative of the Ministry of Environment. The Agricultural Board was successful in attaching a rider to this solution, to the effect that the existing size and organization of the stock breeding industry may not be affected. This limiting condition also created a basis for compensation claims in areas where, as a result of the special protection requirements, stricter manuring standards were enforced than those that applied nationally. The option of reduction of the agricultural sector, advocated by the environmental movement, had to be considered more seriously, when the minister of agriculture imposed a deadline of 1994 to the solution of the problem of the manure surpluses. If this attempt failed, then volume measures (reduction of livestock numbers or closure of certain farms) would have to be taken (Termeer, 1990: 99).

The solution requiring the relocation of manure surpluses is an initiative of the Agricultural Board. This organization insisted that the problem of manure surpluses should not be viewed as a regional, but as a national one. The manure surpluses are then treated as a problem of logistics, which can be solved by transferring the surpluses to areas without surpluses. To support this solution, the Agricultural Board took the initiative of setting up a National Manure Bank, whose task was to establish an infrastructure to deal with the relocation of manure surpluses. Later on, the minister of agriculture gave this National Manure Bank a legal status (Doorewaard, 1990:67, 69).

The solution that mandated the encouragement of innovations, is also an initiative of the Agricultural Board, yielding a varied set of instruments. For example, the Board achieved a progressively stricter phasing of the manuring standards over a time span of thirteen years. In comparison with the target standard, which will be valid in the year 2000, the initial standards are very flexible. The phasing was considered necessary to develop technological options for the prevention, disposal and processing of manure. This solution also resulted in a surplus tax. On the one hand this tax has a regulatory function by encouraging a reduction in manure production; the less manure produced the less tax one has to pay. On the other hand, the proceeds from this tax are used to finance innovations in the agricultural sector intended to solve the manure problem.

The agricultural sector itself has played a very important role in implementing policy in the first two solutions. A manure accounting system has been introduced, to ensure that the rules for manuring are being adhered to. The cooperation of the traditional institutions is required to implement this system. These institutions are the responsibility of the Agricultural Board. To relocate manure surpluses, manure banks have been created, administered by participants in the agricultural network. As a result of personal ties, the administration of these manure banks is firmly in the grip of the agricultural network.

Another point worth mentioning, is the strengthening of the infrastructure in the agricultural sector, which is a result of the third solution. The agrarian industry itself has also gained in strength, as a result of this solution, and it is precisely this group which is exercising political pressure to prevent reductions in the agricultural sector. As the stimulation of innovations has proofed to be successful, it will be more difficult to make reduction of the agricultural sector a policy target in the future.

Third phase (1989-1993): the network tries to recover its balance. Because of the growing criticism concerning the agricultural pollution of the environment, and influenced by a new wave of environmental attention at the end of the 1980s, the agricultural network developed a more offensive strategy. The Ministry of Agriculture is trying to regain the initiative on the formulation of agricultural environmental policy, which means that it wants to play a role at an earlier stage of the policy formulation process. In an interview with us, the head and an officer of the General Environmental Policy Sector stated: 'As Ministry of Agriculture we should stop adopting a reactive and defensive approach and make our input in an earlier phase'. Furthermore, it was observed by the respondents, that the attitude to environmental affairs is shifting within the Ministry of Agriculture. The external pressure creates more opportunities than before for 'reformers' within the organization. Attitudes do not change as much on all environmental subjects. The Ministry of Agriculture has taken more initiative on manure policy than, for instance, the introduction of environmental management systems in farming, according to our respondents.

The year 1989 is a landmark, as important documents were delivered both by the Agricultural Board ('*Integraal Milieu Actieplan voor de Land- en Tuinbouw*') and the Ministry of agriculture ('Structuurnota Landbouw'), in which self-regulation by the network through the development of sustainable production methods was the central element. Both documents advocated the use of such policy instruments such as covenants. The advantage of such instruments for certain agricultural areas would be, that the implementation of environmental policy could be temporized (soft-pedaled) and financially better supported. Besides that, the publicity effect of covenants was recognized, as a lot of criticism concerning agriculture could be parried this way.

Meanwhile, at the end of the 1980s the Ministry of Environment started to focus on the agricultural target group. Of the 800 officials employed in the Directorate General for Environmental Protection, approximately 50 are directly involved in agriculture. A management team was created to implement the target group policy for agriculture. Its task is, in conjunction with the agricultural sector, to seek policy measures which promote ecologically sound production methods (Logemann, 1990). In an interview with us, the 'target group manager' stated, that his task was two-sided: he also had to aim at improving the understanding of the agricultural sector within his own ministry.

The environmental interest groups developed a much more tolerant attitude to the target group. In an interview a spokesman of the environmental movement put it like this:

> *We shouldn't think about it, but nevertheless it remains a real possibility that the entire agricultural sector might suddenly decide to collectively ignore the rules which apply to manuring. If we take a radical standpoint this will only create ill-feeling on a regional level.*

The agricultural target group is viewed as an unruly group in which moderate solutions, which leave the future prospects for agriculture unchanged, have the most chance of success (Van Duinhoven, 1991).

With respect to the policy network in the third phase, we can perceive at least four reactions. At first, the Ministry of Environment became accepted as a network participant by the ministry of agriculture. Since 1984 both ministries consult each other frequently with respect to the manure policy. An interdepartmental manure project group assembles monthly, while about 10 to 20 working groups within the scope of this project group meet much more frequently. Demarcation disputes still exist, especially as the Ministry of Environment initiates the further detailing of the manure policy, and it also happens that two separate policy proposals are presented to both ministers. But in the end, policy making is integrated and, with respect to the relation between agriculture and environment, this will be stated in the same way in separated official documents from both ministries. In an interview with public officials of the Ministry of Agriculture it was observed, that 'sitting at the table, it turns out that both parties agree much more than might have been expected beforehand'. Disputes mostly concern the pace and the scope of the policy. The environmental section wants to go faster and further, which can be explained by the fact that, although the interrelatedness between the Ministry of Environment and the farmers has increased, the commitment is still weak. There is still little resistance of these policy makers to ambitious objectives.

Secondly, the ministry of environment has been accepted as a network participant by the target group. In spite of the reserved attitude of the Ministry of Environment in this respect, a lot of consultation with the Agricultural Board does take place. In an interview with us, the agricultural target group manager at the Ministry of Environment puts this as follows:

> *The farmers realize that they won't make it with the Ministry of Agriculture alone. Their experience is that they don't have enough influence on the environmental minister through their own minister of agriculture. That is why they want to contact the Ministry of Environment directly. A recent image study has shown, that farmers attach a more positive image to the Ministry of Environment than they do to the Ministry of Agriculture. Farmers experience the Ministry of Environment as a more client-oriented and flexible organization.*

Thirdly, the iron triangle still exists, although it functions more openly than it did. Expertise from outside is more welcomed. 'The environment' has become a daily issue, for example in the farmers' daily newspaper. The admittance of external expertise has been made easier because of external pressures on the function of the agricultural network and particularly its ministry. Organizational problems have speeded up the opening of the iron triangle. Environmental as well as agricultural intermediaries initially entered the Ministry of Agriculture separately to do their lobbying. At the moment they are invited to enter the ministry at the same time for consultation. Also, several organizations within the agricultural network are becoming more and more environmentally oriented. The commodity board for cattle fodder, for example, is promoting the introduction of a mineral balance within cattle breeding industries. An important role is played by the Centre for Agriculture and Environment (CLM), an organization started by environmental campaigners who recognized that a 'trench war' against the farmers would not make sense. This center tries to develop solutions to the manure problem which leave room for self-regulation by farmers.

Fourth, we still perceive in this phase, as in the phase before, that the division between the target group and its interest organizations still exists. Because of the one-sided attitude of the agricultural intermediaries, that prevented an early and adequate solution of the manure problem, the farmers are now surprised now by the rapid pace of current policy making (Doorewaard, 1990: 89). The agricultural sector is faced with a dilemma, also known by other target groups of the environmental policy, such as industry and transport: as a result of successful representation by its interest group, the target group itself is given the opportunity to arrange matters and to avoid drastic measures; at the same time, however, the target group runs the risk of achieving too few results in the environmental area, which means that more severe measures will be unavoidable in the future. If the agricultural sector does not succeed in finding a permanent solution for the manure surpluses, then a reduction of the sector will become inevitable. The paradox of this situation is that successful representation of interests in the short term may injure the target group in the long term. Representation of interests will only be successful in the long term, if it is attended by a strong discipline among the target group in respect to self-regulation.

At the beginning of the 1990s, some starting points of the manure policy were already beyond discussion. Consensus had been reached between most actors involved, that the manure problem required strong intervention. Consensus also existed with respect to the of building up of a sustainable cattle-breeding industry, to be realized through technological solutions (reduction of the percentage of minerals through adjustment of cattle fodder, spreading of manure through transportation, and industrial processing up of manure). However, disagreement still existed on the kinds of intervention to be applied, on the pace and intensity of solving the manure problem, and on the division of financial burdens related to interventions. As a result of a strong interrelatedness and a strong commitment, the interest group and some of the policy makers of the Ministry of Agriculture are not prepared in general to make harsh, short-term demands on the target group to solve the manure problem (Klep, 1990; Bressers and Kuks, 1992; Frouws, 1993, Termeer, 1993; Bloemendaal, 1995).

References

Achttienribbe, G.E. (1993) Privatisering van de drinkwatervoorziening, een redelijk alternatief? (Privatization of the water supply sector, a reasonable alternative?) *H2O,* 26(25): 746-752.

Baldock, D., and G. Bennett (1991) *Agriculture and the Polluter pays Principle. A Study of Six EC Countries*. London and Arnhem: Institute for European Environmental Policy.

Bloemendaal, F. (1995) *Het Mestmoeras* (The manure swamp). Den Haag: Sdu Uitgevers.

Brand, E.M.L. (1994) *Groundwater Protection from Pesticides in the Netherlands and Florida. The choice of policy instruments explored.* Enschede: University of Twente, Center for Clean Technology and Environmental Policy.

Brauw, W.M. de, and H.J.M. Naaijkens (1989) *Grondwaterbeheer* (Groundwater management). Zwolle: W.E.J. Tjeenk Willink.

Breejen, E.K. den, (1991) Een bedrijfstak in de zorgen; het Tienjarenplan van de Vewin (An industrial branch in troubles; the ten year plan of Vewin). *Milieu en recht,* 3 (March): 142-150.

Bressers, J.Th.A., and S.M.M. Kuks (1992) *The Agricultural Policy Network and Environmental Policy in the Netherlands. Paper presented at the EGPA Conference in Pisa-Italy, September 1992.* Enschede: University of Twente, Center for Clean Technology and Environmental Policy.

Bressers, J.Th.A., D. Huitema, and S.M.M. Kuks (1995) Policy Networks in Dutch Water Policy. In: Bressers, J.Th.A., L.J. O'Toole, J.J. Richardson (eds.), *Networks for Water Policy. A Comparative Perspective*, pp: 24-51. London: Frank Cass.

Dekker, G. (1988) Het groene web (The green web). *Intermediair,* 24 (31, 5 Aug. 1988): 19-33.

Dietz, F.J., and C.J.A.M. Termeer (1991) Dutch manure policy: the lack of economic instruments. In: D.J. Kraan and R.J. in 't Veld (eds.), *Environmental Protection: Public or Private Choice*. Dordrecht/Boston/London: Kluwer Academic Publishers.

Doorewaard, M.E.M. (1990) *Milieuwetgeving en het bedrijfsleven: de paradoxale rol van belangenorganisaties* (Environmental legislation and the industry: the paradoxical role of interest groups). Groningen: Wolters-Noordhoff.

Duinhoven, G. van (1991) Het heeft geen zin de boeren tegen de haren in te strijken (It does not make sense to trouble the farmers). *WUB Magazine* (bijlage bij het Wagenings Universiteitsblad), 13 June 1991: 47-49.

Frouws, J. (1988) State and society with respect to agriculture and the rural environment in the Netherlands. In: J. Frouws and W.T. de Groot (eds.), *Environment and agriculture in the Netherlands: a socio-political perspective*. Leiden: CLM.

Frouws, J. (1990) Groen front en mestproblematiek: `erosie' van het agrarisch neo-corporatisme (Green front and the manure problem: 'erosion' of the agricultural new corporatism). *Namens*, 5 (10): 30-40.

Frouws, J. (1993) *Mest en Macht. Een politiek-sociologische studie naar belangenbehartiging en beleidsvorming inzake de mestproblematiek in Nederland vanaf 1970 (Manure and power*. A political-sociological study of advocacy groups involved in the manure problem in the Netherlands since 1970). Wageningen: Landbouwuniversiteit Wageningen (Ph.D. Thesis).

Geleuken, B. van (1991) Het grondwaterbeschermingsbeleid en de nitraatverontreiniging: een waterdicht beleid? (Groundwater protection policy and nitrate contamination: a waterproof policy?) *H2O*, 24 (8): 203-207.

Glasbergen, P., M.C. Groenenberg, and F.A. Roorda (1989) *Naar een strategisch grondwaterbeheer* (Towards strategic groundwater management). Den Haag: VUGA.

Jehae, M., and J.P. van Soest (1990) Stratego in waterland; bestuurlijke mogelijkheden voor drinkwaterbeleid beperkt (Strategic games in waterland: governance options for drinking water policy are restricted). *Nieuwe Bèta* 4(4, April/May): 8-9.

Klep, L. (1989) De mest, het vee, de techniek en het geld (Manure, livestock, technology and money). *Nieuwe Bèta*, 3 (1, Jan.): 8-9.

Klep, L. (1990) De centraal geleide landbouw (The central planned agriculture). *Nieuwe Bèta*, 4 (7/8, July/Aug.): 10-11.

Klok, P-J (1995) Classification of Instruments for Environmental Policy. In: B. Dente (ed.), *Environmental Policy in Search of New Instruments*. Dordrecht: Kluwer Academic Publishers.

Knaap, J.W.M. van der (1987) Provinciale invloed boven water; ervaringen met de Waterleidingwet in de provincie Utrecht (The provincial influence on water supply; experiences with the Water Supply Act in the province of Utrecht). *Bestuur*, 10 (Nov.): 300-305.

Kuks, S.M.M. (1988) *Mestbeleid in grondwaterbeschermingsgebied: de effectiviteit van bemestingsnormen en schadevergoeding bestuurskundig onderzocht* (Manure policy in groundwater protection areas: a policy study on the effectiveness of manure standards and compensation payments). Enschede: University of Twente, Center for Clean Technology and Environmental Policy.

Kuks, S.M.M., G.H.J.M. Neelen (1991) 'De vervuiler betaalt': eenzijdig principe in het gebiedsgericht milieubeleid ('The polluter pays': a one-sided principle in regional environmental policies). *Openbaar Bestuur* 1 (5, May): 23-28.

Langendijk, P.W. (1994) Van Gelderse roulette naar een gestuurde ontwikkeling (From Gelderse roulette towards a controlled development). *H2O*, 27 (15): 431-435.

Ministry of Agriculture, Nature Management and Fisheries [LNV] (1991) *Meerjarenplan Gewasbescherming* (Long term crop protection plan). Den Haag: Ministry of Agriculture, Nature Management and Fisheries.

Ministry of Agriculture, Nature Management and Fisheries [LNV] (1993) *Notitie Mest- en Ammoniakbeleid Derde Fase* (Policy document on manure and ammonia third phase). Den Haag: Sdu Uitgeverij. (*Tweede Kamer, vergaderjaar 1992-1993, 19 882, nr.34.*)

Ministry of Agriculture, Nature Management and Fisheries [LNV] (1995) *Integrale Notitie Mest- en Ammoniakbeleid* (Integral policy document on manure and ammonia). Den Haag: Sdu Uitgeverij. (*Tweede Kamer, vergaderjaar 1995-1996, 24 445, nr.1.*)

Ministry of Agriculture, Nature Management and Fisheries [LNV] (1996) *Emissie-Evaluatie Meerjarenplan Gewasbescherming (MJP-G).* (Emission evaluation long term crop protection plan). Den Haag: Ministry of Agriculture, Nature Management and Fisheries.

Logemann, A. (1990) Geen milieujacht op boeren: het groene gezicht van VROM (No environmental hunt on farmers: the green face of VROM). *Boerderij,* 75 (19, 6 Feb. 1990): 22-24.

Meijden, D. van der (1988) De bescherming van het grondwater (The protection of groundwater). *Milieu en Recht,* 2 (Feb.): 46-58.

Nieuwenhof, R. van den (1995) *Vermesting en water. De overlast voor gebruikers* (Manure and water. The nuisance for water users). Amsterdam: Stichting Natuur en Milieu / Babylon-De Geus.

Reus, J. (1991) Meerjarenplan Gewasbescherming biedt agrariërs te weinig stimulans: Effectief gewasbeschermingsbeleid moet boer en tuinder meer motiveren (Longterm crop protection plan offers farmers no incentives enough: Effective crop protection policy needs to motivate farmers more), in: *Landbouwkundig Tijdschrift* 103 (1991/2): 5-7.

Rijksinstituut voor Volksgezondheid en Milieuhygiëne [RIVM] (1991) *Nationale Milieuverkenning 1990-2010* (National report on the state of the environment 1990-2010). Alphen aan den Rijn: Samsom HD Tjeenk Willink.

Schot, J. van der (1995) Waterleidingbedrijven: de stille milieuridders (Water supply companies: the silent knights of the environment). *Natuur en Milieu,* 1995 (July/Aug.): 19-21.

Smeets, I. (1990) Sprekers over uitvoering van MJP-G: 'Bestrijdingsmiddelenhandel moet ook bijdrage leveren' (Experts about longterm crop protection plan: 'Pesticides traders also need to contribute'). *Weekblad Groenten en Fruit,* 30 November 1990: 8-9.

Termeer, C.J.A.M. (1990) Het landbouwnetwerk onder druk (The agricultural network under pressure). In: J.A.M. Hufen, A.B. Ringeling (eds.), *Beleidsnetwerken: overheids-, semi-overheids- en particuliere organisaties in wisselwerking,* pp. 87-103. Den Haag: VUGA.

Termeer, C.J.A.M. (1993) *Dynamiek en inertie rondom mestbeleid. Een studie naar veranderingsprocessen in het varkenshouderijnetwerk* (Dynamics and inertia around manure policy. A study of change in the network of pigfarming). Den Haag: VUGA.

Velema, W., F. Boer, T. Verheul (1989) Aardappelen koken in Spa blauw (Boiling potatoes in Spa blue). *Intermediair,* 25 (February 24, 1989).

Verdonk, A., and C. Velders (1996) Project Emissie-Evaluatie Meerjarenplan Gewasbescherming voor het ijkjaar 1995: 'Minder uitstoot van gewasbeschermingmiddelen dan in de jaren tachtig' (Project emission evaluation longterm crop protection plan for the reference year 1995: Less emission of crop protection substances than in the 1980s). *Platform,* 1996 (May): 26-27.

Vereniging Milieudefensie (1996) *De buik vol van gif. Tussentijdse evaluatie van Meerjarenplan Gewasbescherming (MJP-G) door samenwerkende milieuorganisaties.* (Fed up with poison. Preliminary evaluation of the long term crop protection plan by the cooperating environmental movement). Amsterdam: Vereniging Milieudefensie.

Vermeulen, W.J.V. (1992) *De vervuiler betaald. Onderzoek naar de werking van subsidies op vier terreinen van het milieubeleid* (The polluter paid. A study on the effects of subsidies in four environmental policy sectors). Utrecht: Uitgeverij Jan van Arkel.

Vewin (1989) *Hoofdrapport en bijlagen Tienjarenplan 1989* (Main report and appendices of the Ten years plan 1989). Rijswijk: National Association of Water Supply Companies.

Vewin (1991a) *De openbare watervoorziening in feiten en cijfers* (The public water supply in facts and figures). Rijswijk: National Association of Water Supply Companies.

Vewin (1991b) *Milieuplan* (Environmental plan). Rijswijk: National Association of Water Supply Companies.

Vewin (1993) *Feiten en cijfers in 1992 over de drinkwatervoorziening* (Facts and figures on the water supply in 1992). Rijswijk: National Association of Water Supply Companies.

Vinkers, H.J. (1991) De zorg voor de kwaliteit van het grondwater als drinkwaterbron (The concern about the quality of groundwater as a drinking water source). *H2O*, 24 (21): 595-601.

Ministry of Housing, Physical Planning and Environment [VROM] (1986) *Mestactieprogramma* (Manure action programme). Den Haag: Sdu Uitgeverij. (*Tweede Kamer, vergaderjaar 1986-1987, 19 882.*)

Ministry of Housing, Physical Planning and Environment [VROM] (1990a) *Milieukerngegevens Nederland* (Essential environmental data on the Netherlands). Den Haag: Ministry of Housing, Physical Planning and Environment.

Ministry of Housing, Physical Planning and Environment [VROM] (1990b) *Naar een glasheldere toekomst. Nota ter voorbereiding van het Beleidsplan Drink- en Industriewatervoorziening* (Towards a crystal-clear future. Policy document to prepare a policy document on the supply of drinking water and industrial water). Den Haag: Sdu Uitgeverij. (*Tweede Kamer, vergaderjaar 1989-1990, 21 536, nr.1-2.*)

Ministry of Housing, Physical Planning and Environment [VROM] (1993) *Deel 1 (ontwerp planologische kernbeslissing) van het Beleidsplan Drink- en Industriewatervoorziening* (Part 1 of the policy document on the supply of drinking water and industrial water: design for a land use decision making process). Den Haag: Sdu Uitgeverij.

Ministry of Housing, Physical Planning and Environment [VROM] (1994) *Drinkwater in Nederland* (Drinking water in the Netherlands). Den Haag: Ministry of Housing, Physical Planning and Environment (Directie Voorlichting en Externe Betrekkingen).

Ministry of Housing, Physical Planning and Environment [VROM] (1995a) *Beleidsplan Drink- en Industriewatervoorziening* (Policy document on the supply of drinking water and industrial water). Den Haag: Sdu Uitgeverij.

Ministry of Housing, Physical Planning and Environment [VROM] (1995b) *Zicht op zorg in de Drinkwatervoorziening. Onderzoek naar de rol van de drinkwaterbedrijven bij de winning van de grondstof* (Vision on management in the water supply sector. A study on the role of water supply companies with respect to the extraction of groundwater). Den Haag: Ministry of Housing, Physical Planning and Environment (Inspectie Milieuhygiëne).

Ministry of Housing, Physical Planning and Environment [VROM] (1996) *De kwaliteit van het drinkwater in Nederland in 1994* (The quality of drinking water in the Netherlands in 1994). Den Haag: Ministry of Housing, Physical Planning and Environment (Inspectie Milieuhygiëne).

Ministry of Housing, Physical Planning and Environment [VROM] (1993-1997) *Knipselkrant* (Selected clippings from several newspapers), Den Haag: Ministry of Housing, Physical Planning and Environment.

Ministry of Transport and Public Works [V&W] (1989) *Derde Nota Waterhuishouding* (Third national policy document on water management). Den Haag: Sdu Uitgeverij. (*Tweede Kamer, vergaderjaar 1989-1990, 21 250, nr. 1-2.)*

List of Abbreviations

AID	General Service for Inspection of Agriculture
CLM	Centre for Agriculture and Environment
DLV	Agricultural Extension Service
IPO	Inter-provincial Consultation
KIWA	Institute for Inspection of Water Supplies
LEI	Agriculture-Economic Institute
LNV	Ministry of Agriculture, Nature Management and Fisheries
Nefyto	National Foundation for Phyto-Pharmacy
NCB	Noordbrabantse Christelijke Boerenbond (Christian Farmers League in the province of Noord-Brabant)
RIVM	National Institute of Public Health and Environmental Protection
Vewin	National Association of Water Supply Companies
VROM	Ministry of Housing, Physical Planning and Environment
V&W	Ministry of Transport and Public Works
WMD	Waterleiding Maatschappij Drenthe (Water Supply Company Drenthe)
WML	Waterleiding Maatschappij Limburg (Water Supply Company Limburg)
WMO	Waterleiding Maatschappij Overijssel (Water Supply Company Overijssel)
WOB	Waterleidingmaatschappij Oost Brabant (Water Supply Company Eastern Brabant)
WOG	Waterleidingmaatschappij Oostelijk Gelderland (Water Supply Company Eastern Gelderland)

8. CONTROL CAPACITY - THE NETHERLANDS

STEFAN M.M. KUKS
Center for Clean Technology and Environmental Policy, University of Twente, Enschede, The Netherlands.

8.1 Introduction

In this chapter we discuss the capacity of water suppliers to control the prevention of pollution. This 'control capacity' specifically concerns the relationship between water suppliers and farmers. Although water suppliers have no regulatory competence in this relationship, regulations do play a background role. On the one hand, water suppliers are regulated by drinking water quality standards, which force them to produce drinking water above a specified quality level. On the other hand, farmers are regulated by the Dutch national and provincial government, which impose restrictions on the use of fertilizers and pesticides in agricultural operations. In Chapter 7 we described how developments in the environmental regulation of farmers - in particular the manure legislation since 1987 (regulatory context) - as well as increasing groundwater pollution - in particular an expected rise of nitrate levels (problem context) - provoked dynamics in the network context and the way in which water suppliers and farmers regard each other. Despite the regulatory gap in the relationship between water suppliers and farmers, we want to know whether water suppliers do try to bridge this gap by encouraging the prevention of agricultural water pollution by other means. And if they do so, how successful are they? Or, to put it in other words, what is their control capacity?

In our study of the Netherlands, we assume that an actor - a water supplier - has 'control capacity', if on the one hand he is making efforts to gain control over the behavior of another actor - a farmer - and if, on the other hand, these efforts appear to be successful. An attempt to gain control over the behavior of another actor will be operationalized by the actual transfer of resources from a water supplier to a farmer, the water supplier's aim or motivation being to diminish agricultural pollution of drinking water sources. We are specifically interested in the types of resources that are most successful.

Geerten J.I. Schrama (ed.), Drinking Water Supply and Agricultural Pollution, 191-230.

A measure of success can be found at several levels. At first, we can say that an attempt is successful if it achieves better quality groundwater. This can be proved by measuring the quality of the upper groundwater, just below the ground level, or the deeper groundwater as pumped by a water supplier, and by finding a relationship between the attempt and the improvement in the water quality as an effect.

Secondly, we can say that an attempt is successful if it changes farmers' behavior and moves them to take preventive actions which they would not have undertaken in the absence of the attempt. More generally, a preventive action can be each environmental innovation within a farm (for example, an action that reduces the amount of manure produced), or more precisely speaking, an innovation that is directly related to improving the groundwater (such as an action that reduces the application of nitrogen in a groundwater protection area).

Thirdly, we can say that an attempt is successful if it motivates farmers, causing them to consider preventive action. In other words, an attempt can be seen as successful if it changes the willingness or motivation of a farmer. This kind of success can be measured by looking at the participation of farmers in prevention oriented projects or by looking at farmers' reactions to the water suppliers' invitation to consider preventive action.

In our study the measure of success is a combination of the second and the third criteria. We conclude that an attempt by a water supplier to move a farmer towards prevention is successful, when the farmer shows a willingness to work on prevention, making commitments to specific preventive actions. The water supplier's attempt considered to be the independent variable, with the water supplier's motivation to encourage prevention and to transfer resources as necessary elements. The farmer's behavior is considered to be the dependent variable, with the farmer's willingness or motivation to work on prevention and an investment of his resources in preventive action as necessary elements. We also try to formulate statements on specific types of resources that can be successful and therefore helpful to a water supplier if they are transferred to farmers to encourage prevention. In our opinion, an indicator of the success of a specific resource is not the only proof that it works; this can also appear from the susceptibility of a farmer to that resource and the potential influence of a resource on a farmer's consideration of preventive behavior.

In this chapter we will start in the next section by describing what strategies were undertaken in the five selected provinces by water suppliers in co-operation with provincial authorities, in order to encourage farmers in groundwater protection areas to take preventive action. In another section we will take a more detailed look at the motivation of water suppliers, what they expect from their work on prevention, and how these motives are related to the motives of the other actors involved (provincial authorities, farmers and farmers' organizations). We will consider both the willingness of water suppliers to encourage prevention, as well as the responsioveness of farmers to these attempts. In the following section we will analyze the resources that were transferred during the water suppliers' prevention oriented actions. Again, we will consider each of the resource types in two ways. On the one hand we look at the extent to which water suppliers are dependent on a resource if they want to encourage farmers to take preventive action. On the other hand we look at the extent to which farmers are susceptible to a resource when they consider preventive action. At the end of the chapter we try to formulate conclusions about the control capacity of water suppliers, its relationship to the motives and resources of water suppliers as well as farmers, and on the background role played by the problem context, the regulatory context and the network context, as described in Chapter 7.

8.2 Preventive strategies on the provincial level

8.2.1 PERIOD 1988-1991

The Soil Protection Act, which lays down the basis for groundwater protection policies in the provinces, was passed in 1987. It required the provinces to designate groundwater protection areas before 1989, and to provide legislation for these areas in order to achieve adequate groundwater protection. The provinces have the competence, by means of ordinances, to regulate the use of fertilizers and pesticides and, in doing so, to be more stringent than the standards adopted under the national manure and pesticides policies. Such additional regulations mean that farmers within groundwater protection areas will be more stringently regulated than farmers outside these areas. Provinces have an interest in bringing about an accelerated protection in groundwater protection areas, since groundwater protection policy, over and above a national policy, is their constitutional and political responsibility. On the other hand, provinces do not feel any responsibility to go further than the final standards that were set by the national government as a long term goal for the Netherlands as a whole. The provinces also choose to use the same basis for regulation as the national manure standards, which is the amount of phosphates applied per hectare per year. The specific protection level, created by the provinces, should have a function as long as the general protection level, created by the national government, does not provide sufficient protection to groundwater sources that are used for drinking water production. Because of the present state of the groundwater quality, the national legislators who passed the Soil Protection Act believed they could not effort to wait for a national protection level in order to provide adequate protection to drinking water resources. They wanted to enable the provincial authorities to bring this about s soon as possible.

The Soil Protection Act also created the provision to compensate farmers in groundwater protection areas for the disadvantages they are faced with due to the more stringent regulations. Such disadvantages are considered to be the extra surplus of manure that farmers have to discharge and the extra costs they incur in respect of alternative means to fertilize the soil and protect their crops. Formally speaking, farmers should be compensated for their financial losses by the provinces, and the provinces should in turn be compensated by those who abstract groundwater and who profit from groundwater protection, i.e., the water supply companies, or rather the consumers, who pay the compensation, in the end, as part of the drinking water price.

The creation of this compensation provision was not just a political choice to weaken the resistance of farmers in groundwater protection areas to more stringent standards, it is also considered to be a logical continuation of a practice in some provinces, in which water supply companies were already compensating farmers for the dehydration damage to crops due to the abstraction of groundwater and the lowering of the groundwater level in water abstraction areas. Since the beginning of the 1980s, the water supply companies in some provinces were even giving subsidies to farmers to reduce the amount of manure spread on the land in water abstraction areas. So it seemed to be a logical consequence to continue the paying role being played by the water supply companies.

In the provinces of Drenthe, Overijssel, Gelderland, and Limburg the compensation provision has caused the water supply companies most closely involved to start informal negotiations with farmers organizations (in all cases the regional department of the national Agricultural Board) on direct arrangements for compensation payments, bypassing the formal route via the provincial authorities. These negotiations started around the same time that the provinces were preparing their ordinances (in the period 1987-1989). The provinces had no objection to these informal arrangements. On the contrary, they considered that it would be better for the acceptance of the more stringent standards if the water suppliers and farmers could achieve a satisfactory mutual understanding. So the provinces tried to encourage such negotiations. These were especially concerned with standardizing the types of costs that would be allowed in the compensation of the farmers. Such a standardization also gives the farmers the advantage of having greater certainty in advance about payments, rather than having to wait and see what they would be get paid after action had been taken. Furthermore, an informal arrangement would have the advantage of keeping administrative costs as low as possible. In the informal construction, the water supply companies would not have to pay the administrative costs of the provincial bureaucracy. In 1989, compensation arrangements were available in all four provinces, having been set up by the most closely involved water supply companies. The other smaller companies followed the agreements and signed them, too. Farmers in these provinces appealed to the provincial authorities on less than five percent of the damage claims, because they could not finally agree with the water supply company about their right to compensation payments. In those cases, the provinces were advised by an expert committee to adjudicate on the proposed damage claim. Exceptionally, in the province of Noord-Brabant the relevant water supply company did not wish to negotiate, giving preference to the formal way of handling the farmers' financial claims, which means that farmers will be compensated by the province, and that the province will thereafter be compensated by the water supply company.

In 1991, the water supply companies and the provinces started to come to the conclusion that the specific protection level in groundwater protection areas was having an effect in terms of reducing the agricultural pollution of the soil, but that hardly any effect could be found on the groundwater quality. In 1994, an evaluation study by the Dutch Ministry of Environment on five years of groundwater protection policies confirmed this conclusion. A first explanation for this result is that the manure legislation did not relate standards to the use of nitrogen, but to the use of phosphates, while nitrogen in artificial fertilizers was not included when the standards were set. Secondly, a large amount of historical pollution is still making its way underground and this is distorting the effects of policy measures on the groundwater quality. Improvements are difficult to measure if unknown amounts of underground pollution will affect the quality of groundwater for many decades to come. This obscures the policy makers' short-term intentions to achieve improvements in the groundwater quality. Thirdly, the geographical delimitation of groundwater protection areas was based on geohydrological assumptions which, the passage of time revealed, had not always been correct. It could happen that a water supply company, after some years, had to say to a farmer 'sorry, we made a mistake, your land is not part of a groundwater protection area any more.' So the weakness of some geohydrological assumptions also weakened the effects of the policy (VROM, 1994).

Looking at the policy mix of standard setting and financial compensation, the water supply companies concluded that this mix scarcely encourages farmers to work on prevention of agricultural pollution. Water supply companies do not like to spend their money this way, mainly helping farmers to get rid of their manure surpluses. They prefer to spend this money on environmental improvements in farming practices, with a more sustainable effect on the groundwater quality. On the other hand, water supply companies realize that they gained a lot of confidence among farmers, by being cooperative in compensating farmers, and that this confidence base helped the enforcement of the provincial standards in groundwater protection areas. However, this is only the case in provinces where the water supply company negotiated about a compensation arrangement. The compensation arrangements helped to weaken the resistance to provincial standards, although this resistance was originally massive and turbulent, especially during the period 1987-1989, when the farmers first had to become accustomed to the introduction of national manure legislation and then to more stringent provincial standards which, in their experience, made matters even worse for them. After five years of experience with groundwater protection policies, evaluation studies in the four provinces with a compensation arrangement, showed that about 70% of the farmers involved were satisfied with the way the provinces and the water supply companies handled the groundwater protection measures. These farmers considered the water supply companies to be reliable partners in business.

The water supply companies have also concluded, that the instrument of compensation payments offers them more information about the way in which farmers in groundwater protection areas are operating their farm and how this might affect the groundwater quality. The compensation payments to farmers enable water suppliers to 'buy' information about 'non-point source pollution', and such information is normally hard to get. In the end, this information will help water suppliers to improve the geohydrological models they use to predict the future quality of their drinking water sources.

8.2.2 PERIOD 1991-1995

In 1991, the way in which provinces and water supply companies approach farmers in groundwater protection areas gradually started to shift towards another policy style. It was learned that the 1989 ordinances generated hardly any effects on the groundwater quality. The standards set in these ordinances are not related to the nitrate pollution of groundwater, but only to the phosphate contents of manure. To protect groundwater it is necessary to regulate the application of nitrogen directly. This means that even the use of nitrogen in artificial fertilizers should be regulated. Moreover, the sanitation of harmful pesticides is not proceeding fast enough, mainly because of a delay in the implementation of the national pesticides policy. Most compensation arrangements, which have a term of six years, will end in 1995. In all, about 1,000 farmers in the Netherlands are involved in a compensation arrangement with a water supply company. It is not only the setting of standards that has been criticized by the water supply companies, but also the compensation provisions. Water supply companies think they do not achieve adequate results (in terms of groundwater quality improvements) for the money they spend on compensation payments. They prefer to spend this money on investments in more prevention oriented farming operations, to achieve a more lasting effect, instead of spending the money on the discharge and manufacturing of manure surpluses.

In the interim, the Centre for Agriculture and Environment[1] started a pilot project in the province of Overijssel, in which pilot farms account for their nitrogen losses through minerals book keeping, trying to decrease these losses by optimizing their farming practices. The participating farmers are meeting each other in study groups, in which they discuss and will be informed about possibilities for reducing the use of nitrogen and pesticides. The Agricultural Extension Service (*Dienst Landbouw Voorlichting* or DLV) has also been involved in these study meetings, partly because of their expertise, but also to increase the acceptance of study meetings by involving an information service that farmers are already familiar with. The pilot project turned out to be very successful, partly because the participating water supply company was offering a financial contribution to innovative measures with a demonstrable effect on the groundwater quality. This pilot project was an initiative of the CLM in collaboration with the Vewin, the national association of water supply companies, which since 1991 has tried to encourage other water supply companies to start comparable projects for farmers in groundwater protection areas (Vewin, 1993).

From that time on, the water supply companies in the provinces of Overijssel, Drenthe, and Gelderland, encouraged by the Vewin, started their own stimulation policy, which gradually came to replace the compensation policy. For that matter, the national compensation provision in the Soil Protection Act arranged that compensation payments should be only temporary and should only be offered for as long as there was a difference between the protection levels offered by the national legislation and the

---

1 *Centrum voor Landbouw en Milieu* (CLM), an institute funded by the national government for studies on environmental innovations in farming practices.

provincial legislation. Because of the progression in national standard setting, the difference is gradually reducing to zero (with a target of the year 2000). The water supply companies in Drenthe, Overijssel, and Gelderland decided to restrict the compensation payments to a minimum, in order to give the new stimulation policy a more prominent place. The resulting minimum compensation payments are at least available to farmers that are not willing to do something more, while farmers participating in the projects might get payments for measures with a broader environmental scope than only improving the groundwater quality. A difference from the compensation policy is that the stimulation policy is related to the whole farming process, while the compensation payments are related only to those parts of a farmers' land that are located in a groundwater protection area. Another difference is that the stimulation policy activates farmers to do more than is required by the existing legislation, on a voluntary the basis. Monetary encouragement does not conflict with the polluter-pays-principle, because farmers are encouraged to do more than they need to do, compared with farmers outside groundwater protection areas.

Water supply companies are sending the message to farmers in groundwater protection areas, that they can be leaders in prevention compared to farmers outside these areas by means of temporary subsidies. This message is received as a positive one by the farmers, who are tired of the process of manure legislation that has been ongoing for almost ten years now. The farmers' interest in the stimulation policy is growing and their participation in such projects is estimated by the respondents in the water supply companies to be about 20 to 40 percent of the total farmer population in those groundwater protection areas where a stimulation policy is available.

On the other hand, one could ask why it is still difficult to encourage the rest of the farmers. One explanation mentioned by the respondents is, that after all the discussions about environmental problems in the agricultural sector, a large part of this sector has lost confidence in public policy in general. It is difficult to motivate these farmers to do more. The step from external regulation towards self-regulation is difficult to communicate to this group. Another explanation mentioned by the respondents is, that many farmers calculated that the water supply companies' stimulation policies are unattractive to them. There are more costs than revenues in their case. A third explanation mentioned by the respondents is, that many farmers are reluctant to provide the commercial information about their own firms, as required by the stimulation policy, since such information will be made public. Participation in a stimulation project requires, that a farmer is willing to provide precise data on his farming methods and his minerals accounting. Water supply companies have to guarantee in the negotiated agreement with agricultural representatives, that they will treat this information confidentially, without passing it on to the provincial or national authorities. Water supply companies have to make clear that they are not public controllers for the enforcement authorities.

But even the water supply companies are not completely satisfied with their new policy. After three years of experience with this policy in the provinces of Overijssel, Drenthe, and Gelderland, the respondents from the water supply companies in these provinces conclude, that a lot of study groups do not come up with measures that might be effective for the protection of the groundwater. The agricultural sector is too much interested in subsidies and demonstration projects, and too little in getting results. Besides, many farmers do not realize that the water supply companies are only interested in financing measures that improve the groundwater protection, and not in financing other environmental measures. Intermediary organizations in the agricultural sector and agricultural advisers give the advice to farmers, that it is important to do something, while caring too little for the results of what is being done.

The agricultural sector expects that the support for this stimulation policy can be increased by leaving the implementation of the policy to the intermediary organizations of the agricultural sector itself. That would counter the reservation that farmers have about providing information about their own business. An intermediary of the agricultural sector would not be regarded as part of the government. This could increase the confidence of the farmers in the policy.

On the other hand, the water supply sector does not have enough confidence in the intermediaries of the agricultural sector. Besides that, the acquisition of information from the farmers, as well as the building of confidential relationships with farmers through direct contacts, are essential parts of the policy. In the experience of the water supply companies, individual farmers are often more receptive than intermediate agricultural organizations.

In 1994 the Vewin started to stimulate a new type of policy instrument that might generate more precise effects on groundwater quality. This instrument is termed 'payment by results' (*resultaatbeloning*) and can be characterized as a 'performance subsidy'. Again, this instrument has been developed in collaboration with the Centre for Agriculture and Environment (CLM), and is a more elaborate form of minerals accounting and stimulation policy: only results for the groundwater quality will be subsidized. The Vewin has created this instrument in collaboration with the water supply companies of Drenthe, Overijssel, and Limburg (CLM, 1994).

8.3 Comparison of the five selected provinces

Agricultural water pollution in the Netherlands is especially a problem to water suppliers in the northern, eastern, and southern parts of the country. It shows up in areas where groundwater is used for the production of drinking water and where farmers make intensive use of a vulnerable soil (sandy loam). This combination of circumstances is most evident in five (Drenthe, Overijssel, Gelderland, Noord-Brabant, and Limburg) of the twelve provinces that make up the Netherlands. For the Dutch study we have chosen to focus on these five provinces.

8.3.1 WATER SUPPLY COMPANIES

For this study we looked at one specific water supply company in each selected province (some statistics on these provinces are presented in table 8.1) All the selected companies

TABLE 8.1: Some Statistics

	Total area		Inhabitants		Inhabitants/km^2
The Netherlands	33,900 km^2	100%	14.9 mln.	100%	439
Gelderland	5,000 km^2	15%	1.9 mln.	12%	360
Noord-Brabant	4,900 km^2	14%	2.2 mln.	15%	442
Overijssel	3,300 km^2	10%	1.0 mln.	7%	306
Drenthe	2,700 km^2	8%	0.4 mln.	3%	166
Limburg	2,200 km^2	7%	1.1 mln.	7%	509

Source: EU (1990)

	Agricultural area	Agricultural livestock	Livestock/ha.
Gelderland	257,000 ha.	1,382,000	5.4
Noord-Brabant	275,000 ha.	1,915,000	7.0
Overijssel	210,000 ha.	948,000	4.5
Drenthe	166,000 ha.	275,000	1.7
Limburg	112,000 ha.	564,000	5.0

Source: EU (1990)

Drinking water price per 1,000 liter in Dutch guilders	
Waterleiding Maatschappij Limburg (WML)	Hfl. 2.81
Waterleiding Maatschappij Overijssel (WMO)	Hfl. 2.41
Waterleiding Maatschappij Oostelijk Gelderland (WOG)	Hfl. 2.23
Waterleiding Maatschappij Oost-Brabant (WOB)	Hfl 1.79
Waterleiding Maatschappij Drenthe (WMD)	Hfl. 1.79

Source: Consumentenbond (1997)

use groundwater as their main source for drinking water production. Also they all pump groundwater in predominantly agricultural areas. Given these criteria, we interviewed the largest companies, supposing that they are the leaders in taking preventive actions on agricultural water pollution.

- *Drenthe* has two water supply companies, both of which serve cities as well rural areas. Some communities along the borders of the provinces are served by water supply companies in the neighboring provinces (Groningen and Overijssel). For this study we looked at the *Waterleiding Maatschappij Drenthe* (WMD), which is the largest water supply company in the province. This company has 14 pumping sites.
- *Overijssel* has three water supply companies, one of which serves an urban area (Enschede), the other two serving larger regions. For this study we looked at the *Waterleiding Maatschappij Overijssel* (WMO), which is the largest company in the province. This company has 28 pumping sites.

- *Gelderland* has four water supply companies, which all serve on a regional scale. Two of them are purely water supply companies, while the two others are combined water and energy utilities. For this study we looked at the *Waterleidingmaatschappij Oostelijk Gelderland* (WOG), which is large, but not the largest water supply company. This company is not involved in energy distribution. Gelderland has 54 pumping sites in total.
- *Noord-Brabant* has six water supply companies, three of which serve urban areas (Den Bosch, Eindhoven, Tilburg), the three others serving on a regional scale. For this study we looked at one of the regional companies, the *Waterleidingmaatschappij Oost Brabant* (WOB), which is the largest company in the province and which has 18 pumping sites.
- *Limburg* has two water supply companies, one of which serves an urban area (Maastricht). The other one is called the *Waterleiding Maatschappij Limburg* (WML) and it serves ninety percent of the province of Limburg.

8.3.2 AGRICULTURAL POLLUTION OF WATER SOURCES

Agriculture in the province of *Drenthe* is predominantly potato cultivation, for which large amounts of pesticides are needed. Pesticides constitute the main problem of agricultural water pollution in this province. A lot of groundwater is polluted with a soil disinfectant that has been applied in potato culture since 1968. On the other hand, farming in this province is less intensive than elsewhere. Therefore fertilizers are much less a problem in Drenthe. Groundwater is abstracted from quite shallow levels at moors. Most pumping sites of the WMD (Drenthe) are rather vulnerable.

The provinces of *Overijssel* and *Gelderland* have predominantly intensive livestock farming. Most problems are caused by the spreading of (animal as well as artificial) fertilizers on the sandy grounds. Pesticides are a less serious problem in these provinces. Groundwater is abstracted from sandy soils from quite shallow levels. Most pumping sites of the WMO (Overijssel) and the WOG (Gelderland) are rather vulnerable.

The province of *Noord-Brabant* is also notorious for its intensive livestock farming - the most intensive of all the Dutch provinces - but it has only a few vulnerable areas. Most of the soil in this province consists of clay, which offers a sound protection to the groundwater, that is abstracted from very deep levels. There are only two shallow pumping sites in sandy areas (Boxmeer and Vierlingsbeek) where the groundwater is highly vulnerable to agricultural pollution (both fertilizers and pesticides are a problem there).

The province of *Limburg* also has a culture of intensive cattle farming. Especially the northern and middle part of Limburg have the highest manure production in the country. Groundwater is only withdrawn from the deeper aquifers in these areas, so pollution effects of excessive manuring will not become apparent within 50 to 100 years. Only a few well fields are vulnerable in a short run perspective, because they have to abstract from shallow water layers. Although the nitrate standards are exceeded in some water wells, mixing with water from cleaner wells helps to keep the nitrate level in the water down. It is also planned to close some of these polluted wells and to start using surface water sources instead for the production of drinking water.

Half of the drinking water produced in Limburg originates from groundwater sources in the southern part of this province. These sources are very vulnerable to nitrate pollution, because they are not covered by underground protection layers. Although farming in the south of Limburg is not very intensive, a lot of historical pollution is still on its way underground. It is expected that the first of the twenty water wells in this area will not show too much pollution before 2000. Many more water wells will follow after that. For the time being, problems with nitrates are solved by mixing water instead of using purification techniques. Pesticides play a role in Limburg, although a modest one compared to Drenthe. Limburg has relatively much more arable land and horticulture than Overijssel, Gelderland and Noord-Brabant.

We may conclude that in all the provinces studied, the water supply is highly dependent on groundwater as a source for drinking water production. However, this dependency will diminish in the future, because more surface water will be used for the drinking water production to prevent problems with the dehydration of nature reserves. In Limburg, the provincial authority planned to increase the amount of surface water sources from 10% up to 40% within eight years. In four out of the five province in this study, the groundwater is rather vulnerable to agricultural pollution. Noord-Brabant is an exception, as most of the groundwater is quite well protected by clay layers.

All provinces have some water wells in which nitrate pollution is a problem, since nitrate levels in the pumped water exceeds the nitrate standard for drinking water. Until now this has not occurred on a large scale and water suppliers succeed in bringing down nitrate levels by mixing with water from less polluted wells. Gnly a few water wells had to be closed, because other solutions would be too costly. There are barely any violations of standards in water wells (VROM, 1996).

To conclude, agricultural pollution of groundwater seems to be manageable for the moment and for the near future, especially as the result of avoidance strategies (mixing water or abstracting from deeper layers or surface water). By contrast, in the longer run pollution might leach to the deeper groundwater layers and cause problems on that level. Meanwhile, an unknown amount of historical pollution will show up in the future and will cause problems of an unpredictable extent.

8.3.3 POLICY APPROACHES TOWARDS AGRICULTURAL POLLUTION OF WATER SOURCES

Drenthe. In 1989, the provincial government designated groundwater protection areas and passed an ordinance with standards for these areas to restrict agricultural operations. The ordinance offers the option of compensation payments. Originally the Water Supply Company Drenthe (WMD) rejected the possibility of initiating negotiations with farmers' organizations on compensation arrangements. The WMD hardly had any contacts with farmers, partly because there was no existing practice of compensating farmers for dehydration damage. The first contacts with farmers' organizations were very formal and difficult. Through mediation of the provincial authorities collaboration with the farmers' organization was finally established, negotiating a compensation arrangement in which standard types of costs are identified for reductions in the use of fertilizers and pesticides. The largest part of the compensations paid are related to reductions in pesticides use, mainly in relation with the intensive potato farming. The other water supply companies in Drenthe followed the arrangements negotiated by the WMD. The agricultural sector in Drenthe is very satisfied about the arrangement with the WMD and the way in which it has been implemented.

In 1993, the WMD started to develop a stimulation policy, following the example of the largest water supply company of the neighbor province Overijssel, where a pilot project appeared to be successful in 1991. The Vewin encouraged the WMD to develop its own stimulation policy. At the end of 1994, about 100 out of 250 farmers in groundwater protection areas, were participating in stimulation projects. The regional department of the national Agricultural Board is playing a modest role in the policy implementation by communicating the policy to the farmers, although the WMD recognizes them as an important partner in creating support among farmers for such projects. The province itself prefers to be a partner at a distance and leaves the initiative as much as it can to the water supply company.

Overijssel. In 1989, the provincial government designated groundwater protection areas and passed an ordinance with standards for these areas to restrict agricultural operations. The ordinance offers the option of compensation payments. Shortly after the presentation of a draft of the ordinance, the Water Supply Company Overijssel (WMO) and the regional department of the national Agricultural Board started negotiations on a compensation arrangement, which were very soon successful. Although the arrangement also applies for pesticides, such compounds are not much of a problem in this province, certainly not compared to nitrates. The quick establishment of an agreement on the compensation payments between the water suppliers and the farmers' organization can be explained by the already established practice in Overijssel of compensation payments to farmers for dehydration damage. Both parties actually preferred to keep the provincial government at a distance, since they did not need a mediator due to that earlier relationship. In the end, the other water supply companies in Overijssel also signed the agreement negotiated by the WMO. The agricultural sector in Overijssel is very satisfied about the arrangement with the WMO and the way in which it has been implemented.

In 1990, Overijssel was the first province where a water supply company started a pilot project to experiment with a stimulation policy. This experiment was set up by the Centre for Agriculture and Environment (CLM). The first experiences in this pilot project were very good, and at the end of 1994 about 100 farmers out of 600 farmers in groundwater protection areas, were participating in such projects. Although the regional department of the Agricultural Board would have liked to play a more central role in the implementation of the stimulation policy, the WMO argued that its own contacts with individual farmers were good enough to do most of the implementation work on its own. So the agricultural organization was not needed as much as it was in other provinces to build up a basis of support among farmers for the new policy, although the WMO still recognizes their role in building support if that is needed. At that time, the historically good relationship between WMO and farmers motivated the Vewin and the CLM to start a pilot project in this province. As in Drenthe, the province itself prefers to be a partner at a distance and leaves the initiative as much as it can to the water supply company.

Gelderland. In 1989 the provincial government designated groundwater protection areas and passed an ordinance with standards for these areas to restrict agricultural operations. The ordinance offers the option of compensation payments. Very shortly after the presentation of the draft ordinance, an agreement on compensation payments followed between the Agricultural Board and the Water Supply Company of Eastern Gelderland (WOG). Gelderland was the first province where a water supply company signed a negotiated compensation arrangement. At that time, other water supply companies and the Vewin heavily criticized this agreement, which was too generous in their view.

Initially (in 1987) the Agricultural Board was trying to negotiate with the Vewin on a national model for compensation settlements in the provinces. The Board planned to impose such a model on the regional agricultural boards and the water supply companies in the provinces. However, the Vewin did not want to commit itself to such a model, since this would deny the autonomous position of water supply companies, which is quite characteristic of the water supply sector in the Netherlands. The Agricultural Board concluded that such a model would only be viable if the national associations involved are strong enough to give a lead to their own sectors. For example, negotiations on a national model for a compensation settlement between the agricultural sector and the energy sector were very successful, because the Gas Union, the representative of the energy sector, controls 80 % of Dutch gas distribution.

In 1987, the national Agricultural Board started to negotiate on the provincial level. It chose to start in the province of Gelderland, where a generous compensation scheme for dehydration damage was already in place. The Boards high expectations, based on the reputation of the WOG, proofed to be right when the first negotiated agreement on compensation payments was established in Gelderland. It was so favorable to farmers, that the water supply companies in other provinces were alarmed about the tough negotiators of the Agricultural Board. The WOG, on the other hand, had opted for a favorable settlement to build up a good relationship with the farmers in order to create guarantees for effective groundwater protection. It was aware of its dependence on the farmers. Manure standards are difficult to enforce, and one has to rely a lot on the willingness of the target group to cooperate. Further, the WOG considered it a matter of prestige to be the first one in the country with a compensation settlement for groundwater protection. Not long after, the other provinces followed suit. In all cases the national Agricultural Board assisted the regional boards. Although the presence of a previous compensation settlement for dehydration damage was favorable to the success of the negotiations, the large water supply companies in the other provinces were more reserved during the negotiations.

After the agreement was reached in 1989, it still took two to three years before the other water supply companies in this province agreed to endorse it, which they did in the end. Since then, all the water supply companies of Gelderland and the regional department of the Agricultural Board meet twice a year, with WOG acting as chairman. The compensation arrangement applies to fertilizers as well as pesticides, although, as in Overijssel, fertilizers are much more of a problem in Gelderland.

In 1993 the WOG started a stimulation policy, initiated by the Vewin and following the successful example of the neighbor province of Overijssel. As in Drenthe and Overijssel, the province itself prefers to be a partner at a distance and leaves the initiative as much as it can to the WOG. It only participates through a financial contribution to the stimulation projects. The introduction of the stimulation policy has led to a modification of the compensation agreement, as is the case in Overijssel and Drenthe: compensation payments will be reduced to a minimum, while farmers can receive a larger subsidy, based on their contribution to groundwater quality improvements, instead.

Noord-Brabant. In 1989, the provincial government designated groundwater protection areas and passed an ordinance regulating the use of fertilizers in these areas, but not the use of pesticides. As an experiment, the ordinance was only applied to the two most vulnerable areas. The proposed standards were very strict and they included a total ban on the use of artificial fertilizers. As the standards were not established in agreement with farmers or farmers' organizations, they provoked tremendous resistance among the target group. The heavy protests forced the provincial government to withdraw the ordinance and return with a revised version, which was much weaker in its approach to the groundwater pollution problem. The new standards have been bargained down and the ban on artificial fertilizers has been annulled. In a 1995evaluation of the revised ordinance of 1989, the province concluded that it has had too little an effect on the groundwater quality, partly due to the lack of support for it (Provincie Noord-Brabant, 1995).

The lack of policy support might be explained by the absence of a compensation agreement in Noord-Brabant. The relevant Water Supply Company East Brabant (WOB) refused to negotiate with farmers' organizations on this. The involvement of the WOB is much less than its colleagues in the other provinces under study, especially because the WOB does not depend to any great extent on vulnerable water well fields. The official opinion of WOB is that they do not expect much effect from a protection level in groundwater protection areas additional to the national protection level. Therefore they do not advocate area specific protection. They also advocate that the national government should not walk away from its responsibility to provide an adequate, nation-wide groundwater protection level.

Besides that, the relationship between the WOB and the agricultural sector has always been difficult. The dominating culture inside the WOB organization is to keep the agricultural sector at a distance and they are not pragmatic in that attitude. In a manner of speaking, they are capable of spending 100 guilders to prove that they spent one guilder too much on compensation payments to farmers. For that reason the WOB also disapproves of compensation payments for pesticides, since they don't think in terms of pesticides as crop protection but only in terms of pesticides as despicable polluting materials. Among farmers this attitude provokes reactions like 'water supply companies are our opponents, like governments, creating obstructions that make it even more difficult to survive in these hard times, so we have to get rid of both of them.'

At the beginning of the 1990s, the province realized that it would have to change its policy style towards farmers. The groundwater protection ordinance of 1989 considers groundwater protection too much as a technocratic issue, denying the political aspects. An important motive for changing the provincial policy style is that high level governmental officials and high level representatives of influential agricultural organizations in Noord-Brabant have started consultations to come up with a political reaction to the extensive debate in society on manure policy, which took place at the end of the 1980s. The heat of this debate was especially fierce in Noord-Brabant, since this is the province with the most intensive agriculture, but also with very extended nature conservation areas (forest). The spreading of manure and the resulting ammonia emissions, as well as the withdrawal of water from agricultural lands and the resulting dehydration, are serious problems that threaten the survival of forests in Noord-Brabant. The provincial government realized, that the unique combination of intensive agriculture in an important natural environment needs strong political attention to farmers as well as environmentalists. The consultations that were started, aimed to consider all political matters that relate to the interaction between agriculture and the environment.

Consultations were also strongly advocated by the agricultural organization *Noordbrabantse Christelijke Boerenbond* (NCB), which is very active in the province and which is known in the Netherlands as the best organized and most influential regional farmers association. While in other provinces a regional department of the national Agricultural Board acts as the main agricultural representative, here the strong position of the NCB made it impossible to set them aside. The NCB is heading in its own direction, to a large extent independent of other farmers' organizations.

The strong ties between the provincial government and the NCB are more or less keeping the water supply company WOB out of the game. The province prefers to be the leading player, and the agricultural sector is in favor of a leading role for the province. Therefore, in 1992 it was the province which initiated a policy to encourage preventive actions in the agricultural sector. This policy not only applies to groundwater protection areas, but also to all areas where environmental values need protection. The WOB prefers to be a partner at distance and only provides some financial contributions, although they do not expect much effect from this on the groundwater quality. They see it more as a matter of creating some goodwill.

Just before the province initiated a stimulation policy, the WOB tried to set up a policy itself for the vulnerable groundwater protection areas of Boxmeer and Vierlingsbeek, mainly because Vewin asked them to take such an initiative. However, the resistance among farmers was so great, that only five farmers showed any interest. This might also be explained by the enormous delay in the payment of compensations for dehydration damage in these areas. In some cases farmers have already been waiting for twelve years for compensation. The WOB is much more formal in its handling of such claims. Neither does the WOB incorporate agricultural expertise into its own organization, unlike other water supply companies in our study. In dealing with the agricultural sector, the WOB prefers to leave matters to the province.

Although the agricultural organization NCB is playing an influential role on the level of consultations with high level government officials, their role in the implementation of stimulation policies is more restricted. Provincial officials at the implementation level believe, that they have good direct contact with farmers. They think that direct consultations with farmers are often easier than consultations with their representatives. Many times representatives have opinions that are too much on principle and not sufficiently pragmatic, while individual farmers might be much more flexible and co-operative. Despite that, provincial officials recognize the importance of agricultural mediators in gaining support for provincial projects among farmers.

Limburg. Since 1989 the province of Limburg, like the other provinces, has had an ordinance for groundwater protection areas. In that ordinance a distinction is made between legislation for the southern part of Limburg, where standards are more stringent because of the vulnerability of the whole *Mergelland* region, and legislation for the middle and northern parts of Limburg. Originally, the province made the mistake to present a draft ordinance (although not without consultations with the water supply company and the regional department of the Agricultural Board), without the option of compensation payments. As in Noord-Brabant, this raised strong protests among farmers, who got the idea that the agricultural organizations had already given their commitment to the ordinance without discussing it with the farmers involved. Especially the stricter standards for the south of Limburg provoked a lot of resistance. People were speaking in terms of 'the Margraten war' and 'the battle of Gulpen' (after villages located in Mergelland in the south of Limburg).

This was a signal to the province to get the large Water Supply Company Limburg (WML) and the regional department of the national Agricultural Board around the table, to prepare an agreement on compensation payments. The province took the initiative and not long afterwards the WML and the agricultural representatives had come to an arrangement. The implementation of the arrangement is in the hands of the WML. In practice, 99% of the claims made by farmers are covered by this agreement. The rest is handled by the province, which has an advisory committee to deal with them. The WML adheres strictly to the principle that, if they have to pay, they also want control over the implementation of the agreement, without delegating this to agricultural organizations. The agreement does not apply to pesticides, since the provincial ordinance does not regulate the use of pesticides in groundwater protection areas. In this respect the Limburg ordinance differs from the ones in Drenthe, Overijssel, and Gelderland, although pesticide use in Limburg is higher than in Overijssel and Gelderland. In general the compensation settlement in Limburg has been quite well accepted by the farmers.

In 1995, in an interview, provincial officials stated that they noticed that the style of co-operation between the province and agricultural representatives had changed a lot. They attribute this to new national ideas about policy styles, like 'network steering', which have had an impact in the provinces. Provincial officials think that the opinion about regulations and their feasibility has changed:

Nowadays, everyone talks about network steering and win/win situations. Although many people don't know what it means, they strongly feel that it will bring about a cultural change in the way we deal with each other.

In Limburg the regulatory climate changed around 1990/1991, after the last round of negotiations about final arrangements for standards in Mergelland, the most vulnerable region in the south. The solution there was that the national government was willing to provide a fund with subsidies for farmers who are co-operative in meeting standards that are more stringent than everywhere else in the Netherlands. 'From that moment on, maybe because the last regulatory barrier had been removed, the diametric oppositions were over', as the officials involved concluded.

In 1991, the province started initiatives for a stimulation policy. For that purpose, it created a Task Force on Agricultural Emissions, and invited representatives of the province, the agricultural sector and the water supply company WML to participate. Agricultural organizations are invited to propose projects, or the province can come up with projects itself. Since 1991, about 100 project proposals have been submitted for the province as a whole although, as in Noord-Brabant, these projects are concerned with a much broader variety of environmental improvements than only improvements in the groundwater quality. The WML has not initiated its own stimulation policy, because it is mainly using deeper groundwater, and does not expect groundwater quality to improve due to temporary measures. The results will not be visible within 50 years, and that is too long a period to be an appropriate argument for providing money now. Despite this, the WML has invested in incorporating agricultural expertise within its own organization, and in that respect the WML differs from the WOB in Noord-Brabant. Reacting to an initiative from Vewin and CML, the WML also showed itself to be interested in experimenting with performance subsidies that guarantee a more direct relation between financial incentives to farmers and results in terms of groundwater quality.

8.4 Motives of water suppliers

In this section we will look in greater detail at the motivation of water suppliers, their expectations of working on prevention, and how these motives are related to the motives of the other actors involved (provincial authorities, farmers and farmer's organizations). We will consider both the willingness of water suppliers to encourage prevention, as well as the willingness of farmers to react to the water suppliers' attempts at encouragement. An analysis of motives is important to learn more about the control capacity of water suppliers. In particular this analysis will tell us about the conditions under which water suppliers might be motivated to choose prevention of agricultural water pollution as their strategy and to choose to transfer resources to farmers as part of this strategy.

8.4.1 MOTIVES OF THE WATER SUPPLY COMPANIES

Providing drinking water of good quality is what water supply companies see as their main task. Water abstraction (pumping), water production (purification), and water delivery are their core business. They do not consider preventing agricultural pollution to be their task, and they do not view drinking water directives or standards as an incentive for water supply companies to get more control over groundwater polluting activities. On the other hand, water supply companies are aware that the presence of pollution in drinking water resources is in fact an incentive for them to think about alternatives for water purification and treatment. Thus, agricultural water pollution is forcing them to become active in encouraging prevention, since this might be more efficient than curative solutions. In particular, the extent to which they need to invest in complicated or expensive purification techniques, is seen by them as an indicator of the success or failure of groundwater protection.

A first principle of the water supply sector as a whole concerning groundwater protection, is that the national government should be responsible for the general protection of the groundwater in such a way groundwater should be fit for the production of drinking water throughout the entire country. Secondly, additional protection of groundwater in areas with water wells should always be a temporary measure to solve crisis situations. The general level of protection should meet the specific levels of protection as soon as possible in order to prevent extra costs of additional protection (in terms of financial compensation).

These principles reflect how water supply companies think about governmental responsibility towards groundwater protection: national and provincial governments are not allowed to walk away from their responsibility in providing adequate protection levels. However, since the Soil Protection Act (1987) in the Netherlands created a legal construction in which those who profit from additional groundwater protection have to compensate those who incur losses because of this, water supply companies are prepared to pay for additional protection, on the conditions that additional protection will be temporary and will anticipate provisions for general protection.

The compensation payments in groundwater protection areas are considered to be temporary, and the Soil Protection Act supports this idea since it indicates that compensation will no longer be necessary after 2000, in which year the general protection level should replace the specific level of groundwater protection needed for drinking water purposes throughout the country. In this context, it is quite understandable that water supply companies have problems with the third phase of the Dutch manure policy, which does not guarantee such protection before 2010.

In the end, water supply companies are not completely satisfied with the results of the provincial ordinances of 1989 and the related compensation settlements. Although they realize that it was a learning experience, which made the water supply sector more adaptive towards the way in which the agricultural production sector operates and is affecting drinking water sources, they also think that it was not a very efficient way of handling agricultural groundwater pollution problems.

Water supply companies prefer to apply the compensation money for the encouragement of innovations that optimize the use of minerals and pesticides. Such a policy might relieve the use of excessive amounts of minerals and pesticides. With respect to the use of nitrates, water supply companies think they have succeeded in gaining some reduction with their own strategies to encourage farmers to take preventive actions, in combination with the provincial ordinances, although this has not had any immediate impact on the groundwater. With respect to the use of pesticides, they think they depend heavily on the national pesticides policy, especially the implementation efforts of the 'Long term crop protection plan'. In the short run, regional policies will not achieve the required effects on groundwater quality. These are spoiled by a delay in the implementation of national policies and by the historical pollution that will show up for decades to come. On the other hand, stimulation policies might provoke a better awareness and understanding among farmers of how their farming operations might harm the groundwater quality. It is the experience of the water supply companies, that a considerable group of farmers became interested in the prevention of groundwater pollution. Water supply companies like to achieve a change of mentality in the agricultural sector, through which farmers will see that agriculture and drinking water production can get along hand in hand.

Despite the creation of a better level of support among farmers, water supply companies are still reserved about spending money on innovations in farming operations. It is the requirement to operate the water supply efficiently, as supervised by the provincial governments according to the Water Supply Act, which provokes a paradoxical situation. On the one hand, provinces require that water supply companies deliver good quality for the lowest price. On the other hand, provinces try to shift the burden of additional groundwater protection to the water supply companies. Water supply companies try to avoid subsidizing measures that improve the environment in general without improving the quality of drinking water sources. They keep advocating that national and provincial governments should remain responsible for environmental policy in general. Besides that, they work on the fine-tuning of policy tools for water supply companies to stimulate only agricultural innovations that have a demonstrable effect on the quality of drinking water sources.

Inasmuch as water supply companies want to activate farmers in preventing agricultural water pollution, some companies (in Drenthe, Gelderland, and Overijssel) prefer to contact farmers directly on their own, allocating a very restricted role to intermediate organizations in the agricultural sector. In other provinces (Noord-Brabant and Limburg) the water supply companies do not want spend too much time on activating the farmers, and try to leave this to the provincial authorities and the agricultural sector itself. Where water supply companies are active on their own and build up agricultural expertise within their own organization, in general we can say that these companies have to deal with drinking water sources that are much more vulnerable to agricultural pollution. These companies do not like to rely too much on actions by the provincial government.

8.4.2 MOTIVES OF THE PROVINCIAL AUTHORITIES

Provinces have an interest in bringing about accelerated protection in groundwater protection areas, since groundwater protection policy, additional to a national policy, is their constitutional and political responsibility. On the other hand, provinces do not feel responsible to go beyond the final standards that were set by the national government as a long term goal for the Netherlands as a whole. So, if water supply companies seek more stringent regulation, the provinces do not want to provide it. They consider this as a specific interest that should be striven for by the interested parties, according to the profit principle.

Provinces also consider the costs of additional groundwater protection. They believe that additional protection standards will suffer from severe enforcement problems if they lack support among the target population. Provincial ordinances for groundwater protection areas were only politically feasible in combination with compensation arrangements. Since provinces will not be able to provide the money for this, they rely heavily on the water supply companies (and in fact on the drinking water consumers, who pay the price in the end) to create the required financial support for additional protection measures. Provinces even have problems in finding other partners in society with a financial interest in environmental protection, and therefor they have not come up yet with provincial regulations in other areas, where environmental protection is needed for other reasons than protection of drinking water resources. For example, the Soil Protection Act also asks the provinces to indicate 'soil areas', but there is no social actor with a financial interest in this kind of protection, and therefor no provincial legislation for this type of protection has been passed.

Provincial authorities are very enthusiastic about stimulation policies. They think that these fit quite well with the national tendency to activate 'partners in society' that are able to give environmental policy incentives to target groups (like farmers). Water supply companies are seen as such, since they have a control capacity through their strong interest in groundwater protection, and since they have the financial means to pay for it.

In this context, the provinces see their own role as one of initiating stimulation policies and bringing partners in society together, such as the agricultural sector and the drinking water sector. Provinces try to facilitate such meetings by creating a task force for the integration of environmental issues into agriculture, as has been done in the provinces of Noord-Brabant (*Landbouw-Milieu Overleg*) and Limburg (*Werkgroep Landbouwemissies*). In these provinces, the provincial government created a more central role for itself than the governments in the other provinces (Drenthe, Overijssel, and Gelderland). This is a consequence of the fact that the water supply companies involved in Noord-Brabant and Limburg are less motivated to initiate stimulation policies, since they are less dependent on vulnerable drinking water resources in the short term.

Insofar as provincial governments are active in setting up a stimulation policy, they all are motivated by the results of pilot projects, undertaken through concerted action of the Vewin and the Centre for Agriculture and Environment (CLM), since these results testify to an active participation of farmers in study meetings on environmental innovations. Provinces are convinced that this is the way to activate farmers, and they also participate financially in stimulation projects together with the national Ministry of Environment, which also furnishes money such projects. In general, the province, the Ministry of the Environment, the water supply company, and the participating farmers each contribute 25% of the costs of environmental innovations that are approved in the context of a stimulation project.

8.4.3 MOTIVES OF THE FARMERS

Despite the farmers' strong resistance to the national manure policy in the 1980s and 1990s, and the provincial standards for groundwater protection areas as laid down in the provincial ordinances of 1989, farmers in general feel comfortable with compensation payments and stimulating policy instruments. The national tendency towards tightening manure standards makes them feel relatively lucky to have land in groundwater policy areas, which offers them the advantages of getting payments for environmental innovations, while farmers outside these areas have to pay for environmental innovations themselves, albeit somewhat later.

Also more realism is growing among farmers in general that not all of them will survive in a business climate where environmental innovative actions are becoming inevitable. Environmental policies do have a tremendous impact on how farmers judge their future. Only those will survive who can integrate environmental standards into their farming, and are able to manage and control their mineral streams and water use. Currently, 3% of the Dutch farmers go out of business every year, including those who have no successor or insufficient land to exploit. In ten years 30% will have dropped out and experts predict that only 60% of the farmers will be able to continue farming.

It is expected that stimulation policies will reach not more than one third of the total population of farmers in groundwater protection areas, since they are based on voluntary actions. Participants are mainly the more intensive and larger farms, as well as the more innovative farmers. Innovative farmers are less characterized by age and more by their willingness to accept risks. They are the pioneers within the agricultural community and enjoy authority among their colleagues; farmers about whom others will say: 'if he is doing it, then I might be able to do it too.'

8.4.4 MOTIVES OF THE FARMER'S ORGANIZATIONS

The main principle of the agricultural sector concerning groundwater protection is, that the competitive position of the Dutch agricultural sector in general (compared to other countries), as well as the competitive position of individual farmers (in groundwater protection areas) may not be harmed. As far as the specific and more stringent standards for groundwater protection areas are concerned, they adhere strongly to the principle of financial compensation for the losses due to differences between the general and the specific level. In terms of the international competitive position of the agricultural sector, they are concerned about the - in their view - progressive character of the Dutch legislation. The Dutch government is trying too much to be 'the most decent boy in the international class'.

The Agricultural Board is generally accepted by the farmers in groundwater protection areas as being the sole negotiation partner for the water supply companies. This exclusive position of the Agricultural Board has been established in previous negotiations with the water supply sector on dehydration damage, as well as negotiations with the energy sector on damage due to underground gas mains, in which the board was very successful in making financial deals which were profitable to the farmers.

Although the Agricultural Board initially acted in a very defensive way, nowadays it is increasingly going over to the offensive in being co-operative in thinking about preventive solutions for agricultural (water) pollution. While they still oppose manure and pesticide standards that restrict farming operations, agricultural organizations and their representative managers are also aware, that something has to be done and they are trying to make the best of it. They strongly advocate policy instruments that encourage environmental innovations in agricultural operations. They also strongly advocate a position for the agricultural intermediary organizations in the implementation of environmental policies for farmers. They like to have a key role in the communication between water supply authorities and farmers, for example by running pilot projects, by taking over communicative campaigns and information services, and by taking over the distribution of moneys for subsidies.

8.5 Resources of water suppliers

The control capacity of the water suppliers can not only be explained by their motives, one also has to consider the resources that are available to them for exercise control. In this section we consider different types of resources and we analyze the extent to which they are important for water suppliers who want to control agricultural water pollution. On the other hand, it is important to know under what conditions farmers will be susceptible to steering activities. In other words, the different types of resources can also be considered from the perspective of the farmers and the extent to which they are interested in receiving resources from the water supply companies. So, on the one hand resources play a role in the choice of a preventive strategy by water suppliers, and on the other hand they play a role in the success of this strategy through the susceptibility of farmers.

The first resource we consider is 'organizational capacity'. The question is whether the size of a water supply company matters in the choice or success of a preventive strategy. For example, the size of a company might influence the ability of the company to incorporate agricultural experts and in general the ability of an organization to get on speaking terms with farmers. A hypothesis might be, that smaller water supply companies pay less attention to the behavioral aspects of pollution prevention, and therefore are more oriented to the engineering aspects of water production. Smaller companies might be less able to set up steering activities.

A second resource we consider is 'financial capacity'. If water supply companies want to move farmers towards prevention, they probably will consider whether the costs of preventive action compensate for the alternative of remedial expenses. From a farmer's perspective, we need to know the conditions under which they are willing to change their agricultural operations, if they were to get money for preventive action.

A third resource we look at is 'information'. Do water supply companies have the right information necessary to change farmers' behavior? Are the assumptions about the relation between changing agricultural patterns and possible effects on the groundwater quality stable enough? What information do water suppliers need to be sure, that farmers have been complying with the standards for preventive behavior? Under what conditions are farmers willing to be more open about the way they farm?

A fourth resource that might be worth some consideration, is 'confidence'. For negotiated agreements on preventive action it is important that some basis of confidence exists between farmers and water suppliers. If water suppliers provide money (compensation payments) to farmers, they want to be sure, that they will get results in terms of groundwater quality, while changes in agricultural operations are generally difficult to enforce. Agricultural water pollution is considered to be a kind of non-point-source pollution. Also, farmers in general are critical of the public authorities and environmental regulations. It is difficult to gain their confidence and to be seen by them as a reasonable and reliable stakeholder.

A fifth resource we consider is 'authority'. If one is to make farmers comply with rules or to steer them in some direction, it can be important to have authority. Do water suppliers have formal or informal authority? To what extent do water suppliers depend on public authorities and to what extent are they hindered by a lack of authority? Are farmers susceptible only to authority, or are more resources necessary, in combination with authority, to make farmers change their behavior?

The last resource type we consider is 'time'. The main question from this perspective is whether the water suppliers are in a hurry to get farmers to change their behavior? Does the problem context (the soil type, the size or seriousness of pollution, etc.) require quick action? And the farmers themselves, do they feel the 'heat' of problems of pollution?

8.5.1 ORGANIZATIONAL CAPACITY

Water supply authorities in the Netherlands are in general public utilities, not fully privatized, like in France or the United Kingdom. Provincial governments have to supervise water supply companies to guarantee the delivery of drinking water at a reasonable price and of good quality. In 1997, the Netherlands had 28 water supply companies, some of which operate on a scale of almost an entire provincial territory (such as the WMO in Overijssel, and the WML in Limburg) and others on a regional scale within a provincial territory (such as the WMD in Drenthe, the WOG in

Gelderland, and the WOB in Noord-Brabant). The Netherlands numbers a total of twelve provinces. The smaller water supply companies are in general the older ones, started in times (19th century) when public water delivery was organized only in urban areas. Most of the larger companies started in the first half of this century, serving both urban and rural areas. They have grown because of an increasing connection rate. By now, almost all households in the Netherlands are connected to a water main. By Dutch standards, a viable water supply company serves at least 100,000 connections, and abstracts water from at least two well fields which are mutually connected. The provinces have used their supervisory powers to reorganize the water supply sector by enforcing provincial reorganization plans. The national government is supporting this action, since government policy is in favor of a strong water supply company sector, with a restricted number of viable companies. For reasons of consumer protection and the protection of public health, the government prefers to avoid full privatization (Achttienribbe, 1993; Langendijk, 1994; Van der Schot, 1995).

The efficiency requirements imposed by the provincial governments helped the water supply companies to build strong organizations, able to resist agricultural pollution by investing money in purification techniques. Due to this, scarcely any cases can be found where drinking water standards have been violated in the water delivered to the tap. Such cases as were found, were not so serious as to pose a problem to public health. So, one could say that a viable organization can help a water supply company to prevent violations of drinking water standards.

With respect to water supply companies active in preventing the agricultural pollution of drinking water sources, this study has shown, that those who are heavily dependent on vulnerable sources, are most active in negotiating with farmers on prevention measures, WMD in Drenthe, WMO in Overijssel, and WOG in Gelderland. WOB in Noord-Brabant, and WML in Limburg were less active since they do not feel so much pressure from increasing pollution of drinking water sources in the short term. The active companies do have many well fields in areas with intensive farming, while the groundwater sources in these areas lack the natural presence of covering layers, like the natural clay layers in Noord-Brabant and in the northern and middle parts of Limburg. Unlike the WOB in Noord-Brabant, the WML in Limburg has negotiated with farmers on a compensation arrangement, which cannot be explained by some interest in the short-term protection of groundwater. It is more plausible to search for an explanation in the active role of the province, which wanted to have the water supply company at the table, in order to preserve the credibility of the provincial ordinance.

This study also shows that in general the larger water supply companies were most active in participating in consultations with agricultural organizations, while the smaller ones were more followers, only playing a role at the end of negotiations by signing an agreement as an additional partner on the water supply sector side. Nevertheless, it was not the largest water supply company in each province that lead the negotiations with farmers' organizations. An exception is the WOG in Gelderland which, while it is not the largest, it is the company that is most involved in the problem of agricultural water pollution of drinking water sources.

Further, this study shows that larger companies are better equipped, which enables them to incorporate agricultural expertise within their organization. This is the case for all the water supply companies studied, except for the WOB in Noord-Brabant, where a predominantly production oriented organization culture appears to be a barrier for organizational change in this sense. Although neither the WOB in Noord-Brabant, nor the WML in Limburg have a strong interest in consultations with farmers, the latter shows that, despite this, a water supply company may have an interest in developing its own agricultural expertise. Personal factors, like the personality of the manager, are also important in this context. In general, it is assumed that managers who have already been in service for a long-time are more traditional in their thinking, still emphasizing 'pumping and billing'. They don't feel comfortable with negotiating about changing polluting behavior.

A very important role in motivating water supply companies to consult with farmers, was played by the national union of water supply companies (Vewin) and the Centre for Agriculture and Environment (CLM). Since the Vewin incorporated agricultural experts (some of whom had previously worked for the CLM), this tendency had a considerable impact on the individual water supply companies as well. The ideas developed by the CLM to integrate environmental innovations into agricultural practices were essential demonstrations for the water supply companies that something could be done by them.

8.5.2 FINANCIAL CAPACITY

Water supply companies are financially self-supporting. No tax revenues are needed from the government. The prices charged for drinking water are low throughout the country. The choice for low prices and the reduction of profit is based on a concern for greater efficiency in the water supply sector. Groundwater is considered to be a collective property, and therefore water supply companies may not be regarded as profit making industries selling a marketable product. The provincial governments are authorized by the Water Supply Act to control the efficiency of the water supply sector. As long as water supply companies perform well, no interventions will be made by the authorities. Most water supply companies are organized on a partnership basis, the majority of the shares being in the hands of the provincial authority. Provinces are authorized to impose reorganization plans on the water supply sector.

In general, the national government is very satisfied with the performance of water supply companies, in terms both of finance (i.e., their efficiency) as well as their environmental performance. In the National Environmental Policy Plan, the water supply companies are held up as an example of industries in which environmental thinking is largely internalized. Provinces depend very much on the water supply sector as far as the effectiveness of the provincial ground water policies is concerned. Provinces are aware of the fact that the standards for specific protection can only be enforced with financial investments. The provinces are not able to provide the money on their own, and therefore they treat the water supply companies as their 'policy partners', while farmers' organizations are considered to be representatives of a target group. This offers water supply companies the opportunity to participate in the policy formulation process.

By contrast, the water supply companies believe that the national and provincial governments are shifting the financial burden too far in their direction. Delays in the implementation of pesticides and manure policies force the water suppliers to make large investments in purification techniques, and maneuvers them into a very uncertain position in which they do not know what kinds and amounts of agricultural pollution that will emerge from their wells. As far as the national pesticides policy is concerned, the sanitation of harmful pesticides is not proceeding fast enough, in part due to a delay in the European harmonization of pesticides policies in the Member States. As far as the national manure policy is concerned, there is too little regulation of nitrates use. Until 1995 nitrates were not regulated at all, because all manure standards were related to the use of phosphates, while artificial fertilizers were excluded. Furthermore, the implementation of the European directive for nitrates in groundwater, which is very important to the water companies, has also been delayed, since the government believes that the Netherlands cannot meet the European standard before 2010. The water supply companies are not satisfied with this situation, in which governments are forcing the water suppliers to meet drinking water standards for public health purposes, while on the other hand they do not offer sufficient support to the water suppliers in preventing drinking water sources from being polluted.

Another case in which water suppliers see a shift of the financial burden, is the introduction of a groundwater levy. Since 1995 the Dutch government has imposed a levy to finance nature protection activities, and to make drinking water consumers more aware of water scarcity. Water supply companies pay 34 cents per m^3 of drinking water (1000 liter), while industrial users who extract groundwater from their own sources, pay half this amount. The water supply companies fix the levy on the drinking water price that consumers pay. In 1994, when the Dutch government presented its plan to impose a levy, the water supply sector protested fiercely. The director of Vewin spoke of 'opportunist politics' and considered it to be a veiled taxation measure to create money for environmental policy at the cost of drinking water consumers. The real problem for water supply companies is that the national and provincial governments are on the one hand forcing the water suppliers to operate as efficiently as possible, while on the other hand, they allow drinking water prices to increase in order to finance all kinds of environmental policy measures. This is the case with the compensation payments, with the stimulation policies, and with the imposition of the groundwater levy (VROM, 1993-1997).

According to the water supply companies, in addition to the national and provincial governments, the agricultural sector also regards them too much as an industrial sector able to pay for groundwater protection. Although the water supply sector is willing to pay for extra groundwater protection on a temporary basis, they prefer investments in sustainable protection of groundwater sources. Performance related subsidies are strongly preferred over compensation payments that offer too little encouragement to farmers to act to protect the groundwater. In that context, water supply companies and the Vewin prefer to further develop the idea of 'payment by results' (*resultaatbeloning*), which is a policy instrument that can work as a regulatory incentive, a performance subsidy. It might become a self-supporting instrument if expanded with a regulatory levy system, in which farmers pay money for the amount of pollution in the groundwater and earn money for the amount of pollution they can sanitize (CLM, 1994).

In general water supply companies are anxious to lose control over the money they spend on preventive measures in groundwater protection areas. They prefer to implement the compensation arrangements, as well as the stimulation policies themselves This attitude of the water suppliers can be explained in terms of the strong incentives from the national and provincial governments to operate in an efficient way. 'If we have to pay, then we also need to have the control,' is what the water suppliers say. They fear that provincial agencies will create excessively high bureaucratic costs, if they were to do the implementation work. In the case of the agricultural organizations, they are more concerned about a possible target shift, brought about by inventive farmers coming up with proposals that would optimize farming operations but which would have no result on the groundwater quality. 'They are just trying it on,' stated a spokesmen of a water supply company, that has rejected many proposals for stimulation projects. If the implementation were to be in the hands of agricultural intermediaries, water supply companies fear that these agricultural officials would not be sufficiently critical about the contributions of proposals to the groundwater quality.

The Vewin calculated that the actual price of drinking water in the Netherlands will increase by half within ten years, and may even double in some parts of the country. The main reason for the price increases is the strongly fluctuating quality of the groundwater and surface water input. To be able to deliver drinking water of good quality, water supply companies have to incur greater personnel costs, to expand their expertise and have to invest in research and development for new, complicated and expensive purification methods. Also, the discharge of the pollution removed from the groundwater is quite costly. All these environmental costs are considered to be an important reason why the number of water supply companies is reducing in the Netherlands.

8.5.3 INFORMATION

It is very important for the water supply companies to gain information about the type and amount of pollution that is on its way through the soil to the water wells. Information about historical pollution is hard to get, and water suppliers have to await what is coming up with the groundwater. are hard to predict Predictions are rather uncertain, depending on the sophistication of the geohydrological models used, and the reliability of the data on the biodegradation rates and mobility of nitrates and pesticides. Differences in soil condition, may lead to differences in residence between one year and more than a century. For water supply companies depending on relatively unprotected sources, it is very important to get information about present farming operations, and even to control future farming operations as far as these are affecting the water that is flowing underground into the direction of their well fields.

The compensation arrangements have increased the water supply companies' knowledge about non-point-source pollution. The compensations helped them to learn about the amounts of nitrogen and pesticides applied, since farmers have to submit their bookkeeping in order to receive payments. This has increased the control of water suppliers over the groundwater protection areas, because the information enables them to start consultations with farmers on their farming operations, which encourages farmers at least to become more aware of the direct relationship between their operations and the difficulties of drinking water production. The water supply sector is even coming to regard its stimulation policy as a better way to gain access to the books of individual farmers. A part of the farmer population is prepared to provide the water supply companies with information about their farming in exchange for subsidies. This enables water suppliers to anticipate potential agricultural threats to the groundwater.

On the other hand, many farmers are reserved about providing information to water suppliers, since they fear too much control by these companies, which many farmers experience as an 'extended arm' of the government. As part of the agreements, the water suppliers had to assure that the information would be treated as confidential, and would not be passed on to government or public agencies. Nevertheless, water suppliers still have to explain to farmers, that enforcement of legal standards is nor their task. Gaining confidence is very important for the water suppliers, if they are to get the information they want. Although about 70% of the farmers in groundwater protection areas are satisfied with the way compensation arrangements were handled by the water supply companies, the participation rates in the voluntary stimulation projects are limited to only 20 to 40%.

Water supply companies have to deal with conflicting role expectations. On the one hand, the national and provincial governments want them to act as partners in helping to create a better environment by reporting pollution and cleaning up water sources, which means that - in the farmers' eyes - they are acting as government agencies. On the other hand the water suppliers have to show the farmers, that they are independent producers, like the farmers themselves, who are able to handle confidential information in the same way as partners in business would do. The mutual confidence between water suppliers and farmers is a crucial factor in this.

The difficulty of this mission also depends on the accuracy of the geohydrological assumptions used in the consultations with farmers. For water suppliers it is very difficult to be sure about underground patterns in the soil, the basis of their assessment of the effects of agricultural practices on groundwater quality. It has happened that assumptions used in 1989 to designate groundwater protection areas, had to be changed five years later. Some farmers had to be told that they are no longer in a groundwater protection area, while others had to be told that they were now in one. It is hard to identify the exact borders of a water catchment area, and water supply companies want to feel free to change policies if their models change. Another example: in 1989/1990 a study of the soil in an area in Noord-Brabant showed that the amount of pyrite in the soil was sufficient to allow higher concentrates of nitrogen to be applied in that area, since the pyrite would prevent a large amount of nitrates from leaching into the groundwater.

Although it was good news for farmers to tell them that the manure standards could be relaxed, it did not create a very reliable impression.

Cases like the one described above provoke statements by farmers that the administrative standards are often too unrealistic and too ambitious. Even water suppliers sometimes have problems with the lack of flexibility of the legal standards. They do not like the regulatory rigor of governmental agencies, and prefer a more flexible approach to groundwater protection as their own policy style. Water suppliers think that the way they deal with stimulation projects is valued by farmers, as being more flexible and realistic, since the operationalization of standards in agricultural practices is discussed in study groups. Pilot projects are important for them to demonstrate how practical changes in agricultural operations can bring about visible results and improvements of the groundwater quality.

8.5.4 CONFIDENCE

The degree of mutual confidence between water supply companies and farmers differs from area to area in the Netherlands. Although the parties are still not on speaking terms in a few groundwater protection areas, in most areas mutual confidence has grown due to the compensation settlements. Most farmers (by respondents estimated at about 70%) are satisfied with the compensations paid. This also increased the willingness of farmers to think in terms of the prevention of agricultural pollution. The majority of the farmers is aware, that the water supply sector is serious in its willingness to pay for sustainable solutions. This study shows that the mutual confidence is at higher levels in provinces where water supply companies play an active and leading role in consultations with farmers about preventive measures in groundwater protection areas. This is especially the case in Drenthe, Overijssel, and Gelderland, while in the province of Noord-Brabant and (although to a lesser extent) in the province of Limburg, provincial officials are taking the initiatives, and more distance appears to exist between water suppliers and farmers. Although farmers are also participating in stimulation projects in Noord-Brabant and Limburg, they don't have the specific purpose of improving the groundwater quality, as they rather aim at sustainable farming in general. Where water supply companies are involved in stimulation projects and take a leading role, farmers are more motivated to work on specific improvements for groundwater quality.

The mutual confidence is an important factor for creating receptiveness among farmers to preventive measures. Water supply companies have to make clear, that they are not environmental controllers, but partners who want to do business on a confidential basis. The separation in the Netherlands between the regulatory role of the provinces in setting standards, and a more consulting and consensual role for the water supply companies doing business in buying innovations in farming operations is rather well chosen in that respect. It works as a kind of double play in which the one is carrying the stick, while the other carries the carrot. Water supply companies that were active in negotiating compensation arrangements, like the ones in Drenthe, Overijssel, Gelderland, and Limburg, have broken down a lot of resistance to groundwater protection measures, as they were introduced in provincial ordinances in 1989. The combination of provinces handling compensation payments in a strictly formal and economic way, and the water supply companies handling compensation payments in a more generous way is also important in this respect.

The way in which the parties involved deal with each other has changed since the beginning of the 1990s, after the negotiations on provincial ordinances and compensations arrangements ended. Nowadays, many of them perceive a more consensual approach on each side. Farmers' organizations are tired of all the conflicts that arose during compensation negotiations. They are tired of doing battle for more peanuts. Provincial officials feel that the change to a new policy style, started in the Hague, with ministries increasingly thinking in terms of 'network steering', 'consensual steering', 'working together with target groups and partners in society' and 'win/win situations'. Although many are unaware of the actual meaning of these paradigms, they embrace them as something new and interesting. Representatives of water supply companies are changing their style, since the national union of water supply companies (Vewin) in collaboration with the Centre for Agriculture and Environment (CLM) is encouraging them in initiating more co-operation based projects with the participation of farmers, water suppliers, and provinces. Many of the participants involved also think, that the entrance of a new and younger generation of representatives on all three sides has made it easier for them to talk with each other. Many of the larger water supply companies were represented by newly recruited officers who had just completed their higher education in agricultural and environmental issues. They also meet with younger farmers with higher education and a different attitudes than the older generation. They think less in terms of conflict models, and are more willing to work on a consensual basis.

Water supply companies also perceive among the farmers in groundwater protection areas, that they are beginning to see the advantage of getting subsidies for environmental innovations on their farms, while their colleagues outside these areas have to pay for most innovations themselves. Although there is a different time schedule for farmers inside and outside the protected areas, in the end they both have to end up observing the same standards. For the younger farmers who still have a long time to ahead of them in farming, it is just an investment in the future, while it is more a burden for those older farmers who do not have the perspective of a successor for their farm.

8.5.5 AUTHORITY

The authority to provide regulation for groundwater protection standards lies exclusively in the hands of the provincial governments within national and European regulatory frameworks. They have the competence to issue legal standards for their areas. The water supply sector would not be happy to have such a competence itself. They even interpret the lack of having a regulatory competence as an advantage. Complaints about unclear or severe regulations are now addressed to the provinces. This guards the water supply companies against an unpopular image. The unpopularity of manure standards is attributed to the national and provincial authorities.

Further, the water supply companies have low expectations of solving the problems by regulation alone. Other types of policy instruments should also be used to encourage farmers to prevent groundwater pollution. On the other hand, water supply companies are concerned about the tendency of provinces to rely too much on the stimulation policies which are developed by the water suppliers, meanwhile forgetting that such a policy leaves their own regulatory duty undisturbed. Although provincial governments are declaring that stimulation policies are a central pillar in the environmental protection by the provinces, water suppliers emphasize, that provinces cannot give way to low expectations of regulation and enforcement.

Water supply companies do not consider the prevention of agricultural pollution to be their task, and they will not accept that drinking water directives or standards should be seen as an incentive for the water supply companies to take over control of groundwater polluting activities. On the other hand, the water supply companies are aware that the presence of pollution in drinking water resources is an incentive for them to think about alternatives to water purification and treatment. Thus, agricultural water pollution is forcing them to become active in encouraging prevention, since this might be more efficient than curative solutions.

Besides that, water supply companies emphasize that the national government should be responsible for the general protection of the groundwater in such a way that, nationwide, it should be possible to use groundwater for the production of drinking water. Additional protection of groundwater in areas with water wells should always be temporary to solve crisis situations. The general level of protection should meet the specific levels of protection as soon as possible, in order to prevent extra costs of additional protection (in terms of financial compensation).

8.5.6 TIME

The idea behind the designation of groundwater protection areas was to speed up the protection of groundwater sources that are used for the production of drinking water. The time offered to farmers outside protection areas was considered to be too long to achieve the protection levels that are required in areas were groundwater is used for the production of drinking water. In 1994, a national evaluation of the provincial groundwater ordinances concluded, that they did not produce the improvements in the groundwater quality that were needed. In most areas the standard setting came too late, since a lot of pollution from the past was already moving underground towards the water wells. Although the standards will have an effect in the long run, they will have no meaning in the short run, and they do not prevent water supply companies investing in purification techniques.

Only in areas without considerable historical pollution, the water supply companies may profit from the more stringent standards in the provincial ordinances of 1989. This is, for example, the case in the Mergelland region in the south of Limburg, where the rather extensive agriculture will be prevented from intensifying the application of manure. However, a weakness in the provincial standard setting, as well in the national one, is that nitrogen was not regulated until the mid-1990s, and many pesticides are still not subject to regulation. In the province of Limburg, the provincial groundwater ordinance does not even apply for the use of pesticides in groundwater protection areas. So, in fact the provinces were not really leading in providing a better protection level for groundwater protection areas, and therefor farmers in protected areas did not have the same start as farmers outside these areas as far as improving the groundwater quality is concerned.

8.6 Conclusions on the control capacity of water suppliers

8.6.1 INFLUENCE OF THE PROBLEM CONTEXT

In the Netherlands, the agricultural pollution of drinking water sources is mainly a problem of nitrates and pesticides. Violations of European standards are not found in the drinking water delivered to consumers. And even as far as the deeper groundwater sources are concerned, where the public utilities get their water for drinking water production, still only few violations of the standards are found. Treatment of water from these deeper wells is a matter of engineering by applying purification techniques, which water supply companies do not really see as a problem. Many more problems with nitrates and pesticides appear in the upper groundwater levels, just beneath agricultural fields, especially in the sandy areas, where the upper groundwater is rather unprotected, and where there are no protecting layers of clay. In the Netherlands, 70% of the agricultural fields on sandy grounds are facing a violation of the European groundwater directive in the shallow groundwater. In the long run this pollution might also affect the deeper groundwater, although it is uncertain in what time span and to what extent this can be expected. This complicates the task of water suppliers and provincial authorities as regulators to quantify how great a reduction of pesticides and nitrogen use is required for the protection of drinking water sources. The debate between farmers and water suppliers on the relationship between agricultural operations and their effects on groundwater, sharpens into some sort of scientific discussion on the extent to which nitrogen from manure spreading will be absorbed by the crops that grow on the soil, and the fraction of nitrate that will leach through the soil into the groundwater. Farmers advocate that for good crop growing results, a certain amount of nitrogen is needed, while water suppliers argue that the nitrogen loss during manuring is much greater than the farmers assume. The time span in which effects will occur not only makes it difficult for farmers to come up with strong arguments, it also raises the question of whether it makes sense to encourage the farmers inside protection zones more than farmers outside these zones for a relatively short period of 10 to 15 years.

Water suppliers expect that it will take many more years before they will face much more severe problems with agricultural water pollution. This also means that it might take more than 100 years before present changes in agricultural operations will generate effects on the groundwater quality. A lot of historical pollution is already present and is still on its way. Although pollution levels and the variety of polluting substances might increase, water suppliers also expect that adequate purification techniques will be available to treat polluted groundwater used for drinking water production. However, with an increasing need for purification, the costs of drinking water production will rise , as will the price to the consumer. An important question is whether society will consider this acceptable. Do we want to pay a higher price for drinking water, knowing that the price increase is related to the purification of polluted water? The water supply sector, in common with the Dutch authorities and the European Union, adhere to the polluter-pays-principle, and that prevention should be preferred over the application of remedial techniques.

8.6.2 INFLUENCE OF THE REGULATORY CONTEXT

In the Netherlands, groundwater protection has its origins in the national manure policy. Central to this policy is that a regulatory instrument has been chosen (standards for the maximum allowable levels of manuring). The strength of this instrument, which is still in force, might have been that the government sparked off a national discussion on the role of the agricultural sector as a polluter of the environment, while until that time the sector had mostly been seen as one of the driving forces of the national economy. This discussion paved the way for the gradual acceptance of restrictions on agricultural operations for environmental purposes. At the same time the national ministry for the environment gained access to the traditionally closed 'iron triangle' of the agricultural sector.

Although a weakness of the regulatory instrument seemed to be its enforcement and its dependency on a system of manure bookkeeping by the farmers themselves, the bookkeeping system made farmers accustomed to logging the input and output of minerals on their farm, which is nowadays (1997), ten years after the introduction of the manure legislation, accepted as a normal management practice in the agricultural sector. Filling out a mineral form, which has to been sent to the authorities, has become as familiar as completing a taxation form.

Because the national manure standards, as introduced in 1987, would not provide sufficient protection to drinking water sources underground, a compromise was reached at the national political level, which allowed the provinces to come up with additional standards for specified groundwater protection areas or zones. The farmers in these areas became entitled to compensation payments from the water supply companies for the losses in productivity they experience in comparison with farmers outside the protection zones, who are only regulated by the national and not by provincial standards.

This initially led the water supply companies to place their principal reliance on this strategy. Somewhat reluctantly, they accepted this approach to control. Further, four of the five provinces that we investigated went in the direction of settling claims for damages directly by negotiation with the farmers. Since the water supply companies in general had little experience of engaging with farmers on the groundwater pollution issue, the initial result was not much more than a combination of regulations, drafted by the province, and a system of compensation payments, which some water companies were used to, since they had been paying farmers for drought damage in agricultural areas where water extraction had led to a lower water table, thus damaging the agricultural crops.

8.6.3 INFLUENCE OF THE NETWORK CONTEXT

Right from the start, the coordination of groundwater protection was set at the provincial government level by the Soil Protection Act of 1987. The assignment of groundwater protection areas is a more appropriate task for this tier of government. The autonomy of the water production companies, which commonly operate on a regional or provincial scale, means that the provincial government has much more contact with them than does central government. Furthermore, the water companies had only a weak national organization up to late in the 1980s. The agricultural sector is well organized, both nationally and provincially.

The agricultural organization then first attempted to gain a favorable compensation arrangement for the farmers in one of the provinces, and subsequently to use this as a precedent in order to gain a similar ruling in the other provinces. They succeeded in gaining a favorable regulation in one of the provinces, where the regional water production company was faced with an urgent problem with its drinking water and so adopted a willing attitude. The water companies in the remaining provinces resisted this, at the same time attempting to create a uniform, national regulation in cooperation with the Vewin, the IPO (the representative body for provinces at the national level), and the national farmers' union. This, however, was frustrated by differences between the regional water production companies.

The drinking water sector has no tradition of 'social engineering'. The sector's expertise has always been strongest in the areas of geohydrology, civil engineering and water economics. When the Soil Protection Act went into force in 1987, most of the companies did not have any agricultural expertise in their employ. However, the regulation of groundwater protection areas made the water companies more alert to agricultural issues, leading them to recruit agricultural expertise. The presence of the CLM, too, was important in bringing about a change in the attitude of the water companies. This was because the CLM had already played a role in suggesting national solutions to the manure problem by presenting the environment and agriculture, not as conflicting interests, but rather by indicating that the environment should be regarded as an obvious factor that should be integrated into agricultural practice. The CLM set up pilot projects with the water companies to seek ways to protect groundwater quality in exchange for financial support from the water companies for innovative agricultural practices.

There was already some relationship between the farmers and the water companies in a number of provinces, built up since the mid-1980s due to the compensation paid for drought damage. These relations were quite limited, however: there was indeed contact between the two groups, which was viewed positively by the farmers (affective dimension) since payments were made, but there was no extensive exchange of views (cognitive dimension) on common approaches to innovations in groundwater protection in agricultural practice (preventive measures).

In those provinces where a direct covenant between the farmers and the water companies was agreed for the protection of the groundwater, a greater degree of 'interrelatedness' came into existence, since discussions took place between the parties on possible preventive measures. The affective dimension increased due to the financial support offered by the water companies for environmental innovation. The interrelatedness also increased strongly for those water companies that had recruited agricultural expertise. This relationship is principally cognitive in nature, in the sense of an empathy with each other. The affective aspect is experienced more by the farmers than the water companies, who remain suspicious about whether the farmers actually do apply the financial support they receive to the improvement of groundwater quality. The CLM remains continuously active in suggesting new and innovative ways to collaborate, such as payment by results.

8.6.4 MOTIVES OF WATER SUPPLIERS

The mission of the water supply companies is to produce drinking water of good quality and at a reasonable price. The participation of the public authorities means that water suppliers in the Netherlands feel responsible for the implementation of drinking water quality standards. It also suppresses the profit motive in the drinking water production sector. However, water suppliers are increasingly feeling concerned about how to guarantee the levels of drinking water standards in the future. Although they traditionally have a strong orientation to water production as a process of technical engineering, and although they strongly believe that the state of the art of purification technology makes it possible to remove a lot of pollution, they feel uncertain whether technological developments will be able to resist all kinds of pollution that may emerge in the future. They also realize that increasing pollution will make drinking water production increasingly expensive, due to the need for more purification techniques. This is in competition with the intention to produce drinking water at a reasonable price in an efficient way.

Therefore the water suppliers adhere to the principle that, where pollution is concerned, prevention is better than cure. They are even willing to provide money to encourage prevention, on the condition that in a cost/benefit analysis pollution prevention can compete with the alternative of purification. But they do not like to be the only actor to encourage farmers to take preventive action, and therefore they advocate a strong role for public authorities to force farmers, through regulation, towards prevention, too. On the other hand, for enforcement reasons the public authorities (the national and provincial government) do not want to regulate farmers inside groundwater protection areas more strictly than farmers outside these areas without a compensation provision being paid by the water suppliers. In this way they pass part of the responsibility for groundwater protection on to the water suppliers. So water suppliers are forced to make compensation payments, if they want to have farmers inside groundwater protection areas working harder on pollution prevention than farmers outside these areas. Whether water suppliers are willing to pay for this extra protection depends on what results they expect from it.

In some provinces (Drenthe, Overijssel, Gelderland, and Limburg), where groundwater sources are quite vulnerable, the water suppliers negotiated directly with the farmer's organizations to conclude an agreement on compensation payments. In these provinces, the provincial authorities initiated the negotiations between water suppliers and farmers, but after that they preferred to stay at a distance as far as the compensation payments were concerned. In one province (Noord-Brabant), the largest water supply company involved did not want to negotiate with farmers, mainly because groundwater sources are rather well protected in their catchment area, except for two wellfields only. The small number of farmers involved in groundwater pollution problems in the short run prevented this water supply company from becoming more farmer oriented.

The water supply companies that worked actively on achieving a covenant with the farmers' organizations, have since noticed that a basis of trust has grown between both parties, which appears to facilitate further discussion on preventive measures. In the province where no covenant was signed, it appeared to be far more difficult for the water companies to gain recognition from the farmers as serious discussion partners. In those provinces with a covenant, the relevant water company also commonly takes the initiative in implementing the stimulation policy, that has been common to all provinces since the beginning of the 1990s. In the province of Noord-Brabant, it is the provincial government itself that took the initiative, since attempts by the water company were not accepted by the farmers. The initiation of stimulation policies by water production companies is, by the way, not just the result of the existence of a basis of trust between the companies and the farmers, it is also due to the restrained attitude displayed by the provincial government, emphasizing that the water company must be viewed as a partner in the implementation of provincial environmental policy. The CLM, too, played an important initiating role by starting pilot stimulation projects with pro-active companies in Drenthe, Overijssel, and Gelderland.

Since that time, these water companies enjoy more contacts with the farmers and have a better insight into the relationship between agricultural activity in protected areas and the quality of the groundwater, thanks to their financing of activities that go further, in terms of results, than is demanded by national policy. They are finding an increasing acceptance by farmers, changing from resistance to regulations on fertilizer use to interest in financial support for environmental innovation. The water companies, however, are only interested in environmental innovations, if the can expect that the assessment of costs as against benefits, in terms of future treatment costs as against current prevention costs, comes down on the side of prevention.

8.6.5 RESOURCES NEEDED BY WATER SUPPLIERS

Organizational capacity. The larger water production companies succeed better in guaranteeing the quality of their drinking water. One important advantage of scale is that one has more means to invest in quality research and purification technology. That would appear to ease the pressure to work on prevention. But what is also important, is that the larger companies also have more room to invest in acquiring their own agricultural expertise. While it would appear that a certain company size is a necessary condition (smaller companies are not initiators), it is not a sufficient one. It is also necessary that the importance is acknowledged of working on prevention, which is in turn related to geohydrological conditions, which determine the urgency of the problem.

Financial capacity. The water companies are experiencing an increase in environmental costs, which they justifiably have to carry (scarcity of groundwater, investments in more expensive production from surface water, investments in treatment technology, payments for prevention in agriculture); on the other hand, they are experiencing increased pressure to produce efficiently at the same cost, and to supply at competitive prices. This means that they adopt a critical attitude towards their preparedness to pay for preventive measures. They only wish to support such measures financially, if they yield the maximum possible benefit to groundwater quality and if a preventive strategy appears more attractive to them than a remedial one. They do not want to pay in advance for any prevention, merely because the government has elevated this to one of the cornerstones of its policy. Here, too, it is true that the availability of financial means is a necessary condition, but it is not sufficient for the maintenance of a preventive strategy. Once again, it is the urgency of the problem that is decisive and, in the second place, the possibility to obtain guarantees on the results of innovations from the information made available by the farmers.

Information. The water production companies are highly dependent on the models available for the prediction of the subterranean behavior of potential pollutants, as well as on the information that is input to the models, which has to be obtained from the farmers in groundwater protection areas. The availability of such models governs the degree to which preventive measures are encouraged. There will be support for only those measures that are known to have a clear benefit to groundwater quality. The dependence on information from the farmers is a major driving force behind the pursuit of a stimulation policy. As a result, one acquires information that allows a better prediction of the pollution that might be expected in the future, but the acquisition of such information demands a satisfactory relationship with the farmers.

Confidence. It is not necessary for the water companies to have built up a relationship of trust with the farmers prior to steering activity, but one does observe that matters proceed more smoothly in those provinces where a good relationship with the farmers already existed as a result of the drought damage issue. What is important is that they choose an approach that incorporates the building of such a relation. This is crucial if the farmers are to accept it.

Authority. In the Netherlands the drinking water producers do not themselves have any regulatory competence assigned to them by government. They are dependent in this regard on the provinces, which do have such authority assigned to them, within both national and European frameworks. This does not diminish the fact that regulation and the power of authority are important tools in encouraging farmers to take preventive action. The water companies do not expect regulation alone to have much effect. They expect far more from a dual approach, in which the government lays down rules and the water companies use financial stimuli to provide an extra impetus to the farmers to observe the rules by adopting preventive measures. What this does involve, according to the water companies, is that sufficient regulatory pressure must be exerted by the authorities, and they would like more of it. Whether companies maintain a preventive approach, supplementary to regulation, differs per province. This implies that government action, as the regulatory authority, is certainly necessary but not sufficient in getting the water companies to opt for a preventive strategy.

Time. The water companies are generally in no hurry to adopt a preventive strategy, in the sense of expecting considerable results from short-term actions. The application of treatment technology appears to be unavoidable due to the historical pollution that is still on its way underground. Why, then, do they not wait until the government, via its strategy of gradually tightening up the manuring standards, maneuvers the farmers into taking preventive action? It would appear that a number of companies want to seize the opportunity to build up a relationship of trust with the farmers in groundwater protection areas, which means that they will not only be able to exchange a limited amount of purification in the future for prevention today, and which also means that they can obtain more information about, and possibly more control over agricultural activities in groundwater protection areas as well as possible future threats to groundwater quality.

8.6.6 SUSCEPTIBILITY TO RESOURCES ON THE FARMER'S SIDE

Organizational capacity. The farmers are receptive to innovative expertise, but they are more interested in wider innovations than those specifically concerned with groundwater. What is important is that expertise is coupled with financial support, since expertise can also be gained via their own advisory services.

Financial capacity. A number of farmers are becoming increasingly receptive to financial support, since they believe that innovation will be unavoidable in the long term. They thus have a subsidy advantage over those farmers who are outside protected areas. Financial instruments can thus tilt the farmers' cost/benefit considerations in favor of preventive measures.

Information. The receptiveness to the transfer of information is related to the financial exchange and to the expectation that the water companies will guarantee that, as partners, they will use the information themselves, and not reveal it to third parties, such as government inspectorates.

Confidence. The confidence of farmers in the water companies depends on three factors: Are the water companies prepared to pay up? Can they be counted on in the procedure for allocating financial compensation? And will they preserve the confidentiality of the information that the farmers give them when applying for financial compensation?

This means that financial instruments are not only instruments that will have an influence on the farmers' cost/benefit considerations; they are also a way for the water companies to gain access to important information and whereby they can build up a better relationship with the farmers (investments are made in building up social capital or, in other words, a preparedness to cooperate). Covenants are more readily achieved in areas where the water companies were already active with drought damage. The research has also shown that stimulation policies also achieve a greater measure of penetration in those areas where they water companies were already active with covenants, since a good relationship had already been built up.

Authority. Initially the farmers offered great resistance to authoritative action. Tighter regulations in groundwater protection areas were only accepted with reluctance, in part due to national resistance to the manure regulations. This resistance started to fade in those areas where it became more certain that there would be a coupling with compensation via covenants, and when it became clear that, with time, the regulations would also unavoidably be tightened outside the protected areas, albeit over a longer time frame. Nevertheless, the authority of water production companies and the regulative government agencies is still a matter of discussion, in respect of the connection they seek to assert between necessary regulation and the improvement of ground water quality. This discussion is especially pointed in respect of the definition of nitrogen losses and the behavior of nitrogen in the soil. The farmers argue that a certain nitrogen level is necessary for plants, while the water companies argue that this nitrogen level is too high in view of nitrogen losses to the groundwater. Inadequate or changed information and uncertainty in this area damages authority.

Time. The farmers are in no hurry to adopt measures: indeed they are more inclined to delay any policy approach. Nevertheless, things are changing gradually, since a number of farmers see the attitude of those water companies that are prepared to pay compensation as a temporary advantage in implementing unavoidable innovations with financial assistance, in contrast to other farmers outside the protected areas. These pro-active farmers, as they are called, do thus have a certain degree of haste, since they see that such attractive financial measures will disappear in due course.

References

Achttienribbe, G.E. (1993) Privatisering van de drinkwatervoorziening, een redelijk alternatief? (Privatization of the water supply sector, a reasonable alternative?) *H2O* 26 (25): 746-752.

Centre for Agriculture and Environment [CLM] (1994) *Resultaatbeloning in de Grondwaterbescherming II. Voorstellen voor invoering in de praktijk* (Payment by results to protect groundwater II. Proposals to put it into practice). Utrecht: Centre for Agriculture and Environment.

Consumentenbond (1997) Een probleem van het zuiverste water. (A problem of the most clear water) *Consumentengid*, 1997 (November): 35-43.

EU (1990) *Euro Statistics: data for short term economic analysis*. Luxembourg: Amt fur Amtliche Veroffentlichungen der Europaischen Gemeinschaften.

Langendijk, P.W. (1994) Van Gelderse roulette naar een gestuurde ontwikkeling. (From Gelderse roulette towards a controlled development) *H2O* 27 (15): 431-435.

Nieuwenhof, R. van den (1995) *Vermesting en water. De overlast voor gebruikers.* (Manure and water. The nuisance for water users) Amsterdam: Stichting Natuur en Milieu / Babylon-De Geus.

Provincie Noord-Brabant (1995) *Eindrapportage van Evaluatie Besluit gebruik meststoffen in grondwaterbeschermingsgebieden provincie Noord-Brabant 1989-1993.* (Final report on evaluation of the ordinance on the use of manure in groundwater protection areas in the province of Noord-Brabant 1989-1993). Den Bosch: Provincie Noord-Brabant (Bureau Bodembescherming).

Schot, J. van der (1995) Waterleidingbedrijven: de stille milieuridders. (Water supply companies: the silent knights of the environment) *Natuur en Milieu* (juli/aug 1995): 19-21.

Vewin (1993) *Stimuleringsplan Grondwatervriendelijke Landbouw.* (Plan to stimulate groundwater friendly agriculture). Rijswijk: National Association of Water Supply Companies (Projectgroep Preventief Stimuleringsbeleid).

Ministry of Housing, Physical Planning and Environment [VROM] (1994) *Eindrapportage Evaluatie Provinciaal Grondwaterbeschermingsbeleid.* (Final report on the evaluation of provincial groundwater protection policies). Den Haag: Ministry of Housing, Physical Planning and Environment.

Ministry of Housing, Physical Planning and Environment [VROM] (1996) *De kwaliteit van het drinkwater in Nederland in 1994.* (The quality of drinking water in the Netherlands in 1994). Den Haag: Ministry of Housing, Physical Planning and Environment (Inspectie Milieuhygiëne).

Ministry of Housing, Physical Planning and Environment [VROM] (1993-1997) *Knipselkrant.* (Selected clippings from several newspapers). Den Haag: Ministry of Housing, Physical Planning and Environment.

9. POLICY FRAMEWORK - ENGLAND AND WALES

JEREMY J. RICHARDSON
Department of Government, University of Essex, Colchester, England

9.1 Policy problem[1]

The separate streams of problems, policies, and politics come together at certain critical times. Solutions become joined to problems, and both of them are joined to favorable political forces. This coupling is most likely when policy windows, opportunities for pushing pet proposals or conceptions of problems are open. ....windows are opened either by the appearance of compelling problems or by happenings in the political stream. Thus, agendas are set by problems or politics, and alternatives are generated in the policy stream. (Kingdon, 1984: 21)

9.1.1 BEHAVIOR CHANGING ACTIONS

The quotation above, made as a general observation of the policy process, in fact captures the essence of the policy problem caused by the interaction of the agricultural and water policy sectors in England and Wales. At its most basic, there is an unavoidable and legitimate conflict of goals between these two policy sectors. This conflict - and the need to resolve it - is the core policy problem which an extensive range of actors (both public and private) is now trying to address. On the one hand, the agricultural sector - the farmers and the associated agro-industries such as pesticide manufactures - are legitimate profit maximizes. (However, some small farmers may be farming as a way of life, rather than maximizing profits). They supply an increasingly

---

1 The final version of this study was completed in June 1996. I would like to thank Annabel Kiernan for her assistance in the earlier stages of this project, and the many officials in the water and agrochemical industries who agreed to be interviewed and who commented on earlier drafts of this report. I would like to thank John Brearley for commenting on an earlier draft. The views expressed, and any remaining errors, are, of course, the sole responsibility of the author.

Geerten .J.I. Schrama (ed.), Drinking Water Supply and Agricultural Pollution, 231-251.

sophisticated and demanding group of consumers. On the other hand, the water industry in England and Wales is also a profit maximizing industry since privatization in 1989 (Maloney and Richardson, 1995) facing an even more demanding and knowledgeable set of consumers. Both industries operate in the context of a complex regulatory framework and in an increasingly politicized environment. Thus, the political salience of their activities is already high and is likely to increase. These two sectors are, however, not the only policy 'stakeholders' (Richardson, 1996) in the policy process which has developed to address the core policy problem. Intermittently, other actors claim stakeholder status, such as 'independent' scientists and scientific institutes, environmentalists, the medical profession, the media, and the large food retailers. At times, therefore, detailed issues might be resolved in tightly drawn and restricted 'policy communities' (Richardson and Jordan, 1979), and others in a very loosely coupled (if coupled at all) extended 'issue network' (Heclo, 1978). For example, the two main government departments, the Department of the Environment (DOE) and the Ministry of Agriculture, Fisheries and Food (MAFF), listed no less than 180 organizations to whom they sent a (very minor) Consultation Document mainly concerned with field storage of silage, in September 1995 (DOE/MAFF, 1995). Of even greater importance is the fact of Europeanisation of the policy process. Much of the activity described in this report is in response to *European* rather than British legislation. This reflects the fact that agricultural and environmental matters are areas of European competence, and any British legislation will be implementing directives agreed in Brussels. Both the agenda and the framework policies have been set elsewhere, producing an essentially reactive policy style on the part of all participants.

The 'policy problem', when analyzed more closely, turns out to have at least two key aspects, the physical changes in water resources caused by modern agricultural practices, and the legal standards set by European (and to a lesser extent, national) legislation. In the highly politicized world of regulatory politics, the relationship between actual physical conditions in the water resources and their relevance to human health and to the environment are subject to intense debate. Thus, a central policy dispute is about the actual parameters set for pollutants, whether they are founded on 'good' science or whether they have been generated by a more 'political' process. Changes in the process by which standards are set, changes in the actual levels of parameters, and changes in monitoring procedures can all have very dramatic effects on the definition of the problem to be addressed. For example, if the 50 mg/l parameter for nitrate pollution of water were to be raised significantly, then much of the nitrate problem would 'disappear' overnight. Equally, as many in the agricultural and water industries fear, if it were to be lowered, the problem could worsen overnight. Thus, the policy problem is in part an artifact of the political process and it is impossible for anyone to specify the true nature of the problem. In essence, the definition of the problem is bargainable, with existing standards the outcome of a very complex and unpredictable political process, mostly at the Euro-level (Richardson, 1994).

A third aspect of the policy problem is, of course, the nature of the regulatory system itself. Once pollution parameters have been set, a second category of policy problems then arises, namely, what administrative and regulatory structure should be devised for ensuring compliance with these parameters? A central issue is to identify those structures, institutions and processes which have the maximum leverage on the behavior of individuals. Here, it should be noted that policy analysts would argue that public policy is notoriously ineffective in changing behavior, compared with its success in merely transferring resources from one section of the population to another. A related problem in this case is the need to co-ordinate the actions of a rather diverse set of 'stakeholders' in the two sectors. As we shall see in sections 9.2 and 9.3, the number of public regulatory agencies is quite large and the number of individual private stakeholders very considerable. It is, therefore, unsurprising that the processing of the policy problem in this case is characterized by a multiplicity of policy instruments and by a regulatory style which is largely consensual and incremental. This style is consistent with the long history of public regulation in Britain, essentially avoiding adversarial regulation and relying on persuasion. It is probably also an inevitable outcome of the nature of the problem to be solved.

The key feature of the problem is, of course, the fact that the farming industry (even in Britain with a comparative high proportion of large agricultural units) is atomized, with approximately 186,000 farming units in England and Wales (NRA, 1992: 16). One cannot over-emphasize the importance of this simple fact. Most farms are small and most farmers are very individualistic. They are not easily regulated and are proud of it! This characteristic is of quite fundamental importance in devising an effective 'regulatory' regime. Such a regime must be capable of changing the behavior of thousands of individualistic actors, either by persuasion, incentives, or penalties. The latter option, as all interviewees stressed, is an extremely difficult option except where point source pollution is concerned. (Even then, it may not be sensible to prosecute and thus risk losing the co-operation of farmers in the future). Pollution is caused by increased use of fertilizers, and pesticides and can be extremely difficult (with current technology) to attribute to any one farmer. Moreover, the increased use of these products is, of course, an economic response to demands to increase food production, together with changes in cropping patterns, European subsidy policies and concentrations of livestock.

Another feature of the policy problem is the difficult question of the attribution of costs as between different classes of product group (e.g. between the agricultural and agro-industries sector, and the water supply industry), and between different classes of consumers (e.g. between consumers of food and consumers of water). In Britain as in the EU as a whole, the 'polluter-pays-principle' is said by government to be central to its whole philosophy of pollution control. Yet, clearly, this principle is not being applied in this case in that the costs of agricultural pollution are not borne directly by farmers or the manufacturers and distributors of pesticides. An accurate estimate of the true costs of these externalities is impossible to obtain. However, it appears that the cost to the water supply industry over the past five years is approximately £1 billion in investment and running costs needed to reduce pesticide levels (caused by agricultural and non-agricultural pollution) to EU limits. Ultimately, of course, these costs are borne by water consumers, as the water industry post-privatization has a 'cost-pass-through' mechanism, supervised by the price regulator (see section 9.2). In some ways, this cost-pass-through mechanism, designed to facilitate privatization (see Maloney and Richardson, 1995) and totally unrelated to the policy problem being discussed here, actually facilitates problem solving between the water and agricultural sectors. Providing the technology is available - and, generally, it is - the costs of removing agricultural and other pollutants from drinking water can simply be passed on to water consumers. Moreover, the current heated public and political debate over increased water charges is structured in terms of the alleged profiteering by the privatized water companies and the effects of European legislation. Thus, there is relatively little *public* debate about the 'agricultural problem' although there is now considerable debate and activity amongst the range of policy actors involved, i.e. this is not a case of a problem which is being ignored. On the contrary, there is a vibrant policy process at work, as we will see.

Before discussing this process, however, we need to review the actual effects of modern agricultural production on water resources in England and Wales. In doing so it should be noted that focusing on agriculture is somewhat artificial, as many other sectors are responsible for use of similar polluting products. Taking herbicides as an example, non-agricultural use is significant, as is its effect on water resources. Indeed, in some cases, non-agricultural users of such products were initially the primary target of various actors, such as the National Rivers Authority and water companies, in order to protect water resources. Thus, some pesticides are now banned for non-agricultural use. Data suggest that this initiative has been quite successful. In a sense, these non-agricultural users were the first target because they were responsible for the most exceedances of the 0.1 μg/l EU standard. They were generally public institutions, such as local authorities and British Rail. Although changing their behavior was not an easy task, they are relatively small in number compared with the farming community.

9.1.2 PROBLEM INDICATORS

The NRA report on 'Pollution Incidents in England and Wales 1994' (NRA, 1995a) gives data on the various sources of pollution incidents by sector. Caution needs to be exercised in using these data, however, as the majority of 'incidents' recorded by the NRA are where there has been visible evidence of pollution, e.g. discolored water or dead fish. Pollution by pesticides and nitrates is often detected only by water suppliers when assessing compliance with drinking water standards. During 1994, some 35,291 'incidents' were reported, of which 24,415 were substantiated. Category 1 incidents (the major incidents likely to be environmentally damaging) fell from 331 to 229, although the total number of reported incidents rose slightly. The authority categorized incidents by source and by pollutant. By source, the sewerage and water industry itself accounted for the greatest proportion of incidents, 28%. Next came industry with 21%, agriculture with 13% and transport with 7% (NRA, 1995a). The remaining 31% were from a variety of sources. Of category 1 incidents, agriculture represented16% (mainly dairy farming). Organic waste represented 12% of incidents (16% of category 1 incidents) and chemicals 7% (23% of category 1 incidents). The agriculture sectors share of pollution incidents can be sub-divided as follows: dairy farming (55%), pig farming (7%); arable (6%); mixed (4%); poultry (2%), and sheep (2%). There was also a regional variation, with the largest share of incidents being in the South West region (31%), and the lowest in Thames region (3%). In terms of historical trends, it is difficult to draw reliable conclusions, as annual figures are liable to distortions by special factors (including changes in monitoring techniques). However, the National Audit Office (NAO) concluded from its 1995 review of the NRA's activities in relation to water pollution from agriculture, that the proportion of pollution 'incidents' relating to farms had dropped from 17% in 1985 to 9% in 1993, and the total number of farm pollution incidents had fallen by 12% between 1985 and 1993 (NAO, 1995: 1). However, some unpublished NRA data appear to suggest that incidents related to run-off from land are on the increase as a result of the spreading (rather than storage) of slurry and manure on land.

If we look at particular *types* of pollution, the contribution from agriculture can be more significant. This is particularly the case for organic waste. For example, cattle slurry accounted for 33% of incidents and solid cattle waste (manure) for 9%, and silage liquor for 8% of all organic waste incidents. Again, there are significant regional variations, as might be expected. From the pollution incidents perspective, organic waste is clearly a significant problem for the NRA. At times, too, it can present significant problems for the water companies too, although it is probably lower on their agenda than for the NRA. However, the worrying problem of cryptosporidium (a protozoan parasite) may well be an issue creeping up the political agenda. For example, as a consequence of an outbreak of cryptosporidium in Oxfordshire and Swindon in 1989, the government set up an expert group and a national research program has been instituted. The future of this co-ordinated program of research (which includes one NRA study of the occurrence of cryptosporidium in water draining a catchment area subject to intensive livestock farming) is currently being addressed by the National Cryptosporidium Research Steering Committee (NCRSC). However, MAFF believes that the incidents of cryptosporidiosis do not seem to have been rising and are currently running at 'background' levels. The other two main sources of agricultural pollution of water resources, nitrates and pesticides, are far less important as causes of actual incidents, than they are of long term and diffuse pollution of water resources (nutrients and pesticides are both less than 1% of point source pollution incidents).

The use of nitrate fertilizers (and the use of organic manure) have been important in Britain as elsewhere in Europe and is an essential feature of post-war agricultural productivity gains. Specifically, Britain has experienced difficulties in complying with the 50 mg/l parameter set by the EC. By the mid 1980s, the DOE's Nitrate Co-ordination Group had reviewed long-term data from 25 rivers and shorter term data from 149 sites. It concluded that nitrate concentrations had increased at varying rates, being generally higher in central and south eastern areas, although there were indications that trends were leveling off from the mid 1970s (DOE, 1986). By 1992 the NRA was reporting that, as far as nitrates in groundwater were concerned, those sources with reliable data had indicated a rather mixed situation: some had shown very little increase in nitrates since 1970, a few had shown a decrease, but some displayed a rising trend

(NRA, 1992: 40). Interestingly, the NRA concluded that restrictions on farming practices could be an effective policy instrument. For example, increasing nitrate trends at Batheaston and Monkswood Reservoirs in Wessex Region were causing concern, with predictions that they would eventually exceed EC limits. Financial analysis of these cases had indicated that control of agricultural practices was the cheapest effective option. Between 1985 and 1987, 73% of the relevant catchments were subject to restrictive agreements; where the water companies presumably owned the land and were able to place restrictions on their tenants, essentially changing from cereal production to grassland for sheep. Following this move, winter maxima for nitrates stabilized and nitrate concentrations in springs feeding the reservoir were decreasing (NRA, 1992: 46). As we shall see (section 9.4) the need to comply with the EC's 1991 Nitrate Directive has prompted the strongest regulatory (in the formal legal sense) action against farmers. In fact, the number of water supply zones (relating to treated water, not water flowing through farms) failing to comply with the nitrate standard is low and has been falling gradually. Thus, in 1994, 1.8% of zones did not comply with the requisite standard compared with 3.1% in 1990 (DWI, 1995: 191). The Drinking Water Inspectorate (DWI) noted in 1992 that:

> *the number of water supplies that exceeded 50 mg/l is not as large as would be expected from the increase in number of water sources affected. This reflects the effect of the remedial action taken by water companies in respect of 18 water supplies in 1989 and 17 supplies in 1990.* (DWI, 1992)

In other words, the water supply companies have been effective in managing supplies of raw water through blending and treatment. In that sense, the water companies have been dealing successfully with the externalities of agriculture, although not really through major changes in the actual behavior of farmers.

With regard to pesticide pollution, one can again present two rather contrasting accounts of the severity of the problem. In one sense, there is no major problem, if we take the quality of *drinking water* as an indicator, because the water companies have been making the necessary investments to comply with European legislation. This has involved blending, treating or abandoning water sources rather than, say, reducing pesticide concentrations at source. (However, in some cases, for some parameters, it may be three or four years before full compliance with EU legislation is reached). Yet it is clear that the contamination of groundwater by pesticides is an area of considerable uncertainty, where much more research is needed at both the national and European levels. The key problem for researchers is that pollution of groundwater is subject to very considerable time lags, commonly 25 years or possibly even much longer. As Britain relies on groundwater for approximately 28% of its drinking water supplies - actually, one of the lowest percentages in the EU†- this is clearly an issue which is attracting much attention by the core policy actors. The 'knowledge base' problem is being addressed. For example, the NRA has a long term project (started in 1984) to assess the impact of pesticides on major aquifers. In 1984 the analysis of water samples for pesticides in water was in its infancy and the analysis of aquifer samples had not begun (NRA, 1995a: 3). Thus, much of the current debate might be said to be an artifact of the more recent development of science and technology. Since 1984, more is known about both the use of pesticides and about the flow characteristics of pesticides under certain geological and weather conditions (for a review, see NRA, 1995a). Monitoring suggests that '...relatively few pesticides at low concentrations occur in the groundwater and that the most common are Atriazine, Simazine, and Isoproturon.' (NRA, 1995a: 13) Interestingly, one of the major conclusions of the NRA's research program was the need for more information on non-agricultural use of pesticides (NRA, 1995a: 18). The future of the knowledge base regarding pesticides and water resources may lie in the development of mathematical modeling, although this development is not without its scientific critics. Not for the first time in environmental legislation, the future of public policy may be determined by the trajectory and machinations of scientific research and its use and misuse by different policy actors (see Richardson, 1994). As is often the case with data on environmental pollution, data are interpreted differently by different actors. For example, Ward, et al. cite the 1992 Friends of the Earth (FoE) report, which argued that an analysis of DWI figures suggested that some 14.5 million customers in England and Wales were being supplied with water which breached the EC's standards for pesticides (Ward, et al., 1993: 7). This is in sharp contrast to the conclusions of the DWI itself (see below). Both FoE and DWI use data from the water companies, but use different statistics. FoE scores a zone as 'failing' even if only one pesticide exceeds the 0.1 μg/l standard, for however short a period and however uncertain the result (there may be a thousand pesticide results generated for a single water supply zone in a year). DWI, on the other hand, stresses the large percentage of pesticide results that meet the 0.1 μg/l standard. Whatever the 'truth' may be, two points are clear. First, from the water supply industry's perspective, the use of pesticides (both for agricultural and non-agricultural use) has presented a major technological and investment challenge for it to meet the EU's guidelines. Secondly, the scientific community does have concerns about the *long term* implications of the use of pesticides, in terms of future water supplies. Again, caution needs to be expressed in seeing this as a solely agricultural problem. The NRA study cited above noted that there was a relative lack of agricultural pesticides in groundwater, for example, triazines were used in comparatively small quantities in agriculture and so their appearance in groundwater raised questions of their non-agricultural usage (NRA, 1995a: 7).

In October 1995, the NRA published a major review of pesticides in the aquatic environment. Drawing on the data of two year's extensive regional monitoring of pesticides at 3,500 sites, it indicates the occurrence and distribution of pesticides in surface waters and groundwaters. Between 1992 and 1993, 120 different pesticides were monitored and 450,000 separate analyses were recorded. Some 100 of the pesticides were detected at low concentrations and the remaining 20 were not detected at all (NRA, 1995b: i). The report concluded that in general, the compliance with Environmental Quality Standards (EQS) is very high. As shown in table 9.1, in 1993, over 99% of sites passed the EQSs for List 1 pesticides and over 96% of sites passed for all EQSs (NRA, 1995b: 23). Moreover, the NRA's data suggested that for those 25 pesticides most commonly found in environmental waters and for which the NRA had proposed EQSs, there were very few exceedances of the proposed EQSs (NRA, 1995b: 28). However, the report emphasized the need for more research and development on such matters as

TABLE 9.1: Pesticides in the Aquatic Environment

Pesticide	Sites failing EQS
All pesticides	3.80%
Diazinon	1.25%
PCSD/Eulan	0.92%
Permethrin	0.81%
HCH	0.66%
Dichlorvos	0.34%
Cyfluthrin	0.33%
Total drins	0.26%
Total DDT	0.14%
Pentachlorophenol	0.09%
Azinphos methyl	0.06%
Malathion	0.06%
Total atrazine/simazine	0.06%
pp DDT	0.06%
Fenitrothion	0.03%
Hexachlorobenzene	0.03%

Source: NRA (1995b).

the rates of movement and pathways to surface and ground water, the fate of pesticides entering controlled waters, and the importance of river morphology (NRA, 1995b: 41). This reflects the NRA's view that the significance of long term exposure and the combined effects of mixtures of pesticides are still not fully understood. Similarly, a report from the Foundation for Water Research (FWR) has emphasized the need for more research into such issues as the rate of pesticide degradation in unsaturated and saturated zones and a model of pesticide transport from the soil zones (FWR, 1995: 27).

One key indicator of the policy problem is, of course, the quality of drinking water. Although this is an indicator of threats only to the mammalian population (as opposed to the ecosystem system as a whole), it is one important indicator of the success of the total range of policy instruments in place and of actions by the water companies. For example, the DWI published a report in June 1995 representing the results of some 3,482,941 tests at treatment works, service reservoirs and in water supply zones. As the DWI points out, 99.3% of these tests demonstrated compliance with the specified standards for treated water (DWI, 1995: vii). Of the 0.7% of tests which contravened standards, the DWI was of the view that these contraventions prevented no danger to the health of consumers. Enforcement action is being taken in respect of microbiological standards at 0.9% of the treatment works and 0.2% of the service reservoirs, and in respect of contraventions of standards in water supply zones is no more than 0.4% of the zones for any one parameter (DWI, 1995: viii). (In fact the precise meaning of this figure of 0.4% is somewhat unclear. It appears to refer to enforcement action being taken as a result of data generated in 1994 and may not include long standing problems where remedial action is underway). The long term aim is 100% compliance but, as the DWI points out, there are practical limits to actually achieving 100% compliance. As to the next ten years, DWI considers that investment in water supply will be dominated by the requirement to solve the aesthetic quality problems associated with the poor condition of some distribution systems.

The DWI uses sixteen key parameters in its tests of water quality: coliforms, faecal coliforms, color, turbitity, odor, taste, hydrogen ion, nitrate/nitrite, aluminium, iron, manganese, lead, polycyclic aromatic hydrocarbons, trihalomethanes, pesticides. In terms of coliforms, they were not detected at 84.9% of the companies' treatment works. At 99.2% of the 5,186 sampling points at service reservoirs coliforms were absent from at least 95% of the samples. Of the 2,552 water supply zones in England and Wales, 92% complied fully with the relevant water quality standards, or had breaches which were either trivial or were being remedied by legally enforceable undertakings given by companies. In 8% of the zones, breaches were considered unlikely to recur, although others might result in enforcement action (DWI, 1995: 185).

If we focus on pesticides and nitrates, 31 pesticides were reported in 1994 as detected at concentrations above 0.1 μg/l in compliance samples (DWI, 1995). In every instance, the concentrations corresponded to exposures many thousands (and sometimes millions) of times lower than those known to be harmful or likely to affect health. For 22 of the 31 pesticides, health-based recommendations on maximum advisable concentrations or exposures have been issued variously by the World Health Organization, the UK Scientific Committee on Pesticides, the UK Veterinary Products Committee, and the US Environmental Protection Agency. With one exception, trichloroacetic acid (TCA), the highest reported concentrations and corresponding exposures were within the ranges considered acceptable by these authorities. TCA was detected at concentrations up to 150 μg/l, which is greater than the World Health Organization provisional Guideline Value of 100 μg/l, but lower than the concentration of 300 μg/l deemed by the US Environmental Protection Agency to be protective of adverse human health effects, and with an adequate margin of safety. Although TCA is a pesticide, its presence in drinking water usually results from its incidental formation during disinfection by chlorination. The 1994 figures showed a further reduction in pesticide levels over previous years, the result of the commissioning of improvement programs by the water companies and lower pesticide use by agricultural and non-agricultural uses.

At the time of writing (June 1996), the DWI 1995 report indicated slight further improvements to drinking water quality, although the Commission has decided to take Britain to the European Court of Justice over the DWI's use of derogations during 1989-90 and in 1993, particularly relating to pesticides. The case is to be vigorously contested by the British government, which claims that the enforcement procedure was supported by the British High Court, and was agreed with the EU, and is consistent with the EU's revised Drinking Water Directive published in 1995.[2]

The DWI figures are *outcome* measures, in policy analysis terms. Despite the political controversy (from organizations such as Friends of the Earth), the quality of the data are high. However, there are those in the water industry, though not sharing the views of FoE, who take a less sanguine view of the seriousness of the problem. For example, Thames Water Utilities Ltd. have argued recently that pesticides continue to account for the majority of the failures to meeting drinking water standards. While agreeing with the DWI data that of the over one million analyses for individual and total pesticides in 1993, over 97.6% of these complied with the relevant standards, the Thames scientists argued that this did not give an indication of the importance of

[2] See *Water Bulletin*, 05 July 1995, p. 710.

pesticides as a cause of non-compliance. Their analysis suggested that contraventions for pesticides (both for individual and total substances) were easily the most important cause of contraventions. Thus, for England and Wales the total for pesticides was 73% of all contraventions (see table 9.2).

TABLE 9.2: Contraventions of standards in water supply zones in England and Wales 1990-1993 (values in parentheses are % of total contraventions)

Parameter	1990		1991		1992		1993	
Pesticides	13209	(51%)	27585	(66%)	35679	(75%)	25531	(73%)
Coliforms	3835	(15%)	2709	(6%)	2318	(5%)	1575	(5%)
Iron	2226	(9%)	2515	(6%)	2033	(4%)	1593	(5%)
Nitrite	1743	(7%)	2228	(5%)	2086	(4%)	1876	(5%)
Lead	1598	(6%)	1736	(4%)	1354	(3%)	1263	(4%)
Nitrate	1117	(4%)	1170	(3%)	852	(2%)	364	(1%)
Others	2204	(8%)	3781	(9%)	3317	(7%)	2582	(7%)
Total	25932		41724		47639		34784	

Source: White & Pinkstone (1995: 264).

The difficulty is that there are several ways of interpreting the data. One method is to consider the percentage of tests which failed to comply with each parameter. Another is to compare the number of failures recorded for each parameter. Another is to compare the percentage of failures for each parameter with reference to the total number of failures for all parameters. Another is to look at the number and percentage of zones which fail to comply. Each of these has its merits and disadvantages, primarily because the regulatory requirements for sampling frequencies vary from parameter to parameter, and they allow increased and reduced frequencies in certain circumstances. The DWI report for 1993 (DWI, 1994) gave the number and percentage of tests which failed for each key parameter and the number and percentage of zones which failed to comply in 1993 and earlier years. The Thames scientists give the number of failures and the percentage of total failures for some parameters. The Inspectorate would argue that the Thames data over-estimate the importance of pesticides and that DWI data are a better reflection of importance with respect to compliance, putting pesticides somewhat ahead of iron, with lead next.

The Thames figures emphasized the regional dimension, showing that Thames had a large number of contraventions, relating to the fact that 75% of its raw water comes from lowland rivers draining large catchments with intensive pesticide use, with the remaining 25% coming from groundwater sources (White and Pinkstone, 1995: 263). For Thames alone, the initial capital costs of treatment were estimated to be £400 million. Interestingly, the question of agricultural and non-agricultural use was raised. Thus, they noted that: 'as measures to reduce contamination by industrial and amenity herbicides have taken effect, the relative importance of agricultural pesticides such as Isoproturon has increased...' (White and Pinkstone, 1995: 266) Hence, the agricultural agrochemical sectors might expect further intensification of 'regulatory' pressure on their activities, as other policy actors begin to raise the externalities issue more forcibly.

In fact MAFF, the sponsoring ministry, itself produced worrying data for the farming interests, in its report on pesticides leaching from clay soils, published in 1995. MAFF scientists concluded that the leaching of pesticides from clay soil was greater than that predicted by leaching theory due to rapid movement of water through cracks in the soil.[3] Moreover, they are reported as concluding that there was little prospect of controlling these losses through improved pesticide design, and that the best means of reducing losses to streams will be through controls on application.[4] They also warned that:

> *with approximately one-third of UK soils having a similar hydrological regime (to the test site), this work has significant implications for pesticide risk assessment and management..... improved management of pesticide applications, and not better molecule design, is the key to reducing undesirable pesticide contamination of lead-water streams in many areas.*[5]

9.2 Regulatory framework

In briefly outlining the regulatory framework, it is important to specify what we mean by 'regulation'. The British regulatory style has traditionally emphasized a non-adversarial approach to regulation. Prosecutions are relatively rare and great emphasis is placed upon a consenual 'joint-problem solving' approach. In this particular case, the policy problem may well re-enforce this tendency. Thus, regulatory logic may determine an approach dependent upon persuasion rather than imposition, if only because offenders often cannot be identified. Thus, two important caveats need to be entered when using the word 'regulation'. Firstly, we use the term in a consciously flexible manner. By 'regulation' we mean policy instruments which are designed to change the behavior of the agricultural and agro-industries sectors. These instruments may include formal legal regulation but may also include a whole range of policies and programs (not all introduced or implemented by public authorities) which have no legal sanctions behind them. Thus, we are interested in the 'steering capacity' of the various policy instruments. A second caveat, equally important, is that one should be cautious in seeing agro-industries as, necessarily, offenders. Any future development of policy must recognize that the main cause of the problem is almost certainly not the result of large numbers of farmers deliberately seeking to break the law. Clearly, there are such cases and prosecutions do occur, but the bulk of the problem is the by-product of 'normal' (albeit polluting) economic agricultural practices. Also, it is certainly the case that discontinuation of these practices would not be costless either, though there is evidence that change to alternative biological and organic techniques can sometimes be almost costless, or can actually reduce costs. *Put simply, we use the term 'regulation' to mean*

---

[3] *ENDS* 243, April 1995.

[4] Ibid.

[5] Ibid.

'behavior changing policy instruments, both public and private'. Thus, the term regulation is being used in a deliberately loose way in order to capture the notion that there are many ways of changing behavior. The central focus of our analysis is on those steps which are being taken to change behavior and practices, be they steps taken by public authorities, e.g. the Department of the Environment or steps taken by purely private actors, such as the water companies, or the pesticide manufacturers represented by the British Agrochemicals Association (BAA). As we shall see in chapter 10, much of this behavior changing activity is based upon education, information, and persuasion - joint problem solving - rather than US or continental European style adversarial, rule based, regulation.

Nevertheless, it would be wrong to characterize the regulatory framework and regulatory process as devoid of conventional legal regulation or penalties. In fact, there is a wide range of public laws and regulations (and some prosecutions and recovery of costs) and a rather complex array of regulatory structures. In reviewing these structures and formal processes, it should be remembered that many of them were not designed specifically to address the particular problem which is the central concern of this book. Some of them were designed specifically to address this policy problem, such as the initiatives on 'nitrate sensitive areas' (NSAs) and 'nitrate vulnerable zones' (NVZs), or regulations regarding the construction and use of animal waste slurry pits. Others, such as the economic regulation of the water industry, were not set up to address this problem but do have quite major effects on policy outcomes. Within the complex regulatory structures governing agriculture, the agrochemical industries, and the water industry, we have concentrated on those regulatory authorities which are central to the processing of the policy problem as outlined in section 9.1.

We begin with the regulatory structure governing the water industry. This sector is at the sharp end of the problem and is governed by the most formal set of laws and regulations.

9.2.1 REGULATION OF THE WATER INDUSTRY

For our purposes, we need to focus on three key regulatory structures: the 'sponsoring' department, namely the Department of the Environment, and the Drinking Water Inspectorate (DWI); the Office of Water Services (OFWAT), and the National Rivers Authority (NRA).[6] All of these bodies operate in the context of a range of laws and regulations, many of which now stem from European legislation. Thus, the water sector is heavily Europeanized, even though purely national laws and regulations are still significant (for example the Statutory Water Quality Objectives). There is little doubt that the currently high standards at last being achieved owe much to the need to comply with European legislation, such as the Drinking Water Directive and the Nitrates Directive. However, this European legislation has to be translated into national

---

6 On April 1st, 1996 the NRA became part of the newly created Environment Agency. As most of this study is concerned with the period prior to this administrative change, however, the term NRA, rather than Environment Agency, is used throughout.

legislation, mainly the Water Industry Act 1991 (partly a consolidation of the Water Act 1989 which brought about privatization) and a series of Water Supply Regulations defining 'wholesomeness' under the 1991 Act. These regulations are the Water Supply (Water Quality) Regulations 1989, which have been amended slightly by the Water Supply (Water Quality) (Amendment) Regulations of 1989 and 1991 (DWI, 1995: 2). The Regulations incorporate all of the standards (maximum admissible concentrations and minimum required concentrations) set out in the EC Drinking Water Directive (80/778/EEC) and include 11 national standards. In total, numerical standards are set for 55 parameters and descriptive standards for a further two. The basic monitoring unit is the water supply zone, an area designated by a water company, usually by reference to the source of supply, in which not more than 50,000 people reside. (DWI, 1995: 2). Under Section 18 of the 1991 Act, the Secretary of State for the Environment is required to take enforcement action in cases of contravention unless he/she is satisfied that the contravention is trivial or that the company has made a satisfactory undertaking to comply. (Again, the regulatory style is for an 'undertaking' for specified work to be carried out by an agreed date.)

The main source of data on the quality of drinking water in Britain is the water companies themselves, on which DWI relies. From this data, DWI produces a very comprehensive and reliable annual review of drinking water quality, both nationally and company by company, as discussed earlier in section 9.1. These data are generally regarded in the European water industry as some of the most reliable and transparent in the European Union. The DWI was formed in 1990 and is currently part of the Department of Environment. It was given the task of checking, on behalf of the Secretary of State of the Environment and the Secretary of State for Wales, that Water companies comply with their statutory obligations. It does this via a 'technical audit' of each company, consisting of three elements, as follows:

- an annual assessment based on information provided by companies of the quality of water in each supply zone, compliance with sampling and other requirements, and the progress made on improvement programs;
- inspections of individual companies, covering not only a general check at the time of the inspection on the matters covered above, but also an assessment of the quality of the information collected by the company; and
- interim checks made on aspects of compliance based on information provided periodically by companies (Healey, 1993).

As indicated, the DWI is responsible to the DOE (and to the Welsh Office). In fact, the DOE itself has major responsibilities for the water industry, and is one of the two main government departments directly involved in the processing of the policy problem.[7]

The Secretaries of State for the Environment and Wales have important functions regarding the setting up of water quality objectives and in representing the UK in EU negotiations concerning existing and proposed directives. Their functions are summarized below:

---

7 The other is MAFF. In addition, other government departments have a legitimate interest in the problem - for example the Department of Health.

- appointing companies to act as water and sewerage undertakers, the Director General of Water Services and individuals to serve on the National Rivers Authority;
- elaborating and supplementing the regulatory framework by means of subordinate legislation, for example, to establish standards of performance in relation to the supply of water and sewerage services, to determine standards of wholesomeness of water, to set out criteria for the classification of river quality and to impose requirements to take precautions against pollution;
- approving various codes of practice including conservation and recreational duties, and the exercise of pipe-laying powers; and
- enforcing certain statutory obligations imposed on water and sewerage undertakers, for example, by making enforcement orders and initiating the prosecution of any water undertaker which supplies water unfit for human consumption.

In respect of the National Rivers Authority, the Secretary of State's main functions are:

- paying a 'grant-in-aid', setting financial duties and controlling overall staff numbers and conditions of service;
- designating water protection zones and making drought orders;
- approving charges schemes for water abstraction and discharge consents;
- appointing technical assessors of water quality and setting Statutory Water Quality Objectives; and
- applying, in exceptional circumstances, for a special administration order.

The Secretaries of State receive the reports of the Chief Inspector of the Drinking Water Inspectorate, Director General of Water Services and the National Rivers Authority and, in certain circumstances, can give directions to them and receive information from them related to their functions.

The DWI is, of course, concerned with drinking water, and particularly with the health aspects of drinking water. In contrast, the National Rivers Authority has a much broader remit. The NRA is the 'competent authority' for England and Wales for the implementation of the various EU directives concerning water quality. Indeed, it was set up in 1989 as the direct result of legal complications during the passage of the water privatization process, when it was found that the proposed private water companies could not retain the regulatory functions previously invested in the publicly owned Regional Water Authorities (Maloney and Richardson, 1995). The NRA has wide ranging responsibilities as follows:

- monitoring groundwater;
- river quality and the quality of inland coastal waters;
- flood defense control;
- management of water resources;
- fisheries;
- recreation and conservation;
- navigation;
- water abstraction licensing and enforcement;
- discharge consents and enforcement; and
- the enforcement of statutory water quality objectives to be determined by the Secretary of State.

Essentially, the NRA has responsibility for administering river basins (Britain having adopted a system of integrated river basin management based on single purpose authorities as early as 1973), including deciding and enforcing constraints on all sorts of companies taking water from rivers and putting effluents into them (Kinnersley, 1988: 138). The NRA is responsible for responding to pollution incidents and for taking appropriate action against polluters, following which it may recover costs. It also engages on a variety of activities aimed at preventing pollution from all sources, including industry, sewage, and agriculture (NAO, 1995). The NRA is organized on a regional basis (most of its staff used to be employed in the old RWAs and they therefore have good links with the water companies) which has been retained by the new Environment Agency. The regional structure facilitates local and regional responses (and the development of appropriate policy instruments) to the problems caused by agricultural pollution. Also, the NRA is an important source of data on agricultural pollution, for example, on pollution incidents, as discussed in section 9.1. The NRA is, of course, also very concerned with pollution of water resources from non-point sources.

Regulatory bodies such as DWI, DOE, and NRA are all concerned with environmental and quality considerations. (However, the DOE is also very concerned with cost considerations - see Maloney and Richardson, 1995). OFWAT, in contrast, is primarily concerned with the financial viability of the water companies, the costs of water to consumers, and with the quality of service to consumers.

Its relevance to the processing of the policy problem under discussion here was hinted at above, namely, it is the regulatory authority having the power to determine the specific cost-pass-through arrangements for each of the ten major privatized water companies and the twenty-two smaller private 'water only' water companies. The cost-pass-through mechanism is the so-called 'RPI + K factor', essentially the Retail Price Index plus a factor (K) to cover the necessary costs of the water companies meeting their long-term statutory requirements and ensuring their financial viability. As in all cases of economic regulation, the regulator has to make a fine judgments about possible efficiency gains. As far as the requirements on the water companies to supply clean drinking water are concerned, the costs of removing pollutants, from whatever source, are allowable costs in the setting of the K factor. Cost-pass-through is not automatic of course, and depends upon OFWAT's judgment of the cost calculations for each water company. The combination of new technology, a major post-privatization investment program, and the K factor have enabled the water suppliers to now set a good record (after significant earlier slippage) in meeting EU water quality obligations. OFWAT, particularly its Director General Ian Byatt, has not been disinterested in regulations governing water pollution, however. For example, Mr. Byatt has been vociferous in his criticisms of European legislation and has often clashed with the NRA over the costs of environmental improvements (Maloney and Richardson, 1995; Byatt, 1996).

The whole history of post-privatization regulation (not just of the water industry) has been one of 'regulatory creep', with the regulators in some respects re-writing the (generous) regulatory bargain struck at the time of privatization. OFWAT is no exception to this trend. It is always possible, therefore, that OFWAT might further tighten the regulatory regime making it more difficult for the water companies to pass on the costs of dealing with agricultural pollution. The so-called 'periodic review', first conducted by OFWAT for each company in 1994, did, indeed, tighten the regulatory regime for most of them. There are some signs that water companies are beginning to be more open in their criticism of the agricultural sector, as in the case of Thames Water Utilities Ltd., cited above.

9.2.2 REGULATION OF PESTICIDES

As indicated in section 9.1, the sources of agricultural pollution are threefold: organic waste, nitrates, and pesticides (with sediment from soil erosion of increasing concern). As in all EU Member States, the introduction, marketing and use of pesticides is subject to close legislation and regulation. Much of this legislation, as in the case of water quality, is of European origin. As indicated earlier, it is the level of standards, and the specific parameters derived from these standards, which determine whether or not there is a policy problem at all. Many in the water, agricultural, and agrochemical industries would argue that many of the EU standards are not based upon toxicological findings. For example, they see the 0.1 μg/l for pesticides as a surrogate for zero, not based on toxicology. In Britain, the 'lead' department for the regulation of pesticides is MAFF. In practice, the regulation is conducted by a so-called 'next steps' agency, the Pesticides Safety Directorate (PSD). It is a semi-autonomous agency accountable for its operations to MAFF ministers, but making recommendations to ministers in a number of departments. Before a new pesticide can be used in the UK, it must satisfy safety standards laid down in Part III of the Food and Environment Act 1985. Other relevant UK legislation is the Control of Pesticides Regulations 1986, the Pesticides Maximum Residue Levels (MRL) in Food Regulations, and a number of EU directives and regulations. The central task of the PSD is to license pesticides for use and to review existing licenses should circumstances change or new evidence come to light. Thus, all pesticides used in agriculture, horticulture, forestry, amenity, industrial use, food storage, and in the home garden, are assessed by PSD for their safety and efficiency before approvals are granted.[8] Once a pesticide is approved, controls are exercised over its advertisement, sale, supply, storage and use (ACP, 1994).[9]

In view of the highly technical nature of its work, the PSD works via a hierarchy of expert advisory committees. Its main committee is the Advisory Committee on Pesticides (ACP), consisting of eminent (often academic) experts in the chemistry and toxicology of pesticides, and in soil structures and properties, etcetera. Much of the detailed work is, however, conducted via its sub-committees and working groups, e.g. the Medical and Toxicological Panel; Working Party on Pesticide Residues. The review

[8] Pesticides used in other areas are assessed by the Health and Safety Executive (HSE). Ward et al. suggest that the HSE is 'hopelessly understaffed' and that its inspectors tend to focus on human health and safety risks presented by farm machinery, electricity, grain stores, ladders, and pesticide storage on farms rather than diffuse threats to the water environment (Ward et al, 1993:28).

[9] For a discussion of the nature of these controls, see chapter 10.

procedure is very detailed, though possibly biased towards consideration of the effects on human health of pesticides. For example, the generation of data to satisfy the approval procedure for a new pesticide is likely to cost the proposing company over £3 million. Thus, the documents outlining the Data Requirements for Approval Under the Control of Pesticides Regulations 1986 weigh nearly 3 kg! In a sense, the PSD is at the opposite end of the policy process to the water companies. The water companies have the task of removing pesticides from drinking water, whereas the PSD has the power to approve or not the release and use of pesticides onto the marketplace.[10]

9.2.3 REGULATION OF AGRICULTURE

The PSD is responsible to its major parent Department, MAFF. (However, recommendations from the ACP actually need the approval of no less than five government departments.) As with the DOE, MAFF has much wider responsibilities than those of its Executive Agency, primarily, the sponsorship of agriculture. In terms of the problem under study, MAFF has a number of responsibilities relating to agricultural pollution of water. For example, much of the practical responsibility for the designation of nitrate sensitive areas and nitrate vulnerable zones (see chapter 10) lies with MAFF, even though DOE is formally responsible to the European Commission. Also MAFF was formally responsible for the provision of grants (now phased out) for farmers to set up good waste storage and handling systems, under Regulations governing the containment of silage, slurry and agricultural fuel oil. (Farm waste grants were re-introduced for Nitrate Vulnerable Zones in 1996). The ministry is also responsible for the statutory side of a second agency, the Agricultural Development and Advisory Service (ADAS), which in turn is responsible for advice to farmers. (ADAS is now split into a commercial side and a statutory side). More importantly, perhaps, MAFF is directly involved in promoting the utilization of good practices, both for the use of nitrates and of pesticides. (More details of these initiatives are given in Section 4 below). As part of the encouragement of good practice, MAFF of course, draws the attention of farmers to laws and regulation to which farmers are subject. Thus, in its 1991 Code of Good Agricultural Practice for the Protection of Water, MAFF emphasized the importance of the Control of Pollution (Silage, Slurry and Agricultural Fuel Oil) Regulations 1991, as well as the implications to farmers (fines of up to £20,000) for breaches of Section 107 of the Water Act 1989, governing the discharge of pollution into 'controlled water', and the need for discharge consents from the NRA under Section 108 of the Act (MAFF, 1991). (In practice, the NRA policy is not to grant discharge consents for agricultural waste). In total, as will be seen in Section 4, MAFF is a central actor in terms of attempts to minimize the impact of agricultural practices on water resources.

---

10 The degree of co-ordination between these two sets of actors is discussed in chapter 10.

9.3 Other policy actors

In Section 9.1 we suggested that different characterizations of the nature of the policy process were possible; from, a narrowly defined and stable 'policy community' of actors, to at times, a very loose and unorganized 'issue network' of interests, organizations, and individuals claiming some interest in the policy problem. We have, so far, concentrated on describing the formal state actors, the relevant government departments and agencies. However, the policy problem involves other 'core' private actors and number of peripheral actors.

At the core of any processing of the policy problem are, of course, the private water companies in England and Wales. Originally formed in 1973 as ten Regional Water Authorities (RWAs), they are now fully in the private sector and are subject to normal market pressures. As we have suggested, they are definitely at the 'sharp end' of the policy process. They have onerous statutory responsibilities and have faced a multi-billion pound investment program to meet those obligations. Not surprisingly, they are developing a range of relationships with other policy actors as a means of mitigating the effects (and costs) of agricultural pollution. They are represented nationally by the Water Services Association (WSA). In addition, there are twenty one so-called 'water only' private companies who supply 25% of all drinking water in England and Wales and who are represented by the Water Companies Association. (They existed under the 'pre-privatization' regime, alongside the old RWAs).

A second key actor is, of course, the farming industry, represented by the National Farmers' Union (NFU) and also by the Country Landowners' Association (CLA). The NFU and CLA maintain very closely liaison with MAFF (many would say the relationship is corporatist) and have a very long history of co-operation to defend the agricultural sector. The NFU, particularly, is probably Britain's most effective pressure group and has a very high membership density, i.e. it is very representative of the agricultural sector and has an impressive track record in defending it.

A third important core actor is the pesticides manufacturers, represented by the British Agrochemicals Association, together with the Fertiliser Manufacturers' Association (FMA). It is fair to describe the industry as on the defensive. Indeed the BAA's director commented in the Association's 1994 Annual Review and Handbook that 'we shall have to take the heat if we wish to stay in the kitchen.' Recognizing the likelihood that pressure on the industry will increase rather than decrease, he commented that:

> *over the next ten years or so, we must make sure that the pressures do not build up to such an extent that new legislation or lack of return on investment makes it impossible for farmers to use our products and those of the fertilizer industry to farm efficiently, or for that matter to farm at all.* (BAA, 1994: 5)

Here we see the industry 'waving the shroud' in an attempt to slow down the pace of regulatory reform. Also, we see one aspect of the policy problem placed in sharp relief, the public and political perception of the industry is undoubtedly bad (as it is of the chemical industry more generally, throughout Europe - see Richardson, 1994). In many ways the industry is being forced into a reactive mode in response to agendas set some time ago by the environmentalists. Its policy, similar to the agricultural sector as a whole (including MAFF), has been to be very active in trying to set its own house in order, as a means of avoiding further - and possibly draconian - legislation. Its response is identical to any industry under regulatory threat - to try to make self-regulation more effective. Also, it is part of the advocacy coalition pressing for a greater emphasis on scientific evidence in the making of European level policy.

A fourth category of actors, the environmentalists, presents something of a puzzle as there is little evidence that they are now especially active in the processing the policy problem. Yet their role has been enormously important in setting the agenda to which others respond.[11] The very fact that so much attention is being paid by the core actors to the processing of this policy problem is testament to the power of the environmentalists to put the issues on the agenda in the first place. Thus, Ward, et al. document the role of Friends of the Earth in raising the political salience of pesticides in water, based upon its analysis of Water Authorities' (i.e. pre-privatization) data in the 1980s and of DWI data in the 1990s (Ward, et al., 1993: 6-7). However, we could find no evidence of really significant and systematic 'environmentalist' input to the current processing of the policy problem, although groups such as the Council for Protection of Rural England (CPRE), WWF, the National Trust, the Royal Society for the Conservation of Nature, and the Royal Society for the Protection of Birds are involved in a co-ordinating body called Wildlife and Countryside link. This has a broad ranging concern with the impact of agriculture on the environment (and might be seen as some kind of anti-agricultural lobby by some) in which water concerns appear to be relatively minor. The CPRE, however, has made representations concerning the designation of nitrate sensitive areas and nitrate vulnerable zones. For example, its response to the consultation paper on nitrate vulnerable zones argued that farmers should not be compensated for *not* polluting and that MAFF's proposals for implementing the Nitrates Directive were 'of a minimalist nature' (CPRE, 1994). A key difference between the CPRE and MAFF (and the NFU) is the question of whether farmers should follow economic good practice or environmental good practice. The CPRE argues that it is the latter which should be applied.

A fifth category of actor may be emerging as a significant participant, the large retailing chains. In Britain, approximately 60% of horticultural produce is sold by the leading six or seven food supermarket retail chains. As we shall see, some of these retailing chains have become active on the pesticides issue, not out of direct concern for water resources, but in response to perceived consumer concerns.

Finally, the scientific community and campaign orientated bodies such as the Pesticides Trust (considered by some to be simply an environmental pressure group) are of fundamental importance. These 'professional' actors are possibly the most powerful agenda setters in the policy area. The whole process is driven by science and by the politicization of science by such groups as Greenpeace and Friends of the Earth, who act as some kind of (no doubt distorting!) 'megaphone for science'. Regulation is constantly catching up with science and technology, leaving the central actors on some kind of

[11] See Mazey and Richardson (1992) for a discussion of this phenomenon at the European level.

regulatory ratchet (Richardson, 1994), either trying to meet the demands of scientific findings or trying to discredit them. In such situations, little cost benefit analysis is conducted.[12] Although it is beyond the scope of this study, much more work needs to be done on the process by which knowledge is gained, how it is transmitted, and how it is used in formulating European legislation, and to whose benefit.

References

Advisory Committee on Pesticides [ACP] (1994) *Advisory Committee on Pesticides Annual Report 1994.* London: HMSO.

British Agrochemicals Association [BAA] (1994) *Annual Review and Handbook.* Peterborough: BAA.

Byatt, Ian (1996) The Impact of EC Directives on Water Customers in England and Wales. *Journal of European Public Policy,* 3 (4): 665-674.

Council for the Protection of Rural England [CPRE] (1994) *Designation of Vulnerable Zones in England and Wales Under the EC Nitrate Directives.* London: Council for the Protection of Rural England.

Department of Environment [DOE] (1986) *Nitrate in Water: Pollution Paper 26.* London: Department of Environment.

Department of Environment and Ministry of Agriculture, Fisheries and Food [DOE/MAFF] (1995) *The Control of Pollution (Silage, Slurry and Agricultural Fuel Oil), Relations 1991. A Consultation Paper.* London: Department of Environment and Ministry of Agriculture, Fisheries and Food (September 1995).

Drinking Water Inspectorate [DWI] (1992) *Nitrate, Pesticides and Lead 1989 and 1990.* London: DWI/DOE.

Drinking Water Inspectorate [DWI] (1994) *Drinking Water 1993. A Report by the Chief Inspector, Drinking Water Inspectorate.* London: HMSO.

Drinking Water Inspectorate [DWI] (1995) *Drinking Water 1994. A Report by the Chief Inspector, Drinking Water Inspectorate.* London: HMSO.

Foundation for Water Research [FWR] (1995) *Groundwater Issues.* Marlow: Foundation for Water Research.

Healey, M.G. (1993) *Drinking Water Inspectorate's View on What Has Been Achieved What Needs to be Done.* London: Drinking Water Inspectorate.

Heclo, Hugh (1978) Issue Networks and the Executive Establishment. In: Anthony King (ed.) *The New American Political System.* Washington D.C.: American Enterprise Institute.

Kingdon, John, W. (1984) *Agendas, Alternatives, and Public Policies.* New York: Harper Collins.

Kinnersley, David (1988) *Troubled Waters: Rivers, Politics and Pollution.* London: Hilary Shipman.

Maloney, William and Richardson, Jeremy (1995) *Managing Policy Change in Britain: The Politics of Water.* Edinburgh: Edinburgh University Press.

Mazey, Sonia and Richardson, Jeremy (1992) Environmental Groups in the EC: Challenges and Opportunities. *Environmental Politics,* 1 (4): 109-128.

Ministry of Agriculture, Fisheries and Food [MAFF] (1991) *Code of Good Agricultural Practice for the Protection of Water.* London: Ministry of Agriculture, Fisheries and Food.

Ministry of Agriculture, Fisheries and Food [MAFF] (1995) *Report of the Independent Review Panel on Nitrate Vulnerable Zone 1995.* London: Ministry of Agriculture, Fisheries and Food.

---

[12] Indeed, the focus of this study is itself possibly a distortion of reality in that, taking pesticides as an example, far more pesticides are ingested by humans from food than via water.

National Audit Office [NAO] (1995) *National Rivers Authority: River Pollution from Farms in England: Report of the Controller and Auditor General.* London: National Audit Office.

National Rivers Authority [NRA] (1992) *The Influence of Agricultural Pollution on the Quality of Natural Waters in England and Wales.* London: National Rivers Authority.

National Rivers Authority [NRA] (1995a) *Water Pollution incidents in England and Wales 1994.* London: National Rivers Authority.

National Rivers Authority [NRA] (1995b) *Pesticides on the Aquatic Environment Report of the National Rivers Authority Prepared by the National Centre for Toxic and Persistent Substances (TAPS) (Water Quality Series No.26).* London: HMSO.

Richardson, Jeremy (1994) EU Water Policy: Uncertain Agendas, Shifting Networks and Complex Coalitions. *Environmental Politics*, 3(4), 139-67.

Richardson, Jeremy (1996) Policy Making in the EU: Interests, Ideas and Garbage Cans of Primeval Soup. In: Richardson, Jeremy (ed.) *European Union: Power and Policy Making,* pp. 3-23. London: Routledge.

Richardson, Jeremy, and Jordan, Grant (1979) *Governing Under Pressure: The Policy Process in a Post-Parliamentary Democracy.* Oxford: Martin Robinson.

Ward, Neil, Clark, Judy, Lowe, Philip, Seymour, Susanne (1993) *Water Pollution from Agricultural Pesticides.* Newcastle upon Tyne: Centre for Rural Economy, University of Newcastle upon Tyne.

White, S.L. & Pinkstone, D.C. (1995) The Occurrence of Pesticides in Drinking Water. *BCPC Monograph No.62, Pesticide Movement to Water.*

10. CONTROL CAPACITY - ENGLAND AND WALES

JEREMY J. RICHARDSON
Department of Government, University of Essex, Colchester, England

10.1 'Regulation' in practice: consensus building, joint problem solving and self-regulation?

Despite the existence of a wide range of legislative measures and regulatory agencies with specific powers, the overriding 'policy style' (Richardson, 1982) in this case is, as we suggested earlier, consensus orientated. If we were to choose one label to convey the essential nature of this style it would be 'codes of practice'. These numerous codes are all designed to build a consensual and co-operative approach as a means of changing farmers' behavior, with everyone recognizing that the (relatively) light regulatory regime (as applied directly to farmers, at least) might be replaced by a more tougher, adversarial, style of regulation. Cynics might see the consensual approach as simply agenda management, but it is equally possible to argue that it is the most realistic approach, bearing in mind the nature of the policy problem being addressed.

Some actors, however, have the possibility of other than consensual joint problem solving techniques. For example, the Department of Environment (DOE) can press the environmental case knowing that if Britain fails to act it will again join its European partners - such as the Netherlands and Germany - before the European Court of Justice. Thus, in inter-departmental discussions with the Ministry of Agriculture, Fisheries and Food (MAFF), for example, participants are fully aware of this possibility. Hence, the eventual emergence of a nitrate control policy is noticeably more 'impositional' than the previous regime (see below). Similarly, the National Rivers Authority (NRA) has statutory powers which it can use against farmers in certain circumstances, although it is not very prosecution orientated in practice. In the year up to March 1995, the NRA had brought a total of 237 prosecutions for pollution incidents that had occurred in 1994, 94% of which resulted in convictions. A total of 107 was related to agricultural pollution (NRA, 1995: 49). In addition, the NRA issues formal cautions (193 in total) and warning letters for other than category 1 incidents. The level of fines is also quite low. The largest fine was £30,000, the range in agriculture being £0 to £14,000. Costs can also be recovered under the Water Act. In the agricultural sector the range was £125 to

Geerten J.I. Schrama (ed.), Drinking Water Supply and Agricultural Pollution, 253-277.

£2,000. The level of fines is decided by individual Magistrates or Crown Courts. The maximum fine in a Magistrate's Court is £20,000.

10.1.1 REGULATION OF NITRATES: SHIFTING FROM A CONSENSUAL TO AN IMPOSITIONAL REGULATORY STYLE?

Before discussing the details and implementation of the consensual approach, however, it is important to outline MAFF's nitrate control policy, as this is emerging as a somewhat more 'impositional' regulatory style (Richardson, 1982). As with most other EU Member States, Britain has found itself before the European Court of Justice for breaches of certain EU laws. It is also anxious to avoid further cases, including under the Nitrates Directive. In terms of nitrate policy, Sections 94 and 95 of the Water Resources Act (1991) make provision for the designation of catchment areas where a range of controls on the use of nitrogenous fertilizers can be introduced. Ten pilot projects were established in 1990 and a second phase of the scheme followed in 1994. The so-called 'nitrate sensitive areas' (NSAs) are interesting in terms of our concern with the notion of regulatory style. The nitrate sensitive areas were entirely consistent with the British non-adversarial and non-impositional regulatory style in that they were not compulsory. Indeed, the nitrate sensitive areas were also consistent with the historical tradition of British agricultural policy in that they were underpinned by a system of subsidies to farmers. Thus, they fell into that category of public policy, referred to in chapter 9, which aims to change behavior via resource transfers rather than by simple legal imposition.

Previously, the 'regulatory' regime had been even lighter and rested on the encouragement of good practice. For example, in February 1988, MAFF launched a leaflet designed to increase the awareness of farmers of the consequences of pollution caused by nitrate leaching. The leaflet was a good example of our notion of 'joint problem solving' in that it was jointly sponsored by the Country Landowners Association, MAFF, DOE, the Fertiliser Manufacturers' Association, National Farmers' Union (NFU), UK Agricultural Supply Trade Association, the Water Authorities' Association (representing the ten Regional Water Authorities), and the Welsh Office. The leaflet impressed upon farmers the need to follow MAFF guidelines on the use of fertilizers. It was backed up by a series of local meetings in those areas where nitrate levels were significant and by support and advice from divisional ADAS offices. The central theme of this leaflet and of the previous MAFF advice was that good (economic) farming practices were not inconsistent with action to prevent excessive nitrate levels in water resources. (Indeed, the government has been giving advice on nitrate use to farmers since 1905). Current advice from MAFF and from the Fertiliser Manufacturers Association is based upon three principles:

- the crop's requirement, which depends on the species, the expected yield and (sometimes) the required quality of the crop;

- the soil supply of nitrogen, the nitrogen released from soil organic matter, or left from grazing the lost crop, or from livestock manures; and
- the extent to which the available nitrogen will be lost before the crop takes up.

The thrust of MAFF policy has been to achieve effective self-regulation as a means of avoiding more burdensome (to the farmers) public regulation. For example, in 1988, the Minister, speaking at a Silage Effluent Conference (sponsored by ADAS and the Water Authorities' Association), urged a minority of farmers to change their attitude to disposal of silage effluent '*... if farmers are to maintain their role as custodians of the countryside without burdensome restrictions.*'[1] Thus, the threat of more regulation was always in the background and, indeed, in July of that year the government announced that it was drawing up regulations to ensure proper containment of farm silage and slurry. Again, it was stressed that this would not affect the great majority of farmers who followed MAFF's Codes of 'good agricultural practice' (GAP). Also, again consistent with the general policy style, 'generous' grants were to be provided to help farmers reach the standards required.[2] Yet the gradual toughening of the Ministry's stance was becoming apparent when later that month the Minister announced that it was time to get tough on those farmers who caused serious pollution incidents. Also, he pointed out that £4 million had been provided in grants for farmers setting up good waste disposal systems in 1987.

The designation of nitrate sensitive areas was, therefore, consistent with the general development of MAFF policy to deal with agricultural pollution. As more evidence of possible damage emerged, the government's concern increased. As Haigh (1994) notes, 1970 had seen the government re-assured by the Report of the Royal Commission on Environmental Pollution, which had found the evidence on health risks from nitrates to be 'weak and equivocal'. Following the Report of the Standing Advisory Committee on Water Quality in 1984, however, concern began to increase and policy innovation followed, illustrating our suggestion that scientific findings act as a 'ratchet' for public policy. The Advisory Committee had warned that, if trends in fertilizer application continued, the proportion of abstractions from surface water which exceeded EU limits would rise significantly in the future (Haigh, 1994: 4.4-4.12). Thus, in March 1989 the Minister announced that the first nitrate sensitive areas were to be developed. In July 1989, MAFF announced the selection of candidate areas, following advice from the NRA. Two lists were produced: Candidate Areas for designation as nitrate sensitive areas, and Candidate Areas for intensive advisory campaigns only. Prior to the final selection of areas, an extensive consultation exercise was to be held with farming, water and other interests. The Minister emphasized that the pilot scheme would be on a voluntary basis.[3] Following the consultation exercise, the government announced the nitrate sensitive areas in April 1990 (subject to EC Commission approval under Article 93.3 of the Treaty of Rome), with intensive advisory campaigns in nine other areas. A key element of the NSA scheme was that payments were made to farmers who

1 *MAFF News Release*, February 2, 1988.

2 *MAFF News Release*, July 12, 1988

3 *MAFF News Release*, July 31, 1989.

voluntarily undertook to observe restrictions on their agricultural practices. The payments varied depending on the NSA in line with overall differences in costs of compliance between different areas. All farms in the Advisory Areas were to be visited by ADAS and encouraged to voluntarily reduce the risk of nitrate leaching, at little or no costs to themselves, or even a small benefit.[4] Formal authority for the nitrate sensitive areas was granted via a Statutory Instrument laid before Parliament by the Secretary of State for the Environment and the Secretary of State for Agriculture. By 1993, the policy was being extended, as pressure mounted on the government to comply with EU targets. Up to 30 new nitrate sensitive areas were proposed by MAFF. Again, the nitrate sensitive areas were selected on NRA advice and were to be subject to the usual thorough consultation procedures.

Under the ten pilot nitrate sensitive areas, some 163 farmers had jointed the scheme, entering 9,345 hectares and receiving between £55 and £475 per hectare.[5] In fact the ten pilot schemes received a good response, with 87% of designated agricultural land entering the scheme. As suggested elsewhere in our study, the Europeanisation of the policy area was a major factor leading to policy innovation, albeit very gradual innovation, consistent with historic national regulatory traditions. Thus, the *MAFF News Release*, following a ministerial visit to a pilot NSA in 1990, pointed out that EC Directive 80/778 laid down a mandatory limit of 50 mg/liter for drinking water and that the Commission had made proposals for a directive on water pollution caused by nitrate from diffuse sources. As MAFF commented: *'The results of the UK pilot nitrate scheme will assist the Government in introducing whatever overall measures are agreed as part of the Directive.'*[6]

The first three years of the pilot NSA scheme were assessed by ADAS in 1993, which found that leaching had been reduced in nine out of ten of the nitrate sensitive areas.[7] However, by then the EC's Nitrate Directive (91/676/EEC) was beginning to impinge more directly on the policy process. Thus, in December 1993, the government announced that it would shortly be consulting on the introduction of 'nitrate vulnerable zones' (NVZs) under EC law. Ominously for the farmers, the government also announced that:

> *farmers in these area would be obliged to follow certain rules to reduce nitrate leaching. It is the Government's intention that these rules should be based as far as possible on good agricultural practice (GAP)* and as such compensations will not be payable. (emphasis added)[8]

Thus, although MAFF's policy development had hitherto been incremental, consensual, and often underpinned by subsidies, the nitrate vulnerable zones, though logically a

---

4 *MAFF News Release*, April 3, 1990

5 *MAFF News Release*, April 28, 1993.

6 *MAFF News Release*, August 9, 1990.

7 It should be noted, however, that the first three years of the nitrate sensitive areas scheme coincided with a drought and that the lack of rainfall may have reduced the movement of nitrate into ground and surface water.

8 *MAFF News Release*, December 8, 1993.

development of nitrate sensitive areas, were examples of both policy discontinuity and policy succession. They were a break with the past in the sense that the policy style was about to shift from 'consensual' to 'impositional' (Richardson, 1982). Even so, at the end of 1993, MAFF announced an increase in the subsidies under the old NSA scheme, to £75 per hectare.

The shift in policy style has a simple origin - it is seen by some officials in MAFF and the DOE (backed by the NRA) as the only way that Britain can comply with the 50 mg/l EU parameter. (This view is not shared by the NFU however, who see water treatment and blending as the solution). Along with other member states, Britain was obliged to implement its mandatory measures, following on from the introduction of nitrate sensitive areas, which involved voluntary participation. As ENDS noted in September 1994:

> *There are two important differences between NSAs and NVZs. Participation in the former is voluntary; and participating farmers receive financial compensation, in line with rates specified in the new regulations. However, all 28 new NSAs fall within areas proposed for designation as NVZs, and farmers in these areas will have to comply with future mandatory NVZ rules, whether they participate in the NSA arrangements or not.*[9]

Although a major shift in policy style, MAFF is (perfectly reasonably) trying to graft onto it the preferred consultative approach to policy development. The NVZ proposals went out for consultation (in the summer of 1994) with the usual wide range of affected interests. The proposed 72 nitrate vulnerable zones provoked an extremely strong response from the NFU. This response is too detailed to review fully here, but the main thrust of the its response was as follows:[10]

- It considered the 50 mg/l limit to be over-rigorous and that the government must compensate farmers for the 'unjustified imposition' of the limit. (It should be noted that the NFU is by no means alone in this view).
- It believed that the fully-applied directive would have major consequences for livestock farms.
- It questioned the technical merit of the 72 zones, believing that large scale errors have been made, particularly by the NRA (a view disputed by the NRA).
- The government should compensate farmers if they were required to go beyond 'good agricultural practice', since much of the problem is historic.

In a crucial passage, the NFU stated that:

> *[t]he fact that, in some circumstances of geology or climate, the Government now considers the very activities it has previously promoted to be creating 'pollution' does not transform farmers into culpable polluters. It is for this reason that the 'polluter-pays-principle' cannot be applied fairly to the nitrate situation.* (NFU, 1994)

---

9 *ENDS Report 236*, September 1994.

10. For full details, see NFU (1994).

In response to the strong opposition to nitrate vulnerable zones, the government set up the Nitrate Vulnerable Zone Review Panel under the chairmanship of Terence Etherton QC, which reported at the end of October 1995. The panel reviewed objections to 29 of the 72 proposed zones and recommended boundary changes to 7 zones and the rejection of one small zone which the government had specifically asked the Panel to examine.[11] In total, Mr. Etherton's review panel received 87 submissions by objectors (MAFF, 1995). Clearly, the consensual policy style was important in the government's approach as there was considerable emphasis on consultation throughout and a willingness to make detailed concessions. In the event, it has been decided to introduce 68 zones covering approximately 600,000 hectares.[12]

However, in the end, the zones must go ahead under EU law and it could be extremely costly to maintain the subsidy principle, as demanded by the NFU, although the NFU believes that careful targeting could avoid this. MAFF is under pressure, as are all government departments, to restrain public expenditure. (Indeed, the former Agriculture Minister became the Treasury Minister, having responsibility for improving public expenditure control). Also, the government is still committed to the 'polluter-pays-principle' and it is difficult to see how the farming interests will be able, in the long run, to maintain the position cited above. Thus, many manufacturing enterprises had, long ago, to cease certain practices which were, no doubt, consistent with their industry's equivalent of 'good agricultural practice'. Also, there is some evidence to suggest that the concept of restricting agricultural practices in certain catchment areas does indeed work, i.e. that it is a successful policy instrument. For example, scientists at Thames Water Utility Ltd. report that an NSA set up in the Thames Valley five years ago is beginning to have an effect, compared with geologically similar conditions in non NSA areas.[13] Although other water companies appear to have been less active in monitoring the effect of nitrate sensitive areas on aquifers, the Thames results suggest that restrictions - whether subsidized or not - may well work.[14] Also, results of MAFF's own monitoring are reported to show significant reductions in the quantity of nitrates leaving the soil zone, although some water companies have yet to see the benefit and the NRA has seen no evidence of any reductions in nitrates in groundwater due to nitrate sensitive areas.

---

11 It should be noted that, in response to the government's earlier (1994) consultation document on the proposed 72 nitrate vulnerable zones, to which it received 500 written responses, the proposed boundaries of 31 of the nitrate vulnerable zones had been changed and two zones were dropped).

12 *ENDS Report*, March 1996: 45.

13 It should be noted that NSA requirements go well beyond nitrate vulnerable zones restrictions.

14 *ENDS Report 242*, March 1995.

10.1.2 'EVANGELISING' AS A POLICY INSTRUMENT: MOBILISING THE AFFECTED INTERESTS

As the sponsoring department for agriculture, MAFF is directly involved in a range of activities designed to reduce agricultural pollution of water resources, in addition to the major initiatives in nitrates. For example, it has its own R&D program on the effects of agricultural pollution on water resources, and has set up the National Farm Waste Forum. The terms of reference of the forum are to provide a focus for the exchange of views and ideas on farm waste pollution and related issues, between organizations concerned with the implementation and effects of such issues in England and Wales. The membership of the forum is listed below:[15]

- Agricultural Development and Advisory Service (ADAS);
- Country Landowners' Association (CLA);
- Department of the Environment (DOE);
- Farmers' Union of Wales (FUW);
- Institute of Environmental Health Officers (IEHO);
- Ministry of Agriculture, Fisheries and Food (MAFF);
- National Farmers' Union (NFU);
- National Rivers Authority (NRA);
- Scottish Office Agriculture Department (SOAFD); and
- Welsh Office Agriculture Department (WOAD).

The forum is a good example of a common feature of the policy area, namely, the bringing together of the 'affected interests' as a means of collective problem solving. Thus, the processing of the policy problem is characterized by collaborative arrangements of varying types, from a host of bilateral exchanges between, say the Water Services Association and the British Agrochemicals Association, to broader fora such as the National Farm Waste Forum. However, as we will suggest in our conclusion, few, if any, fora seem to bring together *all* of the core actors, the members of which might be defined as the central policy community, let alone mobilizing the whole issue network. For example, the National Farm Waste Forum does not include the water companies or their representative organizations, yet it is the water industry which bears the cost of many of the agricultural externalities, as we have argued. In effect, the industry is represented by the DOE (its sponsoring department) and by the NRA, each of which has a rather different perspective and different objectives. Neither is OFWAT represented directly, even though it has a central concern with the price of water to consumers.

Underpinning these (usually skewed) structures and processes for consensus building and joint problem solving is an assumption that co-operative action rather than impositional regulation is likely to be the most effective strategy. Hence, MAFF, for example, puts considerable effort into 'evangelizing' good practice. It has produced

---

[15] In addition, other bodies can be invited to participate on an ad hoc basis, by agreement with forum members.

three very comprehensive codes of practice.[16] The legal status of these codes, and the related code on pesticide usage,[17] is interesting. They are not legally binding codes, i.e. they would not be classified by us as 'impositional' regulation, although the water code is statutory. This means that not observing the code is not an offense, but it could be taken into account in any legal action. Moreover, following the code is not a defense against prosecution.[18] As an aid to keeping the Code, farmers can receive free technical advice from ADAS via one farm visit. MAFF has commissioned an independent survey of the efficacy of ADAS's free advice, and recommendations for better targeting are being implemented. In 1994 and 1995, ADAS made approximately 3,000 free pollution advisory visits to farms in England and Wales. This is an agreed level under a three year contract with ADAS (which began in 1994) in which ADAS provides either a free pollution advisory visit or assistance in the preparation of a farm waste management plan. In addition, MAFF and the NRA have targeted six catchments per year for intensive campaigns. These complement the NRA's own more intensive scheme of regular visits to farms (see below). The codes are very detailed indeed. A typical example is information provided on the construction of silos, including even the degree of slope of the floors![19]

Finally, it is important to mention MAFF's work on farm waste management plans. The Ministry conducted a pilot farm waste management plan study in 1992-1993 and has decided to extend the program. Thus ADAS advice is now available free to farmers in the preparation of farm waste management plans, and additional river catchments are being targeted for intensive campaigns to promote the adoption of farm waste management plans (DOE, 1995). However, the government rejected a proposal from the NRA - subsequently backed by the Royal Commission on Environmental Pollution (1992) - that pollution grants to farmers should be conditional on proper farm waste management plans (DOE, 1995). Thus, the government reported that experience from the pilot study suggests that, by implementing the plans on a voluntary basis, the government is more likely to succeed on encouraging farmers to adopt them in a positive way (DOE, 1995: 29). The strong emphasis on encouraging voluntary action is, of course, consistent with the government's policy for reducing the regulatory burden on industry and commerce generally. Thus, wherever possible, measures to secure voluntary action are preferred to the introduction of new regulations, in all policy sectors. For example, in the September 1995 consultation paper on 'The Control of Pollution' DOE and MAFF indicated their wish to rely to the greatest extent possible on

---

16 'Code of Good Agricultural Practice for the Protection of Soil' (1993), 'Code of Good Agricultural Practice for the Protection of Air' (1992), and 'Code of Good Agricultural Practice for the Protection of Water' (1991), the latter an update of the 1985 Code of Practice.

17 'Pesticides: Code of Practice for the Safe Use of Pesticides on Farms and Holdings' (1990), published jointly by MAFF and the Health and Safety Executive.

18 The codes systematically point out those laws which do apply, e.g. that it is an offense, under the Water Resources Act 1991, to cause or knowingly permit a discharge of poisonous, noxious or polluting matter or solid waste into any 'controlled waters'.

19 Again, farmers are reminded of their legal obligations for new silos constructed after 1st March 1991.

advice and guidance to farmers in respect of sites used (for storage of field silage) and to revise existing codes to take account of suggestions for improvements (DOE/MAFF, 1995: 5).

As indicated above, the NRA has also been very active in promoting the consensual approach. Its range of activities in trying to reduce agricultural pollution of water resources has recently been evaluated by the National Audit Office (NAO, 1995). The Report indicates that the NRA estimates that it spent £10million over the past two years on farm pollution activities. The Report itself notes that the probable reasons for the decrease in reported agricultural incidents include the existence of enforceable standards for waste facilities, the publicity given to farm pollution, and the increase in preventive work (NAO, 1995: 12).

The NRA's work on farm pollution is co-ordinated by its Rural Land Use Group, made up of representatives of the NRA regions. The group's report (NRA, 1992) forms the basis of the NRA's approach. The report reviewed available data on water pollution and on the effects of agriculture. For example, it noted that the DOE's Nitrate Co-ordination Group has concluded that nitrate concentrations in surface waters had increased at varying rates, with regional variations. Thus in Scotland, Wales, North West and North East England, the increase was 0.1-0.4 mg/l per year; in the North East of England 0.1-0.7 mg/l per year; Yorkshire, Severn Trent, Thames, Southern, Wessex 0.3-0.8 mg/l; Anglian 0.7-1.1 mg/l per year. Nitrates in groundwater were found to be more difficult to assess. Using the limited amount of reliable data from research boreholes, the report concluded that some showed little increase in nitrate since 1970 and, a few had shown a significant increase. However, it also concluded that there was some evidence that nitrate concentrations in water resources can be reduced by control of farming practices (NRA, 1992: 46). Interestingly, when discussing pesticide contamination of water resources, the report commented that:

> *farmers are incurring costs by using pesticides but they also obtain significant benefits. The water companies incur costs to the detriment of their customers and shareholders. A comprehensive economic analysis would be illuminating.* (NRA, 1992: 66)

Farm visits are regarded by the NRA as the most effective way of promoting positive action by farmers and cheaper than attending an incident or prosecuting a polluter (NAO, 1995: 22). Again, persuasion rather than regulation appears to be a dominant theme. However, the NAO notes that the NRA's role in approving installations under MAFF's Farm and Conservation Grant's Scheme is useful and argues that the NRA's powers under the Control of Pollution (Silage, Slurry, and Agricultural Fuel Oil) Regulations 1991, have been of *'considerable value in improving standards....'* (NAO, 1995: 22) In 1993, the NRA introduced a systematic nation-wide approach to farm visits with 22,000 farm visits planned for the period 1993-1996. (This represents, however, under 12% of the 186,000 farm units in England and Wales). The farms are in catchment areas selected using five criteria: failure to meet set water quality standards; a history of pollution incidents; the risk of water resources; areas of high conservation value; and public opinion (NRA, 1995: 23).

The NRA backs up its farm visits program with leaflets and an extensive publicity program. For example, it has produced leaflets on 'Farm Pollution and How to Prevent It', and 'Farm Waste Management Plans'. The latter is a good example of the combination of persuasion and legal requirements. The leaflet emphasizes the 'good citizen' aspect of pollution control, but also reminds farmers of the legal requirements for the construction of waste storage facilities. Some 1,400 Farm Waste Management Plans had been introduced by the end of 1993. As the NAO audit of the NRA's activities on agricultural pollution pointed out, NRA involvement usually produces the necessary action without prosecution. Where necessary, however, the NRA does resort to legal action. It is also assisted by having a role in the administration of the MAFF scheme for grants payable under the Farm Conservation Grants Scheme. Thus for the period 1989-1994, the NRA had approved some 11,000 installations under the scheme. However, in 1994 MAFF announced that the grants were to be discontinued, although they are likely to be re-introduced in nitrate vulnerable zones (NAO, 1995, p24).

As the guardian of Britain's water resources, the NRA has been strengthening its own analytical capacity. For example, in 1993 it concluded that whilst significant work had been carried out on pesticide contamination, this work needed better co-ordination and that there were significant variations between regions. This problem is now being addressed by the NRA's National Laboratory Service. More importantly, the NRA has set up the National Centre for Toxic and Persistent Substances (TAPS), within the Anglian Region of the Authority. This development could have major implications for future policy development, not just for the NRA but also for the other key actors involved. The new center will conduct a systematic review of the potential impact and likelihood of pesticides entering natural waters and is developing a predictive computer model (Prediction of Pesticide Pollution In the Environment, POPPIE) (NAO, 1995: 27). Thus, although there is still a lack of understanding of the long term effects of the application of pesticides and much more research is needed, there is in fact considerable data within the system already, particularly within the NRA and the water companies. The new center is likely to significantly increase the added value of this data and, therefore, influence the policy process in a more informed way. Pesticides will not be the only focus, as the new center will also assist the NRA to develop a national strategy for identifying the scale, importance and costs of the run off of nutrients. In the medium term, there is no doubt that more data will become available. For example, the NRA is currently developing a major computer project known as the Water Archive Monitoring System, at a capital and revenue cost of £22 million over five years. This will provide better water quality and resource information (NAO, 1995: 29).[20]

The National Audit Office's audit of the NRA's actions concluded with the view that the NRA's policy instruments were effective. Thus the NAO believed that:

> *there is clear evidence that farm pollution prevention work has contributed to overall improvements recorded in water quality and to specific improvements in many of the predominantly agricultural catchments reviewed.* (NAO,1995: 31)

---

20 These facilities could, of course, be utilized by European level policy-makers to better inform deliberations on future European policy.

10.1.3 PRIVATE 'REGULATION'

The NAO study did not, however, include an assessment of the full range of policy instruments and 'regulation' in the loose sense in which we are using the term. Thus, it is important to note that many other actions are being taken, outside the formal regulatory structures, but often in co-ordination with them. Of particular note is the action by the industries associated with agriculture - mainly the chemical industry (but also including the fertilizer industry) and, in one particular sector, the food retailers.

If we take the chemical industry first, the main initiative is being taken by the sectoral association representing the pesticide manufacturers, the British Agrochemicals Association. As we indicated earlier, the BAA is acutely aware that it is operating in an increasingly difficult business environment, and has attempted to ensure best practice on farms. It has produced a range of information material for farmers, for example on 'Pesticides and Water Quality'; 'Pesticide Disposal'; 'Think Water Keep it Clean'; and on 'Crop Production: Working with Nature', the latter concerned with Integrated Crop Management (ICM). Considerable emphasis is being placed on persuading farmers of the benefits of ICM, i.e. the use of the appropriate combination of biological, cultural and chemical methods in farming. The BAA's campaigns, as with the NRA, include very specific and detailed advice to farmers. For example, its leaflet on the rinsing of spraying containers includes advice to the effect that empty containers should be drained for at least thirty seconds, gives guidance on pressure rinsing, and on the burning of pesticide containers. Many of the actions to deal with the policy problem under discussion in this study are micro rather than macro policy instruments, as small changes in farmers' behavior can have very significant beneficial effects on local problems, both for the NRA and the water companies. Thus, we should note that detailed and well targeted policy instruments can have quite big effects on such problems as pesticide pollution because of the localized effects of such pollution and the very high marginal costs of dealing with it.

A central feature of BAA's campaign is support for Integrated Crop Management through the so-called LEAF program, 'Linking Environment and Farming'. In fact, LEAF is not a BAA body but is an independent organization which brings together a wide range of interests including farmers, consumer groups, supermarkets, environmental concern groups, research agencies, the government and the supply industry. Initial funding has been provided by the European Crop Protection Association (ECPA), and BAA has agreed to continue it financial support alongside decreased ECPA support. The BAA has also, since the mid 1980s, introduced 'BAA Crop Protection Certificates' for the sales and advisory staff of BAA member companies. The Fertiliser Manufacturers' Association (FMA) has also distributed leaflets designed to encourage the effective and safe use of its products by farmers and others.

All of these activities are designed to minimize the adverse effects of the use of pesticides and artificial fertilizers, and reflect a belief that they are both necessary and, if used properly, present no serious threat to the environment or to human health. The issue of whether there is a real threat is itself a cause of controversy, however. BAA's position is clear and reflects that of the European agrochemical industry as a whole (see

Richardson, 1994). It is that some of the parameters are wrongly set or are purely arbitrary. For example, the association argues that: *'as a science based industry, we are fundamentally opposed to our carefully researched data being judged against arbitrary standards.'* (BAA, 1994) This position is consistent with the government's position, namely that some aspects of EU water laws should be amended, that greater national flexibility should be allowed, and that in some cases the laws should be repatriated. In a sense, there is now an advocacy coalition, both at the national and European levels, to press for a lessening of the regulatory regime and for a reform of the processes by which European water policy is formed. The British coalition includes the government itself, the NFU, the agrochemical industries, the water companies, and one of the regulators, OFWAT (for the latter see Byatt, 1996).

Before turning to the actions of the water companies and the NFU, it is worth noting that in one sub-sector of agriculture, horticulture, a relatively new entrant to the policy area has emerged. As suggested earlier, some of the large retailers who dominate the market for the sale of horticultural produce have become active in assisting in the introduction of better farming practices. No doubt this has much more to do with the need to be seen as 'green' by their customers, as it has with a concern for water resources. However, it is possible that this concern can eventually have some beneficial side effects on water resources, particularly in those areas of high horticultural production. The fact that the UK has a very concentrated food distribution sector opens up the possibility of policy instruments that might not be available generally in Europe. If we take one example - the supermarket chain Sainsbury's - that company has been developing its own Integrated Crop Management program for its suppliers. This began in 1991, after commercial tomato growers opened their tomato houses to the public in order to demonstrate how few pesticides were being used and the extent to which biological controls had been developed. In 1992, the company produced a policy document which went to all of its suppliers in the UK and abroad. Linked to the NFU in its Retailer Partnership (see below), the percentage of Sainsbury's UK suppliers covered by Integrated Crop Management Protocols has risen from 2% in 1992 to 57% in 1994. The use of pesticides by the company's suppliers has fallen as a result. In practice, an increasingly close relationship is developing with the company's suppliers, some of whom want even higher standards as a means of gaining market advantage. Gradually, the company is also developing links with other relevant actors such as Thames Water Utilities Ltd., the NRA and the Pesticides Directorate. Clearly, the NFU and the retailers have a common interest in improving the effectiveness of the management of pesticide use. Thus, the NFU invited six of the leading multiple retailers to join in developing husbandry protocols (guidelines) for individual crops (Wise, 1995: 1). At the end of 1995 twenty-two protocols had been agreed. The emphasis is on 'self-audit' but it is envisaged that an 'audit-trail' should be in place to enable individual produce to be traced from initial receipt of seed, through propagation, production, harvesting, packing, storage and finally through to the consumer (Wise, 1995: 3). The ultimate sanction behind the NFU-Retailer ICM Protocols is likely to rest in supply agreements between grower and retailer/packhouse.

For the horticultural sector at least, the market power of the big multiples might eventually be a significant policy instrument in protecting water resources. So far, the big retailers seem not to be concerned with extending their initiatives to cereal growers yet it is the cereal herbicides which represent one of the main components of the water pollution problem. Thus, while multiple retailers are clearly influential, their practical influence on the water problem is probably very limited at this stage and needs to be developed further. (Also, water quality monitoring is required before the benefits of the retailers' actions can be assessed).

The closer liaison between growers and retailers in the horticultural sector is, of course, yet another example of joint problem solving and consensus building. At a more general level the government is developing its own 'Action Plan on Pesticide Minimization' - outlined in the government's 'Rural White Paper' earlier in the year - via similar processes. Thus, in October the DOE and MAFF jointly convened a conference on pesticide minimization. The conference brought together a wide range of organizations representing farmers, growers, the agrochemical industry, retailers, consumers and environmentalists, in order to provide their views on the future development of the UK pesticide minimization policy.[21]

We now turn to the actions of the water companies themselves. As we described them earlier, they are at the sharpest end of the whole process in that the statutory controls on them are the most 'impositional' and onerous. In discussing the role of the companies one needs to remember the regional dimension mentioned several times in this study. Action does vary from company to company and it is therefore impossible to make firm generalizations about specific types of actions. Similarly, the type of relationship which companies have with other actors in the policy area varies quite considerably. However, the industry is unusually concentrated and this may have some beneficial consequences for future policy design and implementation, compared with most of Britain's European partners.

At the apex of the industry is its main representative body, the Water Services Association (WSA). The association has an Environment and Quality Committee, consisting of the Directors of Environmental Quality of the ten water companies. The committee has four sub-committees, on water resources, on drinking water, on waste water, and on sludge. The WSA also has a number of liaison groups with specific trade associations, with government departments and with the regulators e.g., with MAFF, ADAS, the Detergent Manufacturers' Association, and has meetings with the NFU and CLA. More importantly, the association has a whole range of issue specific contacts with other actors on problems such as nitrates, pesticides, cryptosporidium, sewage sludge. Thus, if one is looking for the existence of coherent 'policy communities' it may well be at this micro level, where there are fora (either ad hoc or permanent) where the key interests for a particular problem can get together in a process of joint problem solving, on the model suggested earlier.

The companies, however, adopt a very strong public stance that their job is to comply with the legal standards and then to negotiate the cost-pass-through with

[21] *DOE/MAFF, News Release*, October 26, 1995.

OFWAT. In complying with the standards (albeit over an extended timescale), the companies have a number of options. The bottom line for the companies is that they do have the technology and the financial resources to install the necessary equipment for water purification. The immediate concern has been to get the equipment in place, alongside other important measures such as blending, and catchment protection (in liaison with the NRA, the NFU, individual farmers, the BAA, and others). They also apply pressure to the government and the NRA, as the ultimate regulators, to formulate the necessary policies and action plans for the protection of catchment areas. There is possibly a caution in pressing either the NFU or individual farmers in too open a way as there is a clear recognition that the co-operation with farmers will produce more results than a confrontational attitude. Also, it is important to remember that the water companies are not the *enforcers* of legislation. They are not 'authorities', that is the responsibility of the NRA. A number of water companies (again, there are regional variations) have farm liaison schemes of various kinds, all consistent and supportive of the approach of the NRA, BAA, and NFU. There is strong emphasis on persuasion. For example, Severn Trent sponsored a 'Spraysafe conference' for over three hundred delegates from within its region in 1992. The reason for such initiatives is clear. As Dr. R. A Breach, the company's Quality and Environmental Manager pointed out, the costs of bad practice with pesticides can be serious for the company. Thus, he cited the example of two boreholes which had had to be shut down, in each case costing the company £500,000 to replace. For the company as a whole, he estimated that it could cost it £12 million a year to meet the pesticide standard, all of which would be passed on to water consumers in the region. Interestingly, in terms of the focus of our study, Severn Trent had at first targeted non-agricultural use of pesticides as part of its 'Spraysafe campaign', e.g. roads and railways. Results suggest that this has been successful (Breach, et al., undated). However, as with Thames, the Company seems now to be turning more attention to agricultural uses as a source of water pollution, as a reduction in pesticides found in water from agricultural use had not taken place in the Severn Trent region (Breach, et al., undated). New techniques were to be used in the future for tracking the use of pesticides from diffuse sources, as Severn Trent had developed its 'CatchIS system' to plot the relative risk of using any particular pesticide on any particular crop. The implication of better modeling techniques is that closer liaison with the various agricultural interests would need to be developed. As Breach, et al. note *'achieving this objective requires regular and constructive dialogue between all parties concerned about the environmental fate of pesticides.'* (Breach, et al., undated)

The need to develop further links with other actors is a common view amongst the water companies. For example, Southern Water sees itself as ultimately opening communications with all constituencies of interest, in what may turn out to be more formalized contexts, i.e. there may be moves to institutionalize relationships as the problem persists on the agenda. Anglian, which hitherto has concentrated on installing the necessary investment to meet all European standards by 1995, is also in the process of developing closer links with agriculture in its region. For example it has set up a small team under its own pesticides project, to identify particular catchment areas where a targeted campaign might be effective. Again, an important factor is the high marginal cost of removing pesticides from water resources and the fact that small changes in pesticide use can have big effects on this marginal cost. This development is probably typical of many of the ten water companies, as the pressure to control costs increases and as they begin to contemplate the renewal costs of equipment installed in the major program of investments post-privatization.[22] Each water company faces its own particular situation and, therefore, strategies vary. Thus for Wessex Water the problems seems to be relatively slight although triazine is an increasing threat as maize continues to expand as a fodder crop. Also as the occasion arises, the company can be involved in particular initiatives, for example it put funding into an ADAS project on low intensity farming. As suggested earlier, companies such as Severn Trent and Thames have the agricultural problem high on their organizational agendas (together with pollution from non-agricultural sources it should be emphasized). Thus, Severn Trent, in addition to the Spraysafe campaign cited above, also runs a 'Chem-E-Safe campaign' and a 'farmsafe campaign', and has close links with organizations such as the BAA, tenant farmers and local authorities. Again, we see the emergence of essentially regional and local policy networks in response to the problem.

The farming community is almost invariably involved in these local and regional policy networks, either farmers directly or via the NFU regional and local offices. The NFU itself has certainly placed the nitrate and pesticide issue high on its organizational agenda and has developed its own expertise in the problems (e.g. see Wise, 1995). For example, in 1993 it conducted its own survey of current practices by farmers in the disposal of agrochemical packaging. On the nitrates issue, currently of the highest concern to the NFU because of the emergence of the impositional style of regulation cited earlier, it has set up a regional structure, with regional advisers who attend briefings in London, and who liaise with MAFF and the NRA.[23] Of all the interest groups reviewed in this study, the NFU is the most politically active, with a strong lobbying stance regarding EU legislation and the particular development of the NVZ policy.

It is, of course, extremely important to stress that the processing of the whole package of problems associated with agricultural pollution of water resources is a dynamic process. Thus, as we turn to a brief review of possible future development of policy, we need to note that relationships between the various actors are evolving in a continuous process of joint problem solving. A more effective policy community for problem solving seems to be emerging.

---

22 For example, Anglian has invested £130 million in removing pesticides and £80 million on nitrate treatment in the last five years in order to meet the legal requirements for supplying its 3.9 million customers.

23 *NFU Nitrates Bulletin*, May 1994.

10.2 Policy options: policy succession and consensus building or radical policy change?

As we suggested earlier, if one were to select one label to describe the primary characteristic of the process of problem solving it would be 'codes of practice'. The fundamental philosophy underpinning the resort to codes of practice is that policy instruments based upon joint problem solving and consensus building are more likely to be effective than a more impositional or adversarial style of formal 'regulation'. Indeed, we suggested that regulation as a term should be interpreted flexibly to mean 'behavior changing' or 'steering' strategies designed to change the practices of farmers. Perhaps the best indicator of the efficacy of this basic strategy is the National Audit Office's study of the NRA's activities in response to the problems of farm pollution. Whilst recognizing the importance of regulation, the NAO nevertheless recognized the importance of educating farmers in 'good agricultural practice', albeit sometimes backed up by regulation in the pure sense. Also, we ourselves have drawn attention to the importance of 'pure' regulation, for example, the very strict and detailed licensing regulations for the introduction and continued use of pesticides exercised by the PSD and the various water laws enforced by the NRA.

The reason for such emphasis on joint problem solving and consensus building, we suggested, was based upon a combination of national regulatory style and the special characteristics of the regulatory problem under study. Thus, much of the problem is difficult to monitor and to attribute to a particular 'offender'. In any case farmers are very numerous and individualistic. The range of policy responses can be seen, therefore, as a fairly rational response to a particular set of problems.

The picture is also one of increased attention to the various problems, i.e. there is plenty of evidence that the problem is high and rising on organizational agendas. If we take the water companies as an example, the rate of policy innovation is varied - as indeed the NAO concluded it was for the different NRA regions - but they all seem to be moving to a position where they see agricultural pollution as more important, now that they (and the NRA) have made considerable progress in persuading non-agricultural actors to pollute less. In a hierarchy of targets, agriculture is rising up the agenda.

The same is true of the regulatory authorities themselves. Thus there is a sense in which the issues are being institutionalized and, therefore, acquire an institutional dynamic, irrespective of any external pressure. This is probably why the environmental interests are far less active in the policy area now. They can rest assured that the problems have now been 'internalized' by insider organizations, such as the NRA, the NFU, the DOE, MAFF, the ten water companies and the twenty-one 'water only' companies, the BAA, and the PSD. These organizations are fully aware that there is always the threat of 'whistle blowing' in Brussels by British environmentalists. Also, the environmentalists are themselves 'issue entrepreneurs' (Richardson, 1995) and are constantly looking for new issues in order to maintain support and raise funds.

10.2.1 LICENSING PESTICIDES: INCREASED ATTENTION TO WATER ISSUES VIA INCREMENTAL POLICY CHANGE

A typical example of the increasing importance of water pollution problems on organizational agendas, is the increasing attention which the PSD is paying to the problem. Thus, if we go back five years or so, the agency was even more part of the agricultural sector and was primarily concerned with acute studies in the field of eco-toxicology. Environmental as opposed to human health problems were fairly peripheral to PSD. Increasingly, however, the PSD's agenda appears to be more sensitive to environmental problems, and it is possible to see the specific problem of contamination of water resources as increasingly important as scientific knowledge expands. This leaves PSD in an interesting phase in its organizational development. The central focus is still on operator and consumer exposure following the tradition of 'acute' studies. Yet there is clearly an increased awareness of broader environmental considerations. For example, the 1994 ACP Annual Report lists a number of 'new provisional approvals' for products. Quite commonly, the provisional approvals, say, for five years, specify that further studies of possible effects on the water environment are required as a condition of the approval, with the new data to be supplied at specified intervals during the approval period. A typical example was Epoxiconazole, a fungicide used in the treatment of winter wheat and winter and spring barley. The Committee noted, on the basis of aquatic toxicity data, an 'acute and chronic risk' from the proposed use. It therefore recommended an interim buffer zone around water courses and requested further aquatic toxicology data and more data on the transport of Epoxiconazole to adjacent water courses. The ACP does not require, as the condition of a license, that the manufacturers specify the means, and the cost, of possibly having to remove the product from drinking water. It is difficult to avoid the conclusion that such problems tend to be addressed rather later in the policy process, under the procedure for reviewing existing products as new data emerge. Thus, when atrazine emerged as a major problem for drinking water suppliers, it was banned by ACP for certain (non-agricultural) uses and further restrictions on its agricultural use were introduced. More recently, a very widely used herbicide, Isoproturon,[24] has been reviewed by ACP. As a result, its use for pre-emergent treatment of wheat was banned, a maximum dosage per hectare was specified, aerial use was revoked, and various lesser measures were introduced.[25]

Such detailed responses are being taken in the context of a more general awareness, within PSD, of water considerations. For example, the ACP Annual Report for 1994 lists two studies of the problem of pesticides in water, the first being a paper presented on behalf of a working group established to consider the data currently available on pesticide contamination of water sources. The task of the working party was to identify areas needing to be improved within the current regulatory procedures and environmental legislation to protect water. A subsequent paper made a series of recommendations aimed at developing and improving the information received by the

[24] White and Pinkstone (1995) for the problems caused by the product.

[25] *MAFF News Release*, July 25, 1996.

Committee with regard to pesticide contamination of ground and surface water (ACP, 1994b, see also ACP, 1994a: 42-3) The Report produced a number of recommendations consistent with our view, that we are witnessing the increased mobilization of policy networks and policy communities to deal with the problems. For example, it recommended that the NRA should be routinely contacted for data on water contamination to ensure that the latest information is incorporated in the reviews presented to the PSD's expert committees. Similarly, data from the water companies, the WSA, and the NRA should be reviewed annually to determine whether any environmental problems are developing. Specifically, The Environmental Quality Standards used by the NRA should be considered by PSD when toxicological concerns are apparent in drain, ditch, and other surface water problems. Also, it recommended that dialogue between MAFF, PSD, the Health and Safety Executive, water companies, and the NRA should be further developed. Of particular interest in terms of our own review of the nature of the problem, the ACP report commented that DWI data on its own should not be used to evaluate the environmental impact of a pesticide (ACP, 1994b).[26] Thus, it was recognized that drinking water data are but one indicator of the policy problem and that broader environmental indicators were needed.

Before we discuss the question of improved liaison between the different policy actors (as the main *policy succession* option), it is important to note that the PSD itself may need to address the question of the range and scope of participation and representation in its own structures. The water industry appears to be poorly represented in PSD's own machinery, the PSD relying on the DOE and NRA to represent water interests. With the exception of one of its subcommittees, there appears to be no *direct* representation of the water companies and there are doubts (expressed to us by some interviewees) about how effective NRA representation is in the detailed regulatory processes of the PSD.

10.2.2 MORE OF THE SAME: IMPROVING LIAISON AND CO-ORDINATION BETWEEN POLICY ACTORS

The theme of the ACP report was that, generally, better liaison and exchange of data would be beneficial. This leads us to one obvious recommendation for incremental policy development, *namely that more integration and better structuring of national liaison* procedures between policy actors should be instituted. Thus, it might be possible to develop a more broadly based and systematic national *policy forum* to bring together the wide range of interests involved. For example, there could be much more *direct* involvement of the water companies, and possibly of OFWAT (representing water consumers). This would bring into the open the question of externalities and how they should be dealt with and would utilize the expertise and data resources of the water companies more effectively in the design of policy instruments.

---

[26] Now that pesticide treatment facilities have been installed by the water companies, DWI reports reflect only pesticide concentrations at the tap, not in the environment.

A second incremental change would be *to increase both the funding and the co-ordination of research.* Above all the problem under study here is one of uncertainty, particularly relating to the long term prospects for groundwater. Everyone is operating under a high degree of uncertainty and there is always a risk of policy failure, as well as of a maldistribution of resources, if policy is made in the absence of reliable data. All of the public authorities, such as the DOE, MAFF, NRA, and PSD are developing their R&D programs, and the water companies are also developing new monitoring techniques, for example using GIS. Again, it is not clear that this amounts to a *national research strategy.* How such a strategy might be funded would raise difficult questions in a period of public expenditure constraint and in the face of cost pressures on the water companies. Also, very difficult market conditions exist for the fertilizer and pesticide manufacturers.

At the local and regional level, *liaison between the water companies and farmers (and other farming related interests) might be further improved.* There are many excellent examples of close liaison, and it is clear that it is on the increase, but it is difficult to avoid the conclusion that the water companies have been shielded by a relatively generous regulatory bargain struck at the time of privatization. Effectively, this has both enabled them to pass on costs and to turn a low value product into a higher value product, perhaps an ideal scenario for an industry which finds difficulty in expanding its home market, other than by risky diversification. *Greater direct involvement of OFWAT in some of these issues might be one possible option to be examined.* This might encourage water companies to conduct a more thorough options search for dealing with agricultural pollution of water resources. In fact, OFWAT does make allowance in the price limits it sets for water companies for them to reduce nitrate levels where appropriate. However, under the current price cap regime it is for companies to deal with the problem in the most efficient manner. OFWAT's approach is to set the price limits, leaving the companies to choose the most suitable option to solve any problems. Thus, it is firmly committed to the principle of *not* having direct involvement with the companies, so that they are free to manage their businesses and choose the most cost effective solution. Hence, greater OFWAT involvement would be a departure for the agency.

All of these suggestions are essentially building on existing policy instruments in the traditional policy succession model of policy development. They are also probably capable of accommodating the consensual and joint problem solving approach which has characterized the policy area so far. There are, however, some radical options which would be difficult to accommodate in this way.

10.2.3 RADICAL POLICY OPTIONS

The most obvious radical policy option would be *to address the externalities problem directly by applying the polluter-pays-principle even to non-point source pollution.* For example, product taxes are a possible way of achieving this widely accepted principle. There is no reason, in terms of regulatory theory, why product taxes cannot be applied to pesticides and artificial fertilizers. Indeed, such taxes would be consistent with more general recent policy development thinking within the DOE. This radical shift in policy would be opposed by farmers and by the manufacturers of these products. For example, the BAA already objects to the continuation of the levy (approximately 2% of sales turnover) used to fund the operation of the licensing system. It has pointed out that it costs £64,000 to register a product compared with £600 in France (BAA, 1994: 4). Clearly, Britain would need to press for Euro-level policy initiatives of this kind in order to avoid disadvantaging both the farming and the agrochemical industries in Britain (see below). The revenue from product taxes could fund water treatment costs incurred by the water companies. However, the familiar opposition from governments to earmarked taxes would need to be overcome for such a proposal to stand any chance of success.

If economic instruments are unacceptable, then *the alternative may well be tougher 'pure regulation' and tougher monitoring and enforcement procedures.* In fact the PSD procedures for licensing the *introduction and use* of pesticides can be defended as being thorough by international standards, even though water pollution considerations might need to be given even more emphasis. In contrast, the actual *monitoring* of the use of the licensed products commands fewer resources. The PSD's view is that it already spends considerable resources on monitoring wildlife incidents, monitoring for residues in food and carrying out the pesticide usage survey. It believes that data from these surveys enables the agency to estimate the extent of compliance. However, not all observers would be so sanguine. For example, in April 1996, The Financial Times reported a survey conducted by a market research group which suggested that *'British farmers' profitability is being threatened and land unnecessarily polluted because of poor advice from agricultural retailers.'*[27] Part of the cause of the problem, the report suggested, was due to the fact that national crop trials had been phased out progressively during the 1980s, leaving pesticide retailers as the main source of information for farmers. Indeed, in the case of organophosphorous (OP) sheep dips, MAFF's own figures suggest that a very high proportion of farmers are failing to comply with the regulations which make it illegal for them to purchase OP sheep dips without a certificate of competence. Of 90,000 sheep farmers, only 12,000 are reported to have registered for the test and 7,500 had passed.[28] The data are worrying in terms of our study, as OP dips are known to expose the *farmers themselves* to serious health risks and it might have been expected that high compliance would have been relatively easy to achieve. Thus regulating actual use of products raises very difficult regulatory problems, as we suggested in our introduction. Even in the more adversarial regulatory systems of continental Europe, there are believed to be far more serious problems of misuse and illegal use of chemicals than in Britain; due to a thriving black market in illegal pesticides and herbicides in continental Europe, largely absent in Britain. In the OP case, even the distributors' Association is reported as believing that controls on sale rather than on use are ineffective.[29] Easy solutions to this fundamental problem are not likely. One radical

27 *The Financial Times*, April 10, 1996, p.11.

28 *The Guardian*, April 3, 1995.

29 *The Guardian*, April 3, 1995.

solution, proposed by Ward, et al., is that the equivalent of the 'prescription' system used for drugs in the National Health Service should be introduced for pesticides. The system would be run by a body independent of the manufacturers or merchants. They propose a completely regulated market in pesticides with so-called 'crop doctors' responsible for responding to calls from farmers and for 'prescribing' the most appropriate treatment for the specific conditions (Ward, et al., 1993: 50). The likelihood of such a proposal being adopted by the British government is extremely low. A variant of this proposal would be to make the manufacturers or suppliers legally responsible for application, on the familiar agricultural contractor model where some farmers already buy in certain services during peak periods. Less radical toughening of 'pure' regulation might include requirements that manufacturers of pesticides should specify the techniques for, and costs of removing their products from water or give an assurance that existing treatment methods will remove them at no extra costs.

If we take a long-term view of the problems caused by agricultural pollution of water resources, the risk to the farming community is, of course, that more impositional regulation will emerge over time if a more integrated approach to pollution control were to emerge, say in the form of an Integrated Pollution Directive. If self-regulation is found not to be effective in the long run, or if subsidies (such as in nitrate sensitive areas) are thought to be too costly for the taxpayer, then governments may simply resort to tougher regulation and force the cost of that onto the agricultural industry (as in the case of the nitrate vulnerable zones). Regulation is a very cheap public resource, as it simply transfers costs to the regulated, although those who are regulated quite reasonably demand proper cost-benefit analyses of proposed and existing regulations.

Whether existing 'regulation' is thought to be effective will continue to be the focus of intense debate. Previously, the mechanisms that led to water contamination were poorly understood. However, there seems little doubt that the more sophisticated our knowledge base becomes, the greater the pressure for tougher controls over agricultural pollution will be. In a sense, regulation, of whatever type, will always be trying to catch up with the advance of knowledge and with changing farming practices and new products. Thus, *we re-emphasize the importance of constructing a more integrated and broadly based network of national, regional, and local policy actors in order to maximize the potential for joint problem solving under conditions of extreme uncertainty.* Without this, tougher legal regulation is likely to emerge.

10.2.4 RELEVANCE OF EUROPE

As suggested earlier, a number of British and European level actors are pressing for some form of relaxation in European laws. A change in European standards can solve or worsen 'the problem' overnight. Whether a problem exists is in part an artifact of definitional techniques. Hence, some British interests see a change in the rules as the best policy option. However, many of the existing standards can be regarded as 'peace treaties' agreed after long and difficult negations in the past. Rewriting these peace treaties is going to be a very difficult task, as there are strong advocacy coalitions in place at the Euro-level determined to preserve the core of existing regulations. At best, some marginal adjustments might be secured. Moreover, decisions in the water sector cannot be isolated from the impact of bargains to be struck in other policy areas. Expecting the British government to 'deliver' on the repatriation or weakening of European water laws is, therefore, a high risk strategy. A more rapid and co-ordinated national system of policy development in the implementation of European laws seems likely to be a better use of resources in the short run. Moreover, Britain may be able to develop a significant comparative advantage because of certain structural factors. For example, it now has a very concentrated industrial structure in water supply, it operates river basin management, and has moved to a system of integrated pollution control under the new Environment Agency. It also has a rather concentrated farming industry compared with many other EU states, and has a very concentrated food distribution and retail system. Thus, there may be a greater potential for co-ordinated and effective action in Britain than elsewhere in Europe, if only because the number of core actors is smaller. If this comparative advantage were to be fully exploited, Britain could begin to set the pace of Euro-regulation rather than reacting to it; an appropriate task for the new Environmental Agency which has absorbed the NRA.

The key task at the European level, as many British policy actors see it, is *to develop a European level policy process which is more firmly grounded on robust scientific knowledge and research.* There is a very strong belief that Europe policy is not founded on 'good science', that the standards are set arbitrarily, cost considerations are either absent or ignored, and unrealistic time scales are set for meeting new standards. Thus, the demand is for reforms of the European level processes for formulating water policy so that greater emphasis is placed on scientific fundings. This might suggest the need for a major Euro-level R&D program on pollution of water resources, in order to better inform policy-makers at the European level.

A second key task is seen as *the need for much more effective monitoring of the implementation of existing European water laws in the member states.* Many British observers resent the accusation that Britain is the 'dirty man of Europe' and believe that this reputation is in part based upon the fact that Britain's European partners are less effective on collecting and publishing pollution data. Thus, some British observers believe that an important European initiative would be to standardize and enforce measurement and monitoring procedures, so that reliable comparisons can actually be made. Though not relating to the focus of this study - namely agricultural pollution of water resources - the European Commission's own data lend some support to this view. For example, as the Commission itself reported in the case of bathing waters, trends were difficult to establish, because of the sharp increase in the number of beaches insufficiently sampled.[30] For example, it singled out the Dutch figures for criticism, noting that for the Netherlands 'the sea water bathing areas are generally of high quality when only the areas sampled sufficiently are taken into consideration. However, this conclusion cannot be applied to all identified bathing areas because a large amount of

30 Quality of Bathing Water 1994. *EUR 15976 EN.*

data are missing.'[31] This explains why the Netherlands' position in the bathing water 'league table' slipped so dramatically in the Commission's 1994 report on the quality of bathing water and why the Netherlands were *'requested to take all appropriate measures in order to reduce the high percentage of insufficiently sampled points.'*[32] In the specific case of pesticides in water, the Council of Europe draws similar conclusions for the Netherlands:

> *In summary a considerable number of the investigated pesticides and related compounds have been detected in these surface water studies. Ten of the compounds occurred in more than 50% of the analyzed samples and another 30 substances occurred in more than 10% of the samples. Regarding the ecotoxicological risks of these findings the measured maximum concentrations exceeded estimated ecologically safe levels for more than 50% of the compounds. For a considerable number of compounds these values were substantially exceeded.* (Council of Europe, 1995: 31)

Thus, in discussing comparative data on water pollution, *extreme* caution is needed in drawing conclusions about the relative success of different member states. For example, the Council of Europe concluded that:

> *In Germany a considerable number of monitoring activities and surveillance of pesticides in different water companies have been achieved during the last decade. However, there is to our knowledge up to this time no comprehensive review available on the occurrence of pesticides in surface waters.* (Council of Europe, 1995: 33)

Finally, a third key task at the European level is seen as *a need for better policy co-ordination between different policy sector, particularly between the Common Agricultural Policy (CAP) and water policy.* The increased use of atrazine (and therefore, the increased incidence of atrazine in water resources) in Britain is a good (or bad!) example of the cross-sectoral effects of CAP. Thus, maize growing has increased considerably, in part because of changes in the European subsidy system. Essentially, the CAP needs to take greater account of the environmental effects of subsidies via changes in cropping patterns. The task of increasing policy co-ordination presents, however by far the greatest European level policy challenge, yet could have quite major beneficial effects on the problems addressed in this study. We conclude, therefore, with the comment that what is being discussed here is the implementation of *European* law. As is common with all implementation problems, they are closely linked to problems with policy formulation, problems which can be addressed only at the European level.

---

[31] *EUR, 15978.EN.97.*

[32] Quality of Bathing Water 1994. *EUR 15976 EN.*

References

Advisory Committee on Pesticides [ACP] (1994a) *Advisory Committee on Pesticides Annual Report 1994.* London: HMSO.

Advisory Committee on Pesticides [ACP] (1994b) *ACP Review of Pesticides in Ground and Surface Water* [ACP 140 (234/94) & ACP 86 (232/94)]. London: ACP.

Breach, R.A., Porter, M.J. Court, A. (undated) *CatchIS - A New Computer Based Catchment Planning and Information System to Assess the Vulnerability of Surface and Groundwater Catchments to Contamination.* Coventry: Severn Trent Water Ltd.

British Agrochemicals Association [BAA] (1994) *Annual Review and Handbook.* Peterborough: BAA.

Byatt, Ian (1996) The Impact of EC Directives on Water Customers in England and Wales. *Journal of European Public Policy,* 3 (4): 665-674.

Council for the Protection of Rural England [CPRE] (1994) *Designation of Vulnerable Zones in England and Wales Under the EC Nitrate Directives.* London: Council for the Protection of Rural England.

Council of Europe (1995) *Pesticides in Surface Waters. A Review of Pesticide Residues in Surface Waters in Nordic Countries, Germany and the Netherlands and Problems Related to Pesticide Contamination.* Strasbourg: Council of Europe.

Department of Environment [DOE] (1986) *Nitrate in Water: Pollution Paper 26.* London: Department of Environment.

Department of Environment [DOE] (1995) *Freshwater Quality: Government Response to the Sixteenth Report of the Royal Commission on Environmental Pollution.* London: Department of Environment (February 1995).

Department of Environment and Ministry of Agriculture, Fisheries and Food [DOE/MAFF] (1995) *The Control of Pollution (Silage, Slurry and Agricultural Fuel Oil), Relations 1991. A Consultation Paper.* London: Department of Environment and Ministry of Agriculture, Fisheries and Food (September 1995).

Drinking Water Inspectorate [DWI] (1992) *Nitrate, Pesticides and Lead 1989 and 1990.* London: DWI/DOE.

Drinking Water Inspectorate [DWI] (1994) *Drinking Water 1993. A Report by the Chief Inspector, Drinking Water Inspectorate.* London: HMSO.

Drinking Water Inspectorate [DWI] (1995) *Drinking Water 1994. A Report by the Chief Inspector, Drinking Water Inspectorate.* London: HMSO.

Foundation for Water Research [FWR] (1995) *Groundwater Issues.* Marlow: Foundation for Water Research.

Haigh, Nigel (1994) *Manual of Environmental Policy: The EC and Britain.* London: Longman/IEEP.

Healey, M.G. (1993) *Drinking Water Inspectorate's View on What Has Been Achieved What Needs to be Done.* London: Drinking Water Inspectorate.

Heclo, Hugh (1978) Issue Networks and the Executive Establishment. In: Anthony King (ed.) *The New American Political System.* Washington D.C.: American Enterprise Institute.

Kingdon, John, W. (1984) *Agendas, Alternatives, and Public Policies.* New York: Harper Collins.

Kinnersley, David (1988) *Troubled Waters: Rivers, Politics and Pollution.* London: Hilary Shipman.

Maloney, William and Richardson, Jeremy (1995) *Managing Policy Change in Britain: The Politics of Water.* Edinburgh: Edinburgh University Press.

Mazey, Sonia and Richardson, Jeremy (1992) Environmental Groups in the EC: Challenges and Opportunities. *Environmental Politics,* 1 (4): 109-128.

Ministry of Agriculture, Fisheries and Food [MAFF] (1991) *Code of Good Agricultural Practice for the Protection of Water.* London: Ministry of Agriculture, Fisheries and Food.

Ministry of Agriculture, Fisheries and Food [MAFF] (1995) *Report of the Independent Review Panel on Nitrate Vulnerable Zone 1995*. London: Ministry of Agriculture, Fisheries and Food.

National Audit Office [NAO] (1995) *National Rivers Authority: River Pollution from Farms in England: Report of the Controller and Auditor General.* London: National Audit Office.

National Farmers' Union [NFU] (1994) *Designation of Nitrate Vulnerable Zones in England and Wales Under the EC Nitrate Directive (91/676). Response to the Government Consultation Document.* London: National Farmers' Union.

National Rivers Authority [NRA] (1992) *The Influence of Agricultural Pollution on the Quality of Natural Waters in England and Wales*. London: National Rivers Authority.

National Rivers Authority [NRA] (1995) *Water Pollution incidents in England and Wales 1994*. London: National Rivers Authority.

Richardson, Jeremy (1982) (ed.) *Policy Styles in Western Europe*. London: George Allen & Unwin.

Richardson, Jeremy (1994) EU Water Policy: Uncertain Agendas, Shifting Networks and Complex Coalitions. *Environmental Politics*, 3(4), 139-67.

Richardson, Jeremy (1995) The Market for Political Activism: Interest Groups as a Challenge to Political Parties. *West European Politics*, 18 (1):116-39.

Richardson, Jeremy (1996) Policy Making in the EU: Interests, Ideas and Garbage Cans of Primeval Soup. In Richardson, Jeremy (ed.) *European Union: Power and Policy Making*, pp. 3-23. London: Routledge.

Richardson, Jeremy, and Jordan, Grant (1979) *Governing Under Pressure: The Policy Process in a Post-Parliamentary Democracy*. Oxford: Martin Robinson.

Royal Commission on Environmental Pollution (1992) *Freshwater Quality, Sixteenth Report*. London: HMSO.

Ward, Neil, Clark, Judy, Lowe, Philip, Seymour, Susanne (1993) *Water Pollution from Agricultural Pesticides*. Newcastle upon Tyne: Centre for Rural Economy, University of Newcastle upon Tyne.

White, S.L. & Pinkstone, D.C. (1995) The Occurrence of Pesticides in Drinking Water. *BCPC Monograph No.62, Pesticide Movement to Water*, 263-268.

Wise, C.J.C. (1995) *The NFU-Retailer Partnership's Integrated Crop Management Protocols*. London: National Farmers' Union.

PART THREE

11. COMPARATIVE ANALYSIS

GEERTEN J.I. SCHRAMA
Center for Clean Technology and Environmental Policy, University of Twente, Enschede, The Netherlands

11.1 Introduction

The comparative analysis undertaken in this chapter considers together the outcomes of the three national case studies, as well as those of the EU case study. In three distinctive steps the analysis is focused to the subject of the control capacity of the water supply sector - the dependent variable in the project. The first step is to present an overview of the policy networks and policy styles at the European and national level. The relevant policy issues concern the curbing of agricultural pollution in general, and water pollution by nitrates and pesticides in particular. The next step concentrates on the national policies intended to influence farmers' behavior, in particular farmers in or near water catchment areas, who may affect the drinking water resources. The final step is the assessment of the actual control capacity of the water supply sector. The latter is the designation used for the water supply industry and the regional water authorities. Control capacity is the sum total of: (1) the mission and orientation of the organizations in the water supply sector; (2) the available organizational resources; (3) the selected steering strategies. The three steps are also summarized in table 11.1.

11.2 Policy issues, policy networks, and policy styles

The first step of this comparative analysis is to present an overview of the policy networks and policy styles at the national level. The relevant policy issues concern the curbing of agricultural pollution in general, and water pollution by nitrates and pesticides in particular. The arenas in which these policies are brought about are mainly the traditional networks, constituted by the respective Ministries of Agriculture and agricultural sectors. The following issues will be addressed:

- the relevant policy issues;

Geerten J.I. Schrama (ed.), Drinking Water Supply and Agricultural Pollution, 281-304.

TABLE 11.1: Comparative analysis in three steps

General issue: curbing of agricultural water pollution, in particular with nitrates and pesticides		
Step 1	Level of analysis: European and national	
	Categories	Theoretical concepts
Policy networks	• agriculture • water management	• policy communities versus issue networks • advocacy coalitions
Policy issues	• agricultural pollution in general • water quality in general • protection of drinking water resources	
Policy styles		• consensual versus impositional
Step 2	Level of analysis: national	
	Categories	Theoretical concepts
Policy instruments	• directly aimed at farmers • a role for the water supply sector in the implementation • in support of the control capacity of the water supply sector	• (a) moral appeal - legal compliance; (b) economic rationality; (c) self-regulation - social responsibility • enlarging versus restrictive
Step 3	Level of analysis: regional and local	
	Aspects	Categories
Control capacity of the water supply sector	• mission and orientation	
	• organizational resources	• financial capacity • organizational capacity • legitimacy and authority • information
	• steering strategies	• economic • juridical • communicative

- the characterization of the policy networks in terms of 'policy communities' versus 'issue networks', and the predominant policy style, notably the extent of the consensus base;

- the position of the Ministries of Environment, and the way they have succeeded in getting the environmental issue on the agricultural policy agenda;
- special attention will be paid to the specific policy efforts directed at farmers within (ground) water protection zones, and to the roles of the organizations from the water supply sector or their representatives at the national level.

11.2.1 AGRICULTURAL SECTOR

The three EU member states offer some fine examples of disintegrating policy communities. All of them have known true *agricultural policy communities* with 'iron triangles' at their core for a long time. Developments within the agricultural sector - with the reconstruction of the EU agricultural policy, and the pressure evoked by the general concern about agriculture's impact on the environment as major driving forces - have eroded the bases of the policy communities.

Germany. In Germany, the agricultural policy network at the federal level can be characterized as a policy community with a relatively simple structure around the core, constituted by the Ministry of Agriculture and the national farmers' interest organizations. Institutionalized links exist with the chemical industry, while environmental NGO's have more difficulty in gaining access to the network. The Ministry of Environment has no significant position. In Germany, the agricultural sector seems to be the least affected by environmental issues. The door to the outside world is still closed in this respect.

England and Wales. In Great Britain, too, the agricultural sector is still a policy community. Central actors are the Ministry of Agriculture and the National Farmers' Union, Britain's most important farmers' association. As everywhere else in the European Union, this policy community is under heavy pressure, but it is still closed to outsiders. The level of interrelatedness is beyond question, but mutual commitment is no longer taken for granted. The Ministry of Agriculture has assumed the role of guardian of environmental interests within the network, probably out of strategic considerations in order to keep the policy community closed and to maintain the consensual policy style.

The Ministry of Environment belongs to the periphery of the agricultural policy network, together with several other stakeholders who have some influence, such as the agrochemical industry, the environmentalists, the large retailing chains (in response to consumer pressure), and the scientific community.

The Netherlands. In the Netherlands, the disintegration of the policy community has proceeded furthest. The policy network is no longer a policy community, because the introduction of the environmental issue has resulted in a conflict of interests between government and the farmers and their interest organizations.

The Ministry of Environment has successfully sought access to the agricultural policy network, in order to get environmental issues - manure policy in particular - on the agenda. As a matter of fact, the Ministries of Environment and Agriculture together have created an 'advocacy coalition'. Although both ministries are still in favor of a self-regulation approach for the agricultural sector, policy formulation based on consensus has given way to conflict management during the last decade.

11.2.2 POLICY ISSUES AND POLICY STYLES

After the description of the policy networks that constitute the arenas for the topic of agricultural pollution, we turn to the specific policy issues that have emerged on the pollution of water by nitrates and pesticides, and the policy styles applied.

European Union. The protection of water from pollution and the maintenance of good quality of water resources have been key issues of the EU environmental policy. In the broadest sense, water quality policy of the EU includes more than the specific pieces of water legislation. Like other policy actions dealing with sectors of the environment - or with specific environmental issues and sources of environmental pollution - water quality management policy is based upon the more fundamental principles and objectives underlying the EU's environmental policy. These are contained in the relevant articles of the Single Europe Act and the EU Treaty. In addition, the more specific policy principles and objectives, as well as the instrumental strategies to be applied, have been formulated for the medium term in the Fifth Environmental Action Program, which is the authoritative statement outlining the course of action and general measures to be taken in pursuit of the EU's environmental (water quality) objectives. In the specific case of pollution of drinking water resources by agricultural activities, EU policy with regard to agriculture is important. Of particular interest in this regard are the measures be contemplated to integrate environmental concerns into the reform of agricultural practices in the Union.

The EU deals with the protection of the water resources in several ways. The EU has been regulating the quality of drinking water supplies in Europe since 1980 by setting standards for maximum concentrations of nitrate and pesticides in drinking water. Via the Groundwater Directive and the Directive on dangerous substances in water, the EU is regulating certain dangerous substances among which pesticides. Because of a lack of concrete measures, the impact of these directives on agricultural practices (and thereby the protection of the water resources) is limited.

Protection of the water against nitrate pollution should follow from the implementation by the member states of the Nitrate Directive. Applying 'good agricultural practices' by the farmers is supposed to prevent nitrate from ending up in water. More can be expected from the action programs which had to be developed by the member states before the end of 1995.

The EU has developed legislation to protect the water against agricultural pollution - although sometimes it is difficult to point out the impact of the directive on agricultural practices. Moreover, several developments like the Groundwater Action Program and the Proposal concerning the Directive on Ecological Water Quality, indicate an even more active role of the EU in the future. Also, the changes of the position and contents of the common agricultural policy (CAP) should be taken into consideration. It is very likely that pesticide use and fertilizer use will decrease because of the developments within CAP.

England and Wales. The environmental issue has become an integral part of the agricultural policy. Agricultural water pollution (including drinking water resources) by nitrates and pesticides ranks high on the agendas of all core actors. A remarkable fact is that the Ministry of Agriculture has introduced the environmental issue to the agricultural policy network, and not an outsider, such as the Ministry of Environment. The Ministry of Agriculture is trying to solve these problems in a consensual way. Meanwhile the National Farmers' Union is fiercely combating EU legislation and the Ministry's policy on 'nitrate vulnerable zones'.

The most urgent policy issue is nitrates, or rather the implementation of the EU Nitrate Directive. The overriding national policy style is (or rather, used to be) consensus orientated, aimed at persuasion of and self-regulation by the farming community. Under pressure from the European Union, the nitrate control policy of the Ministry of Environment is emerging as a somewhat more top-down regulatory style, aimed at enforcement of the legal standards. The Ministry of Agriculture is obviously trying to maintain the agricultural policy network through a consensual policy style. This approach is under pressure from the European Union, which is urging more direct regulation, which will affect the nature of the policy network.

The Netherlands. The national policy on agricultural pollution comprises policy issues such as phosphates, nitrates and pesticides. Combating nitrate pollution is the most urgent one, pursued notably through the manure policy. This policy has resulted in two acts. First, the 1987 Soil Protection Act, mainly formulated by the Ministry of Environment, which focuses on legal standards concerning the amounts of manure farmers are allowed to spread on the land. A scheme of stepwise incrementation of legal standards was established, with its final stage at the end of 1995, when nation-wide manure standards came into force to secure compliance with EU directives. This act also contains the possibility of establishing ground water protection zones. This involves extra restrictions on farmers within these zones and the right to compensation payments. The second act is the 1987 Fertilizer Act, mainly formulated by the Ministry of Agriculture, which is focused on financial and communicative incentives (it deals with the disposal problem: manure bookkeeping, manure banks, fixed quotas, levies on surpluses). The manure policy and the severe legal standards were greeted with fierce resistance by the farmers. Enforcement of these norms, especially the monitoring of individual farmers, will be very difficult, if not impossible. The national government is failing to make a moral appeal to the farmers (through addressing their social responsibility); even the state's legal authority is now being questioned by some farmers.

Pesticides. Pesticides are being regulated by the EU through several regulations. Since 1979, the EU has prohibited a number of active ingredients from use in the EU. Until 1991, the regulation of specific pesticides was left to the member states. In 1991, the Directive on the placing of pesticides on the market was adopted in order to harmonize this situation. Protection of the water from pesticides should flow from the Uniform Principles, recently adopted by the Council. The Uniform Principles use the drinking water standards for pesticides and nitrate as criteria for maximum allowable concentrations in the ground- and surface water. Although one might expect that with the Uniform Principles the protection of ground- and surface water is guaranteed, several aspects undermine this. First, it will take years before effects might be expected because: (1) the member states need to implement these Uniform Principles in their national legislation which will take some time, (2) a transition period exists for those pesticides that do not comply with the Uniform Principles. Moreover, it is possible that pesticides currently used in agriculture will show up in ground water in 20 or 30 years. Second, on basis of the directive on the marketing of it pesticides is possible for member states to temporarily authorize pesticides that do not comply with the Uniform Principles.

Nevertheless, the comprehensive European harmonization leaves little discretion to the member states with respect to pesticides policy, so all the countries under study show similar approaches. To the extent that national pesticides policy networks exist, they are created around the Ministry of Agriculture. The interests of producers (agrochemical industry) and users (farmers) are well represented. Water managers, however, have only a marginal role in this process, while the water supply industry is virtually absent. As a result, the risks of water pollution are not adequately examined, and licensed pesticides are eventually quite often shown to be detrimental to water quality as more research data become available. The policy style can be characterized as 'indirect regulation' of farmers through regulation of the market. There is also an aspect of direct regulation, through prohibition of the use of blacklisted pesticides. Nevertheless, illegal trading of blacklisted pesticides occurs throughout the Union, and offenses of illegal pesticides use by farmers are regularly reported.

Apart from legal regulation of pesticides use, there are also policies aimed at the reduction of the amounts of (legal) pesticides used. Typical policy instruments are communicative. The Dutch 1991 'Long term crop protection plan', for instance, involves a whole range of voluntary and stimulating policy instruments, including a voluntary agreement and a subsidy scheme. In England and Wales, the national system of pesticides registration is financed by a small levy on the sale of pesticides. (The levy is probably too small to function as a regulative economic instrument.)

Conclusion. The interesting question is whether 'old', disintegrating policy communities are similar to 'new' issue networks in terms of the typical policy styles. In the Netherlands and England and Wales alike, there is a strong external pressure from the European Union to use more direct regulation, now the old ways of the (former) agricultural policy communities have failed to solve the problem of excessively high nitrate concentrations in the ground water. Perhaps the Netherlands is the most distant from compliance with EU standards, and Government is most desperate (although the EU directive as such plays no significant role in the publicity about the manure policy). Changes in policy styles are not only in the direction of more regulation (away from ineffective self-regulation), the consensus base is also diminishing. The big problem is that a new policy must be implemented and enforced within a sector that is traditionally accustomed to a consensual approach, and where support for the new policy is absent, at least among the most vocal parts of the sector. In the Netherlands in particular, it remains to be seen whether the new top-down approach is feasible, given the resistance and obstruction by certain farmers' groups.

11.2.3 WATER POLICY NETWORKS

The issue of agricultural water pollution resides in the overlap of two major policy fields: agriculture and water management. The agricultural sector is almost the archetype of a policy community, while the water management sector also has comprehensive policy networks in the European Union, as well as in the countries under study (cf. Bressers, O'Toole and Richardson, 1995).

European Union. In the EU water policy process - at least as far as the initial formulation of policy proposals is concerned - there is little doubt that the Commission is the pivotal actor. In subsequent phases of the EU policy process, it shares center stage with other institutional actors. The key role of the Commission is the result of both its formal power of policy initiation and its pivotal position - and skill - as 'broker of interests'. Sub-networks, concerned with various aspects of water management policy are, for these reasons, centered around the Commission.

The long, complex multi-layered nature of the community processes provides many points of access for interests, and so many opportunities for them to keep informed and to press their cases. Groups organized along both national and Europe-wide lines are actively involved in trying to get their interests and views incorporated into community policy. At the level of the European Commission there has been a trend towards the development of close relationships between groups and the Commission. With regard to water policy, the number and range of actors has increased over time, moving the system from a relative private and limited range of participants to more open, conflictual and public system of decision making.

Germany. A complex water management network exists, with a 'hard core' constituted by the Ministry of Environment, an unofficial but very influential working group constituted by representatives of the States (LAWA), the national association of the water supply companies (BGW), and three standard-setting associations in water management. Through LAWA the States have an important stake in this federal network. Several NGO's are among the peripheral participants. The network is characterized by a high level of interrelatedness, although mutual commitment is not taken for granted.

England and Wales. The network on the field of water quality management can also be characterized as a policy community. Drinking water quality is an important issue in this field. The central actor is the National Rivers Authority (NRA), the governmental agency responsible for the quality of the surface waters (and now part of a new integrated Environmental Agency). Other participants from the water supply sector are the ten large-scale, privatized water supply and sewage companies of England and Wales, their national association (WSA), as well as two regulatory agencies (DWI and OFWAT). Participation seems to be based on common professional and economic interests. The network is characterized by a high level of interrelatedness, and - similar to Germany - mutual commitment is not taken for granted.

The Netherlands. The Netherlands has an extensive policy network in the field of water management, concerning water quality, as well as water quantity. The Ministry of Transport and Water Works is at the center of this network, while the Ministries of Environment and Agriculture are also participants, as is the national association of the water supply companies (Vewin). This particular policy network, however, is not the initiator of the policy concerning agricultural water pollution, as the agricultural policy network, with the Ministry of Environment at its core, has assumed this role.

Conclusion. In Germany and Great Britain, the water supply sector is more or less part of a larger water management sector, which is organized at the national or federal level in all kinds of discussion forums that deal with water pollution - not just from agricultural sources - and the affected interests. Such forums are less evident in the Netherlands, where the emphasis is on the regional level. Here the water supply sector is relatively well organized around the issue of agricultural pollution of drinking water resources.

11.2.4 AGRICULTURAL POLLUTION OF DRINKING WATER RESOURCES

As mentioned above, the issue of agricultural water pollution resides in the overlap of two major policy fields. Although it clearly is an issue within the EU water quality policy, it cannot be taken for granted, that it is an actual and distinctive policy issue at the national level, nor that there is an (issue) network that deals with it. A second point concerns the attention paid explicitly to the pollution of drinking water resources. The situations in the three countries under study are very dissimilar. Interaction between the water supply sector and the agricultural sector is concentrated on the national level in England and Wales, on the regional level in the Netherlands, and on the local level in Germany.

European Union. Interactions between the drinking water sector and the agricultural sector take place within different networks and during different stages of policy making. Several types of interactions within the formal network are mentioned in the EU case study:

1. interactions via initiatives of the Commission, like the conference on the revision of the Drinking Water Directive;
2. consultations between DG XI of Environment and DG VI of Agriculture which are required on relevant issues, the so-called 'inter-service consultations;
3. committees and working groups that prepare Commission proposals, which are another stage where the sectors meet; and
4. interactions of the actors of the drinking water and the agricultural interests in the member states as part of the process of deciding on national positions on EU policy proposals.

Besides all that, interactions take place in the (more) informal circuit. DG XI talks to the interest groups that represent the agricultural sector, notably the European associations, COPA, ECPA, and EFMA. It seems that DG XI is more 'open' in this respect than DG

VI. The intention of DG XI is to stimulate discussions on environmental issues within the agricultural sector. Relationships also exist between the interest groups. The chemical industry, especially ECPA, is very active in establishing relationships with everybody involved with an issue. The European association of the water industry, EUREAU, also maintains relationships with individual chemical companies. All respondents to the EU case study stressed that coalitions between interest organizations are always flexible and vary depending on the (sub)issue.

England and Wales. England and Wales has a real policy network for the issue of agricultural water pollution that can be characterized as an 'issue network of the second order' (see section 2.2). It is constituted by the core of the agricultural policy community with the NRA as additional core actor. The NRA is a member of several discussion forums, which gives it the opportunity to draw attention to their specific interests. The water supply companies hold only marginal positions. Protection of drinking water resources is an important, but not the only issue for this network.

Germany. In Germany, there is hardly any interaction between the water sector and the agricultural sector, as the Ministries of Environment and Agriculture do not get along with each other very well. On the national level their is no linkage between the two structures. *Ad hoc* collaboration does exist, such as the joint declaration by the BWG and the DBV on the EU Drinking Water Directive. At the local level, however, contacts between the water supply sector and the agricultural sector do occur frequently.

The Netherlands. In the Netherlands, agricultural pollution of drinking water resources is not a specific policy issue at the national level and there is no identifiable national issue network that joins the water management and agricultural policy networks together. Several parties with stakes in the protection of drinking water resources are *de facto* represented in the agricultural policy network by the Ministry of Environment, while the water supply sector has no access to it. The Dutch case study shows the most elaborate regional policy networks, characterized by a consensual policy style aimed at persuasion (use of more efficient and less polluting pesticides) and negotiation (e.g. farmers need economic arguments to accept additional restrictions).

Conclusion. There are no specific national policies concerning the issue of protection of drinking water resources. National policies concerning agricultural pollution in general have been developed in the Netherlands and England and Wales, while the situation in Germany is differentiated, as this concerns the competence of the individual States. The introduction of new and more stringent regulations on manure practices, evoked by the need to implement EU directives, is disrupting the traditional consensual policy styles, most significantly in the Netherlands.

11.2.5 CONCLUSIONS

As the subject of this study concerns a problem of affecting farmers' behavior, the efforts of policy makers are focused on agricultural policy and the corresponding policy networks, notably in England and Wales and the Netherlands. In these countries the traditional consensual policy styles have changed towards more regulatory ones, as a result of external pressures exercised by environmental interests and the need to implement EU directives. As a result the characters of the policy networks have changed too, and they are no longer the tight policy communities they used to be. In Germany the policy on protection of drinking water resources from agricultural pollution has been developed within the water management network. This policy certainly entails restrictions on farmers' behavior, but it did not affect the agricultural policy nor the nature of the agricultural policy network.

11.3 Policy instruments

This section concentrates on the national policies intended to influence farmers' behavior, in particular farmers in or near water catchment areas, who may affect the drinking water resources. Three policy instrument characteristics are selected, as discussed in section 2.4:

1. *Direct-indirect.* The first distinction concerns policy instruments with farmers as direct and principle addressees versus those aimed at the role of the water supply sector as intermediary. In fact, three classes can be distinguished:
   - policy instruments directly aimed at farmers, with the water supply sector not involved in the implementation;
   - policy instruments directly aimed at farmers, with a role for the water supply sector in the implementation; and
   - policy instruments not directly aimed at farmers, which may contribute to the control capacity of the water supply sector.
2. *Moral appeal.* The second instrument characteristic - whether the instrument involves a moral appeal issued to the target group - is related to the extent of self-regulation as a characterization of the prevailing policy style. Again, three classes can be distinguished:
   - the one extreme is a moral appeal on legal compliance (including the option of sanctions in case of non-compliance);
   - the other extreme is a hint at (rather than an appeal to) economic rationality (self-regulation without any form of legal regulation or sanctions); and
   - the third form is self-regulation based on social responsibility, where government can either choose to apply truly neutral communicative steering without any moral appeal, or to make a moral appeal to the target group to adjust its behavior for the sake of some general interest, bur without legal regulation or sanctions.
3. *Enlarging-restrictive.* The third policy instrument characteristic is a combination of provision or extraction of organizational resources (enlarging versus restrictive instruments) and the discretion left to the target group (obligatory versus non-obligatory instruments). All types of policy instruments, legal, economic, and communicative, can be either restrictive and obligatory, or enlarging and non-obligatory.

11.3.1 POLICY INSTRUMENTS AIMED DIRECTLY AT FARMERS

Of all policy instruments directly aimed at farmers, it is hard to tell which are used to curb environmental deterioration in general, and which are used for the protection of drinking water resources in particular (although the former are not extensively studied in the present project). All countries apply rather strict legal regulation concerning nitrates (or manure) and pesticides, aimed at curbing the pollution of surface and groundwater, although not aimed at the protection of drinking water resources in particular. These are predominantly restrictive and obligatory instruments of a legal nature, or economic to the extent that regulative levies are imposed. With respect to drinking water catchment areas, all countries have complementary policies, which will be discussed in the next section when the water supply sector is involved. There are also examples of stimulation policy, aimed at the protection of drinking water resources, in which the water supply sector is not involved. This concerns information campaigns to show farmers sustainable alternatives to the prevailing agricultural practices, and to convince them that these will have no negative effects on their incomes. The information is often supplemented by economic stimuli, such as subsidies.

With respect to governmental attempts to influence farmers' behavior, the most important policy instruments - which are exemplary of the consensual policy style and fitted to the traditional policy community - are the non-binding codes of 'good agricultural practice', applied in England and Wales. These and other communicative policy instruments (such as information campaigns and the creation of discussion forums) are voluntary instruments involving a certain moral appeal to farmers' social responsibility. There is an indirect appeal to legal compliance, as it is made clear to the target group that non-responsiveness may lead to a tougher and more adversarial style of regulation.

The policy instruments applied in the three countries involve no explicit moral appeal to farmers, although the British system of 'good agricultural practice', just mentioned, may also be conceived as a way to institutionalize a moderate form of moral appeal.

11.3.2 WATER SUPPLY SECTOR INVOLVED IN IMPLEMENTATION

In all countries the most powerful tool created at the national level seems to be the possibility of establishing some kind of (ground) water protection zones. In the Netherlands and Germany this option has existed for a relatively long time, the authority being delegated to the provinces and the States, where it has been applied on a large scale. In both countries it involves mandatory additional restrictions on farmers. In England and Wales the measures concerned are more recent, they are not widely applied, and originally were on a voluntary and only later on a mandatory basis (obligatory instrument), while the executive power has been retained at the national level, with the Ministry of Agriculture.

Farmers in the groundwater protection zones are entitled to financial compensation to the extent that they are subject to more restrictive regulation than other farmers in the country. In Germany, the compensation payment schemes still involve large sums of money. In the Netherlands, the differentiation in legal standards has almost been leveled out in 1995, as was the right to compensations. In both cases the money was extracted from the water supply companies, and finally from their customers. In England and Wales, only the first cohort of pilot projects was supported by a subsidy scheme, additional legal restrictions being imposed upon farmers later on, without compensation.

The national policies reviewed are contributing to the protection of drinking water resources in many ways, but the ground water protection zones are the only direct contributions to the control capacity of the water supply sector.

England and Wales. The need to implement the EU Nitrate Directive led the Ministry of Agriculture first to introduce the 'nitrate sensitive areas' (1990), still on a voluntary basis, and later on the 'nitrate vulnerable zones' (1994), which involved mandatory restrictions on farming practices. The latter was perceived by the farmers' association as a departure from the consensual approach in the direction of more legal regulation, and was strongly opposed by them.

Germany. The most important policy instruments applied at the national level to protect drinking water resources are the establishment of groundwater protection zones and the compensation payments to farmers. According to the 1986 amendment of the federal Water Management Act, water protection zones can be established around catchment areas, enabling several legal restrictions to be placed on agricultural practices and other types of land use. Germany has established some 20,000 groundwater protection zones, covering over 10% of the whole territory. However, great differences occur between the States. In general, the zones are established by specialized water authorities upon powers delegated from the *State* Ministry of Environment. The unofficial but habitual procedures involve consultations with the municipalities and the individual land users affected.

Compensation payments must be paid by the 'beneficiary'; usually the water supply companies, but sometimes the State or private industrial users. Generally speaking, however, two models are developed. The first is the co-operative model, according to which farmers and water suppliers must reach mutual agreements about information exchange, probable further restrictions, and the compensation payments. The actual negotiations are conducted at the State level between the branch associations. The role of the State is a facilitating one, without forcing any actor to engage in any particular 'exchange relationship'. This model is adapted in North Rhine-Westphalia, Bavaria and seven other States. In North Rhine-Westphalia the agreement involved a supporting program, according to which the agricultural associations are to employ farming experts who will advise on 'extensification' (less intensive) agriculture. The cost of employing these experts will be borne by water suppliers. The second model was pioneered in Baden-Württemberg, where compensation is paid to farmers directly by the state on a lump sum basis.

The Netherlands. The provinces have been granted important executive authority with respect to the establishment of groundwater protection zones. This instrument is applied all over the country, and it is in fact the most visible national policy on the issue of agricultural pollution of drinking water resources. Farmers within these areas are subject to more stringent restrictions than elsewhere. They are also entitled to financial compensation to the extent of the relative disadvantages suffered. The latter policy instrument matched well with previous practices of compensation payments to farmers for the drought caused by ground water abstraction and subsidies for restrictive use of manure, both paid by water supply companies.

Economic incentives and other policy instruments suited for situations of high levels of interrelatedness and mutual commitment are not used at the national level, but the water supply sector is encouraged to employ them at the regional level.

11.3.3 POLICY INSTRUMENTS AIMED AT THE WATER SUPPLY SECTOR

Apart from the tasks and responsibilities attributed to the regional water authorities, there are few - if any - policy instruments aimed at the water supply sector that may support its control capacity over agricultural pollution. The water supply industry is made responsible for the quality of the drinking water at the tap, without much legal equipment for living up to this responsibility. In general, the constitution of the respective 'cost-pass-through' mechanisms determine the scope of the options of investing in preventive action, but they may also contain stimuli to abstain from preventive action, as can be said of the British system.

More specific policy instruments aimed at the water supply industry are the groundwater abstraction charges in Germany and the Netherlands, but these are rather meant as a regulative levy on the use of groundwater and not to support to the control capacity of the water supply sector. However, the revenues of the so-called 'water penny', adopted by an increasing number of German States, are used for compensation payments, but also for other purposes, including advice to farmers in respect of groundwater resource protection.

11.3.4 CONCLUSIONS

There are no specific national policies concerning the issue of protection of drinking water resources. National policies concerning agricultural pollution in general have been developed in the Netherlands and England and Wales, while the situation in Germany is differentiated, as this concerns the authority of the individual States. Introduction of new and more stringent regulations of manure practices, evoked by the need to implement EU directives, are disrupting the traditional consensual policies, most significantly in the Netherlands. Pesticides policies have been characterized by indirect regulation of pesticides use through regulation of the market and by harmonization throughout the whole of Europe.

In general the choice of the policy instruments matches the policy network characteristics (cf. Bressers, 1993, see also section 2.4). To the extent that the traditional policy communities are still functioning, communicative instruments are predominant. The typical policy instruments mix involves information exchange to show farmers sustainable alternatives for the prevailing agricultural practices, and to convince them that these will have no negative effects on their incomes. These communicative instruments are often supported by enlarging economic ones, such as subsidies for investment or transition costs. The policy instruments applied usually contain no explicit moral appeals to farmers, although the British system of 'good agricultural practice' can be conceived as a way to institutionalize a moderate form of moral appeal.

In all countries the most powerful tool created at the national level seems to be the possibility of establishing some kind of ground water protection zones. In the Netherlands and Germany this option has existed for a relatively long time, the authority being delegated to the provinces and the States, where it has been applied on large scale. In both countries it involved mandatory additional restrictions on farmers. More recent measures are concerned in England and Wales, which have not been widely applied, originally on voluntary and only later on a mandatory base, while the executive power has been retained at the national level, with the Ministry of Agriculture.

Farmers in the groundwater protection zones are entitled to financial compensation to the extent that they are subject to more restrictive regulation than other farmers. In Germany the compensation payment schemes still involve large sums of money. In the Netherlands the differentiation in legal standards has almost been leveled out in 1995, as was the right to compensation. In both cases the money was extracted from the water supply companies, and finally from their customers. In England and Wales only the first cohort of pilot projects was supported by a subsidy scheme, additional legal restrictions being imposed later on, without compensation.

The national policies reviewed are contributing to the protection of drinking water resources in many ways, but the ground water protection zones are the only direct contributions to the control capacity of the water supply sector.

11.4 Actual control capacity of the water supply sector

The third analytical step - after national policy issues, networks, and styles, and after individual policy instruments - is the assessment of the actual control capacity of the water supply sector. The latter is the designation used for the water supply industry and the regional water authorities. Control capacity is the sum total of:

- the mission and orientation of the organizations of the water supply sector;
- the available organizational resources;
- the selected steering strategies.

Mission and orientation. The assessment of the control capacity of some organizations with respect to a certain issue should start with their attitudes towards this issue. Here, two kinds of organizations are involved: water supply companies and regional water authorities. For the design of the present project, it is assumed that actual control capacity requires a certain amount of co-operation between these organization, which is the justification for this specific meaning of the term 'water supply sector'.

To water supply companies the agricultural pollution of drinking water resources is an obstruction in the proper execution of their basic tasks. Their orientation towards this problem varies from a strict technical approach - the inclination to take appropriate measures after pollution has been detected - to a preventive orientation, which requires them to accomplish changes in farmers' behavior. Although the technical measures can be very costly, it is in particular the latter case on which critical questions are raised about the extent to which water supply companies are willing to take action and to invest organizational resources for this purpose, as pollution prevention is assumed to be the task of other agencies, notably the regional water authorities.

Regional water authorities are relevant to the present project to the extent that they have a (formal) task concerning the protection of drinking water resources. On the one hand, their mission and orientation is being determined by the institutional context within which they operate and the policies - established by higher levels of government - that they have to implement. On the other hand, they may have some discretion in exploring options to influence farmers' behavior and to act according to their own vision and initiative.

Organizational resources. As discussed in section 2.4, the exercise of control capacity - social steering - requires the application of organizational resources by the water supply sector. The main categories distinguished are:

- financial capacity (money or capital);
- organizational capacity (manpower);
- legitimacy and authority, including confidence gained from the target group; and
- information.

Steering strategies. The completion of control capacity is the selection of effective steering strategies. Here the steering strategies of the water supply sector are complementary to public policies. The basic categories distinguished in section 2.5 are economic, legal, and communicative strategies, all of them with restrictive and obligatory as well as enlarging and non-obligatory varieties (as summarized in table 2.1). In this case restrictive and obligatory approaches are likely to be pursued with legal strategies, more particularly legal regulation enacted at the national level. On the other hand, enlarging and non-obligatory approaches match better with economic or communicative strategies, which can also be enacted at regional or local levels.

11.4.1 MISSION AND ORIENTATION

Water supply industry. The water supply industries in the countries under study have different constitutions. In England and Wales the industry consists of ten large-scale, privatized water supply companies. In the Netherlands there are about forty companies, including a number of relatively large, dominant ones. In Germany, the industry is organized at the municipal level and consists of a large number of relatively small companies. In all cases, the water supply industry is faced with increasing levels of (nitrate) pollution. Short-term measures are required in any case, such as *ad hoc* measures (mixing of raw waters from different wells, or temporary closure of inlets) or technical ones (purification). In the long run water supply companies may be forced to close contaminated wells. Most water supply companies, together with the rest of the water supply sector, endorse the principle of pollution prevention, but they do not consider the protection of drinking water resources as part of their core mission. In general, the orientation of the water supply companies has changed over the years. Originally, a technical orientation predominated, nourished by an engineering culture. Those water supply companies that have been faced with problems of agricultural pollution have come to appreciate the value of prevention. Involvement in preventive actions is no natural activity for them, but all case studies show that they actually do so, but only under certain conditions and up to a certain level. The Dutch case study shows some examples of extremely active water supply companies. This is not a nation-wide phenomenon and it can be explained by the seriousness of nitrate contamination in particular regions, and by factors such as company size, organizational culture - in particular the attitude towards polluters - and the political climate in the province.

Regional water authorities. The regional water authorities are part of the water supply sector by definition: the (semi-) governmental bodies charged with quality management ground and surface water. The NRA in England and Wales is the semi-governmental agency charged with the management of surface waters. In Germany, the formal responsibility and legal authority with respect to water management lies with the States. Sometimes authority is delegated to the districts. The Dutch provinces are the legal authorities with respect to ground water management. They are also the main shareholders as well as the legally appointed supervisors of the water supply companies.

Protection of drinking water resources is part of the core mission of each of the regional water authorities. The organizational resources applied by them for this purpose are discussed below. However, the relative priority attached to this issue, compared to other water management tasks, is not dealt with extensively in this study. The Dutch case study contains some indications that the provinces are mainly oriented towards legal standards, and that they allocate a lower priority to measures aimed to encourage getting farmers to go 'beyond legal compliance'.

There is a certain skepticism, in Great Britain more than on the continent, about the legal standards concerning the quality of drinking water and drinking water resources, as these standards are often arbitrary and not always based on 'good science'. This is part of the context of the missions and orientations of all parties involved.

11.4.2 ORGANIZATIONAL RESOURCES

Financial capacity. Most water supply companies have considerable financial resources at their disposal. The case studies contain a number of remarkable examples of these companies, mostly in the Netherlands and Germany, which are prepared to apply these resources to prevention activities, by way of investments and with good prospects of a return on their investment. Usually these are small scale projects, while - this is mentioned notably in connection with the Netherlands - the water supply companies strongly prefer temporary projects for triggering changes in agricultural practices over permanent schemes involving the payment of compensation or subsidies for the relative economic disadvantages of more sustainable agricultural practices. Apart from considerations of cost effectiveness, such schemes are also in conflict with the 'polluter-pays-principle'.

The legally established compensation payments schemes are certainly large scale matters, but they lack any incentives to farmers to change their behavior. The Dutch case study shows that in some cases water supply companies have assumed the positions of the provinces (the regional water authorities) in the negotiations about these compensation payments. Their aim was to build in some incentives for behavioral change, which often required the application of additional financial means for additional arrangements, on top of the legal restriction imposed on farmers within the groundwater protection zones. During the previous years, in four out of the five provinces under study, considerable fractions of the financial resources and organizational capacity of the water supply companies have been employed for compensation payments. This involved not only the payments as such, but also the administrative and transaction costs involved in the negotiation and execution of the voluntary agreements between water supply companies and representatives of the farmers (1987-1989). In the three provinces where encouragement programs have been developed - originally in addition to the compensation payment schemes - the water supply companies consider such programs - which encompass training and advice, and also financial rewards for actual pollution reduction - as a more effective way to employ their means than the compensation payment schemes.

The situation in England and Wales, where no compensation payment schemes exist, is completely different. The tight regulatory regime governing the quality of drinking water supplies has required huge investments by the water supply companies, in order to meet European and British legal standards. Moreover, despite the political sensitivity of rising water prices - due to high profits and the high salaries paid to the executives of the privatized companies - the regulatory regime allows water supply companies to pass on these costs to consumers. Indeed, it allows them to turn a low value product into a high value product. Thus, the regime for economic regulation does not really encourage long term preventive measures by the water supply industry. The cost-pass-through implications of such measures are, therefore, not clear, and need to be negotiated with the regulator.

The NRA - and to a lesser extent some of the water supply companies - has invested considerable financial resources in R&D, aimed at the gathering of knowledge and information about agricultural pollution in general and the state of water resources. The knowledge and information acquired can be conceived as resources in their own right, useful for planning purposes, but it is not clear whether these resources are also applied as feedback in the relationships with farmers.

Organizational capacity. All national case studies show several examples of projects by the water supply sector in which considerable organizational as well as financial capacities are involved.

In England and Wales, the NRA is conducting an intensive farm visits program aimed at information transfer to individual farmers. The water supply companies and their national association are also involved in similar activities, but to a lesser extent. All these programs, however, are predominantly focused on pesticides use.

In Germany most of the effort of the water supply sector is invested in the establishment of ground water protection zones and the accompanying compensation payments. The costs are borne directly or indirectly by ground water users, notably the water supply companies. Some States have imposed water abstraction charges for this kind of purpose. Encouragement programs, aimed at training and advice - often promoting organic farming - are usually initiated by the States and are additional homes for the revenues raised by the water abstraction charges.

The rather extensive involvement of Dutch water supply companies in encouragement activities has been mentioned above. This cannot be explained only by the seriousness of the problem, as the size of the water supply companies seems to be a relevant factor too. In fact, these initiatives have been developed by a few large water supply companies, who can act on behalf of the water supply industry for the whole province. Size as such may be no organizational resource, but larger companies are able to build up and employ sufficient resources (financial, organizational, legitimacy, expertise) to actively deal with the issue. In comparison, most of the German water supply companies are smaller than the Dutch ones mentioned above, which may be a partial explanation for the comparatively smaller number of encouragement projects in Germany. On the other hand, the British water supply companies are much larger than the Dutch ones, but their limited involvement in prevention programs has been explained above.

Legitimacy and authority. Within the water supply sector, formal or legal authority (competencies) is a prerequisite of the regional water authorities. Legal measures, notably those aimed at the protection of drinking water resources, are usually based on a special legal status of certain areas, such as groundwater protection zones in the Netherlands and Germany, or the nitrate vulnerable areas and nitrate sensitive zones in England and Wales. The relevant competencies are the authority to establish these zones, to impose special restrictions to land use - in this case agricultural practices - and the authority to control and enforce these restrictions.

In Germany, the competence to establish groundwater protection zones lies with the States. Within the relevant legal confines, they also have the authority to impose further restrictions on agricultural practices.[1]

---

1 There is a remarkable distribution of formal and informal authority in this respect. The States have the formal competence to establish these zones, while the acknowledged expertise resides with some national organizations, such as LAWA and DVGW, in which the States and the water supply industry are represented. They play an important role in the determination of the norms and measures required for an adequate level of protection within the groundwater

In the Netherlands, groundwater protection zones are established by the provinces, by means of provincial ordinances from 1989 on. Additional restrictions on land use are announced through these ordinances, but the provinces have little discretion in this respect, as they are based on uniform national standards. The ordinances also call for negotiations with the affected farmers about the compensation payments, which are sometimes conducted by supply companies rather than the provinces. This situation came to an end in 1995, when the more stringent legal regulation of manure terminated the exceptional status of the groundwater protection zones in this respect, together with the rights to compensation payments.

The nitrate vulnerable areas and nitrate sensitive zones in England and Wales are established by the Ministry of Agriculture, as are the legal restrictions on farmers within these zones.

In all three cases the regional water authorities have the legal authority to control and to enforce restrictions on farming practices in places where drinking water resources are vulnerable. The option of establishing special zones has been applied on a large scale in the Netherlands and Germany; in England and Wales this happened only at a later stage, and on a smaller scale. However, in none of the cases do the regional water authorities extensively exercise their competencies in practice.

Water supply companies have no legal authority over farmers, and they do not want to receive it.[2] The Dutch case shows the best examples of effective co-operation within the water supply sector, with the provinces contributing legal authority and the water supply companies financial and organizational capacity, into combined efforts aimed at farmers in the groundwater protection zones. Many of these activities found their origin in the negotiations about the compensations payments.

In certain cases, however, water supply companies have acquired some legal authority, when they gained control over land use by buying parcels of land in endangered zones and leasing them (back) to farmers, notably in Germany and the Netherlands.

Successful stimulation programs - involving contacts with farmers - have contributed to the creation of informal legitimacy, notably acknowledgment of the agricultural expertise of the organizations involved and confidence in the eyes of the farming communities. These side effects may be beneficial in later stages. Especially in the Dutch situation, these forms of informal legitimacy, acquired through activities related to the compensation payments, have been helpful in maintaining the contacts with farmers and making them receptive to the new voluntary encouragement programs.

Information. Information as an organizational resource has been defined in terms of relevant facts, as well as applicable expertise. With respect to the former, the activities undertaken by the water supply sector, notably those involving actual contacts with individual farmers, have often yielded valuable information about farming practices in

---

protection zones. Their strategic positions within the national policy network also contribute to the control capacity of the water supply sector.

2 This is affirmed by the respondent of the European association of the water industry (EUREAU) in the EU case study.

specific areas and the actual condition of specific aquifers. Especially in England and Wales, the application of financial and organizational capacity of the water supply industry was aimed at the generation of relevant information about the condition of the natural resources.

11.4.3 STEERING STRATEGIES

To the extent that the water supply sector is engaged in preventive activities, in all of the three EU member states under study, they follow a consensual approach, involving direct contacts with the local and regional farming communities. Apart from legal requirements, the farmers' co-operation is achieved by persuasion (communicative strategies) and to a lesser extent by material incentives (economic strategies). Persuasion is to be achieved through information about the ecological effects of common farming methods and about less harmful alternatives. In any case the economic order remains unchallenged; farmers are not urged to suffer economic disadvantage for the sake of protection of drinking water resources.

However, the most important stimuli are, of course, the compensation payments. This is no strategic choice by the water supply sector, as the farmers are entitled to them under to the national legislation on groundwater protection zones (in Germany and the Netherlands). Inasmuch as the water supply sector is involved in these schemes, they follow the same consensual approach, trying to make the most out of it through negotiations with farmers.

Germany. The dominant strategy of the German water supply sector is the establishment of ground water protection zones. This can be conceived as a manifestation of a regulatory approach, although it involves consultations with municipalities and individual farmers.

In addition, a consensual approach is chosen in local initiatives. Numerous examples of stimulation programs can be found, aimed at the voluntary adjustment of farming practices, initiated by water supply companies, often in co-operation with the sustainable farming NGO, or agricultural associations. In certain places municipal water supply companies and farmers have set up successful organic farming programs, involving voluntary agreements and compensation payments. Important explanatory factors seem to be the agricultural sub-sector involved - switching to organic farming is easier in the case of crop growing than in the case of livestock farming - the tradition and attitude of the farmers, and the approach adopted by the water supply company.

England and Wales. The NRA is organized on a regional basis and therefore has good links with the water supply companies. Although a number of water supply companies are very active in dealing with the issue of agricultural pollution of drinking water resources, no clear network structures have yet emerged at the regional level. The case study shows that some water supply companies have taken initiatives to institutionalize relationships with farming communities, now that the problem of agricultural pollution is perceived as a persistent one, and regional and local 'policy networks' are emerging.

These networks are aimed at voluntary co-operation, sometimes backed up by subsidy schemes. Such networks are rather effective, although they are limited in scale. The water supply sector is certainly not operating on its own account. To a large extent it is following - and relying on - the national policy, which is consensual in nature, and structured around the codes of 'good agricultural practice' and the establishment of nitrate vulnerable zones. In addition, the sector is supported through several activities by third parties, such as the information and training programs on pesticides use by the chemical industry, and the preconditions on pesticides use imposed by large retailer chains - under consumer pressure - on the suppliers of horticultural products. Further, the water supply industry exercises pressure on regulators, such as government and the NRA, for more support for the specific cause of protecting drinking water resources.

The NRA has a true legal authority with respect to control and enforcement towards individual farmers, which is not exercised in a very active way. Nevertheless, the option to switch over to a more regulatory approach is retained as a last resort, should the consensual approach fail. However, it is not clear whether the NRA can do so on its own account, or whether it would need the consent of higher levels of government to do so.

The Netherlands. The five provinces of the Dutch case study demonstrate a remarkable distinction in the way farmers within the groundwater protection zones were approached. Four provinces chose the consensual approach of direct negotiations between the largest water supply companies and the farmers. This was the most efficient way to deal with large numbers of claims. In Noord Brabant, where a small number of farmers was involved, the official route of direct regulation was followed. Here individual farmers had to claim damages against the province.

In addition to successful negotiations about compensation payments, the large water supply companies in three provinces also developed stimulation programs - started in 1990-1993, and continued thereafter - based on direct contacts with farmers in the groundwater protection zones. A number of advantages compared to the compensation payment schemes can be mentioned:

- while the legal regulations covered only manure application within the groundwater protection zones, the stimulation programs are aimed at influencing farming practices beyond legal compliance on a voluntary base;
- incentives are schemes for subsidies as rewards for performances (concerning, for instance, manure, fertilizer and pesticides use, investment projects, and nature development);
- in addition, communicative instruments are employed: training and advice (for instance through group sessions with farmers from particular groundwater protection zones); and
- secondary targets are the establishment of confidence and good working relations with individual farmers and gaining access to valuable information about polluting activities (such as manure administrations).

Preventive action in the four provinces can be conceived as the concerted application of mutually reinforcing resources by the province and the water supply companies. The contributions from the latter enabled the provinces to develop and implement their policies of ground water protection zones. In fact, the provinces commissioned the negotiations about the voluntary agreements to the water supply companies. They were very eager to retain control of the compensation payment schemes since they were the ones to bear the costs. In later years, when water supply companies started to develop stimulation programs, some discontent grew with the water supply companies, since they feared that the provinces were becoming less active and were relying too much on their (the water supply companies') efforts.

On the regional level the initiative lay with the water supply companies. Of course, the national manure policy and its legal restrictions supported their targets, but there is a remarkable distinction between the national government - whose legitimacy *vis-à-vis* the agricultural sector is very much under pressure - while the water supply companies succeeded in building a strong base of confidence. The national government is now planning to support the water supply companies' stimulation programs in the near future.

Conclusion. In all of the countries under study, there is a growing distinction between the consensual approach of the water supply sector at the regional and local level and the national trends towards a more regulative policy style concerning agricultural pollution at the national level. In so far as the national policies are effective, this contributes to the protection of the drinking water resources, but at the same time the legitimacy of environmental policy to the farmers and their confidence in all agencies involved, including the water supply sector, is under pressure. This may have a negative effect on their control capacity.

11.4.4 CONCLUSIONS

The situations in the three EU member states under study show remarkable similarities. Although no alarming situations have been found, the overall picture is that the water supply industry has to comply with severe demands concerning their output (drinking water), while they have hardly any control over the quality of their input (raw water). Everywhere, the control capacity exercised by the water supply sector at the regional and local level is limited. This encompasses the water supply companies, as well as the water management authorities - the German States, the Dutch provinces, and the British NRA.

Germany. The main problem is that water supplier companies have to act within in a framework - i.e. the EU's Common Agricultural Policy - that does not allow them to tackle the actual causes of the problem. The most promising option for a lasting solution seems to be co-operation with the agricultural sector in support of organic farming. This option requires considerable financial capacity on the part of water suppliers, because they are urged to compensate for economic losses brought about by the legal restrictions on farming. On the other hand, it is the only measure which effectively reduces the pollution of our groundwater resources.

In the majority of the cases the environmental effectiveness appears to be limited. It is unclear whether reductions alone are sufficient to achieve the objective of clean groundwater, or whether the time lag between changes in agricultural practices and any measurable change in groundwater quality is too long for positive effects to be discerned.

The economic efficiency of the prevention projects is difficult to judge, and depends on the criteria employed. At present it is not possible to assess the total cost of the schemes in relation to their environmental effectiveness. Even if such an approach were possible, it could not take account of the fact that the measures are not designed to be economically efficient according to objective criteria, but are the outcome of a political bargaining process. Nevertheless, there has been intensive debate in Germany on the compatibility of the first model, Baden-Württemberg's 'water penny', with the polluter-pays-principle.

In general terms it can be said that the German water industry has become increasingly politicized over the last twenty years. The adoption of favorable positions toward protective water resource management is not a new phenomenon. Recently, however, such positions have fallen more and more in line the stance of the environmental movement, at least where water resource protection and the control of toxic substances are concerned. Increasingly therefore, the conditions are good for the formation of (issue) alliances between the water industry and segments of the environmental movement. Given the widespread concern within the environmental movement about food quality, and its rejection of many aspects of current agro-industrial practices, such alliances may well focus on finding alternatives to the use of chemicals. Co-operation between water suppliers and organic farming organizations is already taking place, as in Osnabrück, a region with severe groundwater pollution levels.

England and Wales. The British water supply sector feels no urgent need for reinforcement of its control capacity. The protection of drinking water resources depends mainly on developments at higher levels, such as the implementation of the EU Nitrate Directive, governmental efforts to control agricultural pollution, and new technological developments.

The picture is one of increased attention to the various problems; the problem is high and rising on the agendas of the water supply companies and the NRA. Thus there is a sense in which the issues are being institutionalized and, therefore, acquire an institutional dynamic, irrespective of any external pressure. Meanwhile, the codes of 'good agricultural practice' are pivotal in the present approach. The fundamental philosophy underpinning the resort to these codes is that policy instruments based upon joint problem solving and consensus building are more likely to be effective than a more impositional or adversarial style of formal 'regulation'. Perhaps the best indicator of the efficacy of this basic strategy is the National Audit Office's study of the NRA's activities in response to the problems of farm pollution. Whilst underlining the importance of regulation, the National Audit Office nevertheless recognized the importance of educating farmers in 'good agricultural practice', albeit sometimes backed up by regulation in the pure sense.

The Netherlands. In the Netherlands the conditions for practicing control capacity by the water supply sector are almost optimum: motivated partners who are willing to contribute the necessary resources in concerted action. Nevertheless, the effectiveness of all the efforts invested has been limited. On the average, farmers within the groundwater protection zones have complied with the additional regulations rather well, and positive effects have been observed with respect to the amounts of polluting substances emitted into the soil. Two major restrictions apply:

1. Dutch legal regulation is focused on the phosphate contents of animal manure - which is no major threat to the drinking water supply - and the achievements of effective changes in agricultural practices are relatively larger with respect to phosphates than nitrates.
2. Due to large amounts of historical pollution of aquifers, reductions of new emissions do not directly translate into less polluted groundwater at the wells.

As the extend of the legally based compensation payments has been diminished, in some provinces the water supply sector has developed stimulation programs aimed at additional restrictions beyond legal compliance. This must be accomplished on a voluntary base, possibly with subsidies and compensation payments, and even then one cannot take it for granted that the farmers will be willing to co-operate. The organizational resource of confidence, built up by some water supply companies in earlier negotiations, will be put to the test.

The farming community is very dissatisfied with the national manure policy. Nevertheless, some 70% of the farmers involved were content with the scheme of compensation payments, which was based on legal regulation. A smaller proportion (20% to 40%) of the farmers involved are willing to participate in the voluntary programs of the water supply companies (which does not imply that they will all adopt sustainable practices). Stated reasons for non-participation are: lack of confidence in public policy in general, costs in excess of revenues or subsidies, reluctance to submit corporate information to inspection.

References

Bressers, J.Th.A. (1993) Beleidsnetwerken en instrumentenkeuze. *Beleidswetenschap*, 7: 309-330.

Bressers, J.Th.A., L.J. O'Toole, Jr., and J.J. Richardson (eds.) (1994a) Networks for water policy. A comparative perspective. Special issue of *Environmental Politics*, 3 (4).

INTERMEZZO

12. POLICY FRAMEWORK - UNITED STATES

LAURENCE J. O'TOOLE, JR.
Department of Political Science, The University of Georgia, Athens, USA

12.1 Introduction[1]

This analysis addresses the case of the United States and the interaction of the water supply and agriculture sectors to encourage pollution prevention from agricultural chemicals of drinking water supplies. The term agricultural chemicals is employed here to refer to both fertilizers and also other chemicals - principally pesticides - used in agricultural production. The focus is on nonpoint source threats, and the investigation is designed to determine:

1. To what extent can water supply authorities in the US act as environmental regulators to prevent agricultural pollution of drinking water resources?
2. By which means can the US and state-level environmental policy makers strengthen the control capacity of water supply authorities in this regard?
3. What lessons can be learned for environmental policy makers who want to affect the behavior of target groups by means of indirect regulation and 'network management'?

However, in at least three respects the US case is different from others in this comparative inquiry. First, the actual threat to drinking water supplies in the US from agricultural chemicals is lower than that in the European cases under examination. The case does not constitute a 'most different' instance, since the overall level of economic development is roughly similar and, indeed, especially regarding the pesticide challenge, some problems have developed in the US that are roughly parallel to those in northern

[1] Support for this research was provided by the Center for Clean Technology and Environmental Policy, University of Twente, The Netherlands. Additional support was provided by the Institute of Community and Area Development (ICAD), University of Georgia (USA). This assistance is gratefully acknowledged. Helpful suggestions from William Blomquist of Indiana University, Gail Cowie of ICAD, Paul Sabatier of the University of California, Davis, as well as many of those interviewed for this research, have improved the analysis. Needless to say, the interpretations offered are those of the author alone.

Geerten J.I. Schrama (ed.), Drinking Water Supply and Agricultural Pollution, 307-339.

Europe. Second, in terms of both population and geographical scope, the US case is very large. Consequently, the work reported here is designed to cover the research questions at the national level and also at the subnational - primarily state - level. (Some of the field work discussed later focuses on more localized instances.) The scope of the overall case means that a broad overview, with illustrative but highly selective details, is the aim. Third, the intra-case variability is huge.

Three dimensions are of particular note:

Physical and demographic conditions. The United States includes climatological conditions ranging from deserts to high-precipitation zones, coasts to continental interiors, groundwater-reliant regions to surface water zones, and relatively agriculture-free locales to almost exclusively farming sections. Population density also ranges greatly. In large parts of the West, water supply *for* agriculture is by far the main water-agriculture issue. Indeed, much more water is used nationwide for agriculture (primarily via irrigation) than is consumed in drinking water and other domestic uses (Benjamin and Belluck, 1994: 9). Furthermore, in some regions drinking water-agriculture interactions are complicated further by the interests of those concerned with species habitat and ecological protection. The Bay-Delta controversy in California, analyzed briefly later, illustrates this triadic pattern.

Policies. There is a huge array of relevant policies. Two dimensions of variation, in particular, can be mentioned. First, the policies on agricultural chemicals and drinking water are spread among a number of sub-sectors: drinking water policy *per se*, plus so-called clean water policy, groundwater policy, pesticide policy, and agricultural policy. Behavior is influenced from these multiple and partially inconsistent policy directions. And second, within several of these sub-sectors, the range of applicable policies is vast. Some determinations are made at the national level. However, the states themselves are substantial - for certain issues primary - actors. As a result, almost no policy generalization applies to the full US The resulting range of policy instruments, implementation arrangements, resource commitments, and sectoral interactions defies summary, let alone careful analysis.

Even on the most basic dimensions of water policy, there are striking differences across the United States. The clearest instance has to do with the framework for water law. The eastern portion of the country operates under a very different set of legal doctrines regarding water rights than does the west. The east uses the riparian rights perspective derived from England: streams and lakes are typically considered private waters, with bordering landowners able to make reasonable use of them. In recent years, some states have placed further restrictions on these users. In the west, water law is based on the doctrine of prior appropriation, which gives preference to those who show early intent to put the water to beneficial use. This latter set of principles benefits large users and deters conservation. Although some modifications to these principles have been made in recent years, the basic framework continues to constrain problem solving regarding water supply issues. Groundwater rights are less individualistic, but here too states vary in their approaches.

Network interactions. The issues of scope and policy variation help to encourage another difference: a huge range in network interactions, both within sectors and between them. As explained in detail elsewhere, there is no American 'water policy network' but an imposing multitude of them (Heilman, Johnson, Morris, and O'Toole, 1994). The agriculture sector is more coherent but still varies somewhat by commodity. With regard to drinking water, the US operates with a mind-boggling array of mostly-small suppliers. While these units are linked to a small extent within states and, to an even lesser extent, through national associations, there is no single, coherent drinking water supply network, even among supply companies alone.

In this analysis, accordingly, the following design choices are implemented:

1. Developments at both national and state levels are included. The former consists of a broad overview of the policy and actor/network context. State developments are also reviewed in exceedingly abbreviated and selective fashion.
2. Four state/regional cases are included for special emphasis. These have been selected on the basis of the following criteria: representation of pesticide and fertilizer threats, strongly regulatory and less regulatory contexts, regional diversity, and an emphasis on innovation.

Data gathering focused on examining documents, analyzing relevant data bases, and (especially) conducting interviews with officials, members of the relevant policy networks, experts, and other interested parties in Washington, D.C. and states, and at local authorities, regional offices of federal agencies, and offices of other significant participants.

12.2 Problem of agricultural pollution of drinking water resources

Threats to water quality from agricultural chemicals have come to be treated seriously in the US, particularly for groundwater. Nevertheless, a difference in salience between the US and Western Europe must be noted. Although a number of significant policy initiatives have been undertaken, surface water quality is generally regarded as stabilized or improving in many regions during the last couple of decades. Groundwater, while a matter of growing concern, does not face the near-crisis atmosphere evident in some countries. The EPA has judged the US groundwater supply as good to excellent. Degradation is considered primarily a localized problem (EPA, 1992c; Benjamin and Belluck, 1994: 10). A recent review of water quality trends in the US notes that systematic information is notable by its scarcity (OTA, 1995: 19). The US Geological Survey monitors a relatively small number of sites around the country. Some states also do detailed monitoring. However, it is difficult to reach unambiguous conclusions. Some pollutants do seem to have been reduced in concentration. However, national government data display no clear trend for nitrates (Knopman and Smith, 1993). Of course, data on groundwater are necessarily uncertain. Thus the actual state of affairs and the trends are likely to remain unclear for the foreseeable future.

The most recent systematic information is provided in the *National Water Quality Inventory* (EPA, 1994), a report reviewing information supplied by the states. Agriculture was named by the states as the leading source of water quality impairment for US rivers and lakes. Not all agricultural threats derive from pesticides and fertilizers, but the states identified nutrients as the second-leading cause of impairment for rivers and lakes. (Nutrients include roughly equal contributions from wastewater treatment works and agriculture.) Pesticides rank as the fourth-leading cause of quality impairment for rivers. In fact:

> *... agriculture is the leading source of impairment in the Nation's rivers affecting 72% of the impaired river miles.... States attribute 56% of problems in lakes to agriculture.*[2]

For groundwater, agriculture was also one of the most frequently-mentioned sources of water quality contamination.

> *The most frequently cited pollutants in ground water include nitrates, identified as a ground water problem by 49 States, volatile organic chemicals (48 States), petroleum products (46 States), metals (45 States), and pesticides (43 States).*[3]

A recent review of the status of groundwater in the US paints a disturbing picture. The report, published by the Interstate Ground Water Management Network, first notes the context of uncertainty in which data must be considered; and then offers an assessment.

> *[I]n excess of 14 million people in the United States, in 1993, may have been impacted by ground water supplied drinking water which exceeds drinking water standards for at least one contaminant. Seventy-nine percent of the 'significant noncompliers' (defined by the EPA as public water supply systems 'with serious, frequent or persistent violations') in 1993 were in very small systems serving less than 500 people. Ninety-five percent of these very small systems are supplied by ground water.* (Jehn, 1995: ii)

On a nationwide scale, the problem of fertilizers and water is a less serious and salient issue than it is in parts of northern Europe. The national nitrate standard in drinking water is 10 milligrams per liter, measured as nitrogen. This figure translates into a slightly more stringent limit than the EU standard. Few areas of the US face serious compliance challenges. Approximately 1.2 percent of community water system wells and 2.4 percent of rural domestic wells exceed this standard (Benjamin and Belluck, 1994: 6). However, some evidence indicates that the issue is more important than these figures suggest.

> *Only recently have California and other States recognized that nitrate contamination causes the annual abandonment of significantly more drinking water supplies than does toxic chemical contamination.* (Jehn, 1995: 42)

In California in recent years, 14 percent of tested wells exceed the maximum contaminant level (MCL) for nitrate. Two points regarding nitrate threats and

---

2 *EPA, Fact Sheet for 1994*, p 2.

3 *EPA, Fact Sheet for 1994*, p 5.

groundwater, in particular, can be made. First, concentrations in groundwater continue to rise in a number of locations. Second, the problem is very unevenly distributed (for a tabulation of various locations, results, and researchers, see Jehn, 1995: 46-48).

The problem of pesticides in US drinking water is primarily an issue of concern for groundwater. Here, the problem is substantial and growing. Pesticide use in agriculture has increased in recent years, at least into the late 1980s (Reichelderfer and Hinkle, 1989: 155-56). As of 1988, the US Environmental Protection Agency (EPA) had found residues of 74 pesticides in groundwater of 38 states (EPA, 1988). Most derived from agricultural use. It was estimated several years ago that more than 50 million Americans are directly affected, or have the potential to be, through their drinking water consumption (Neilsen and Lee, 1987).

A variety of federal agencies now gather data regarding pesticides in groundwater. A number of states also conduct independent investigations. Data on pesticides in groundwater, in particular, suggest the potential for problems in many locales. The 1990 nationwide pesticide survey conducted by EPA showed pesticides or pesticide degradate in 10.4 percent of rural domestic wells.

> *[P]esticide contamination is present in urban water systems at twice the rate of rural systems, despite earlier beliefs that greater problems exist in outlying, underpopulated areas.* (EPA, 1990)

A number of state surveys have shown results exceeding the national figures regarding contamination.

The pesticide issue was not treated as a serious environmental question for many years. Until the 1970s, the issue was dominated almost completely by agricultural interests. As of the late 1980s, pesticides had emerged as a significant issue, and increased attention was being directed at it by a variety of interests, the national government, and many states.

12.3 Regulatory framework

The regulatory framework for the issue of agricultural chemicals and drinking water supplies must be considered in terms of national policy as well as state and local initiatives.

12.3.1 NATIONAL REGULATION

At the national level, regulatory policy consists of a set of independently enacted, and unco-ordinated, efforts. Regulatory instruments of one sort or another are in place for pesticides, clean water (including surface water), drinking water, and toxic substances control. A discussion of national regulatory policy also necessitates some treatment of budgets and institutional arrangements. Public and private spending for pollution control and prevention in the US is roughly $100 billion annually. One to three percent of this total is directed at drinking water, pesticides, and other toxic chemical programs (OTA,

1995: 18). Of the drinking water portion, only three percent is federal in origin; the great majority, 79 percent, is financed by local government. Seventeen percent of the spending is private (households, private providers) (OTA, 1995). These totals omit spending to implement the Clean Water Act (CWA), to be discussed shortly. Approximately one-third of the nation's total spending on pollution is aimed at CWA-related issues, basically surface water questions.

> *The vast majority of this expenditure is to clean up wastewater from identifiable municipal and industrial sources.... Another source of water pollution - nonpoint source pollution from agricultural and urban runoff - is ranked among the very top of remaining risks to ecosystems.... Relatively little has been spent on controlling agricultural non-point source pollution.* (OTA, 1995: 19)

In general, with regard to surface water, 'agriculture is thought to be the single largest source of remaining river and lake water quality problems.' (OTA, 1995: 20)

The Federal Insecticide, Fungicide, and Rodenticide Act, or FIFRA, is the primary national regulatory policy dealing explicitly with pesticides. It authorizes the EPA to establish criteria for registering pesticides and pesticide applicators for sale and use. Pesticides deemed not suitable cannot be sold or applied legally within the country. Another feature of the law authorizes state regulation. This form of policy instrument, a partial preemption, sets requirements and then allows substantial room for state initiatives beyond the national baseline. The discretion granted to the states regarding pesticides regulation has been used strongly in some jurisdictions, little in others. FIFRA has its weaknesses but marks a tightening of pesticide regulation beyond what was in place in earlier periods, when agricultural policy and administration was handled in 'iron triangles' of mutually supportive industry representatives, USDA, and Congressional committees on agriculture (Bosso, 1987). FIFRA creates some incentives for pesticide producers to emphasize the development of safe products. But it also authorizes a pesticide review process sufficiently time consuming and labor intensive so as to preclude effective consideration of the full set of pesticides on the market. Many have not been reviewed at the national level and continue to be used. And as of 1995, EPA has not denied registration or reregistration to any pesticide primarily because of threats to groundwater (Pease, et al., 1995: 5-6).

The primary regulatory policy in the United States addressed specifically to drinking water is contained in the Safe Drinking Water Act (SDWA). This statute specifies standards for publicly-supplied drinking water throughout the country, including MCLs for 87 contaminants, many of which are found in fertilizers, pesticides, or their derivatives. Thirty-five relate to pesticide ingredients or by-products. Only a small minority of pesticides in use now have MCLs developed under the SDWA (see Pease et al., 1995: 6). Drinking water supply organizations, usually local public utilities, are responsible for monitoring water quality and meeting standards. The EPA and state environmental or health agencies oversee compliance and assess sanctions. The SDWA was made significantly more stringent in 1986, such that local drinking water suppliers, especially the smaller organizations, were frequently ill-equipped in terms of expertise and instrumentation. From the perspective of these units, the SDWA is seen as a regulatory monitoring burden.

Some information about the standards is useful in understanding differences between European and American developments. The MCL for nitrates has been set at 10 milligrams N per liter, as mentioned earlier. Those for pesticides vary by chemical, but virtually all are considerably higher than the 0.1 microgram per liter per chemical/0.5 microgram per liter cumulative standard in the EU. For some chemicals, the MCLs are hundreds of times higher in the US. Thus the US nitrates maximum is close to the European standard, while the pesticides limitations are considerably more lenient. Since the US problem incidence for nitrates is not severe nationwide, the issue has not provoked general concern. On the other hand, pesticide levels have been documented at levels much higher than EU standards, but these concentrations often do not violate American standards. A result is - once again - no major outcry over the issue, albeit sporadic regional concern. It should be added that the nitrate limit in the US applies to any time during a year, while pesticides are not in violation unless the yearly average of four quarterly measurements exceeds the MCL. The effect of this rule is to ease further the pesticide concern in places like the midwestern Farm Belt, which experiences seasonally high concentrations of, for instance, herbicides.

Those critical of the SDWA point to a number of limitations of this regulatory approach:

> *Most pesticides in current use have no applicable standards. Monitoring requirements for pesticides are limited.... Moreover, systems with compliance problems can receive waivers from ... standards if the costs of providing a clean water supply are substantial. Many smaller systems need not comply with MCLs at all. Finally, the limited coverage of small water systems and complete exemptions for private domestic wells ... means that water consumers at the highest health risks are offered the least protection.* (Pease, et al., 1995: 6)

In 1994 the SDWA was scheduled for reauthorization, and some actors, including water suppliers, pressed Congress for an easing of the existing standards. Partisan conflicts regarding revisions delayed legislative changes until 1996, after the field work for this research was completed. Among the most significant changes was a shift from merely including ever more contaminants under regulatory control to an emphasis on the most serious problems. The new law also established a $9.6 billion state revolving loan drinking water program, primarily to support community water system improvements, but also with opportunities to use a fraction of these funds for preventive efforts.

A few provisions of the SDWA also aim to protect groundwater for drinking purposes. The law requires states to submit wellhead protection programs to EPA for approval. States are also required to specify how they will delineate areas for wellhead protection.

The Clean Water Act (CWA) is the nation's most far-reaching regulatory water policy. The emphasis is on surface water; groundwater connected to surface water systems is also subject to some protection. The CWA has been the most generously funded national environmental policy. Most of the expenditures have been aimed at subsidizing the development of wastewater infrastructure at other levels of government. The regulatory emphasis of the CWA was, until 1987, focused completely on point-source threats to surface waters. The policy instrument is a permitting system designating site-specific criteria for any point-source pollution entering a water body. Permits issued by the EPA - or, in 38 states, by the state agency authorized to administer the program - stipulate the criteria to be applied to the point source. The regulatory target, for instance a business firm, is held responsible for adhering to the standards and self-reporting any violations. The regulatory agency oversees monitoring and applies sanctions in cases of violations. The same scheme applies to publicly-owned point-source polluters, including municipalities that treat and discharge wastewater.

In the 1987 revisions to the CWA, some features were added that pertain more directly to nonpoint source threats, primarily to surface water. Under current legislation, states are required to establish strategies to co-ordinate and execute 'best practices' for reducing nonpoint pollution and set up efforts to control nonpoint sources. However, EPA is given no formal authority to disapprove state efforts deemed unsatisfactory. The law also authorizes some grants to be given to states to implement groundwater protection activities that aid in executing comprehensive nonpoint source control. The level of funding for nonpoint source efforts, however, is very small.

One other important possible subject of national regulatory policy is groundwater. Here the most important point is that there is no national regulatory program. Several efforts to enact one have been unsuccessful. Therefore, the issue of water supply threats from agricultural chemicals, as well as from other sources, has been dealt with primarily through other policies and/or by the states. EPA has sought to influence others, particularly the states, via official program 'guidance' that carries some of the style of regulation without actual authority to enforce. This kind of instrument might be referred to as quasi-regulatory. It is technically informational and exhortative but has been framed to give the appearance of national leadership. The most detailed national version of this effort is EPA's Comprehensive State Ground Water Protection Program, or CSGWPP, Guidance (EPA, 1992b). Until now, EPA could exhort state action but had to rely ultimately on state decisions. As a consequence, planning processes in many states have been lengthy, and funding for such key activities as prevention has generally been regarded as inadequate. EPA has begun to endorse CSGWPPs in states. Endorsement signifies that 'an agency has demonstrated at least a minimum level of commitment and is implementing a unified approach towards comprehensive groundwater management'.[4]

One other regulatory measure under development by EPA can be mentioned: the Pesticides & Groundwater State Management Plan Regulation, proposed in May 1995. Under the current draft, states develop state management plans (SMPs) to limit the use of designated pesticides deemed to threaten groundwater. The proposal identifies five pesticides for this form of control. Any state unable or unwilling to develop a plan meeting EPA approval would see the relevant pesticide(s) registration(s) canceled within the state. EPA is now negotiating with the states regarding certain details of the proposed instrument. Unlike much of the regulatory policy pertaining to agricultural chemicals at the national level, this proposal emphasizes prevention. The policy effort relies on states as the primary actors.

---

4 *Water Policy Report*, August 2, 1995, p. 13.

There is no tax imposed nationally on pesticide or fertilizer sale, for purposes of either helping to support the cost of the relevant regulatory programs or altering agricultural behavior by shifting incentives by a significant degree. Some states do impose fees or taxes on the sale of agricultural chemicals.

Other federal agencies also play parts in water policy in ways that can influence the drinking water supply. These include the USDA and the US Geological Survey. The latter is involved mainly in data gathering. The USDA subsidizes the production of commodities and administers efforts to require water quality protection via Best Management Practices (BMPs) under the subsidy programs it supports, as mandated by the 1990 Farm Bill.

Although not regulatory, US agricultural programs deserve brief summary here. Agricultural policy in the United States has consisted for decades of substantial subsidies and technical assistance. By the mid-1990s, approximately $10 billion was spent per year on agricultural subsidies. The main vehicle has been a series of commodity programs: subsidies organized by crop and paid to individual farmers or agribusinesses. Farmers are guaranteed a 'target price' for their commodities and are paid the difference if the market does not price the crops at the targeted level. In exchange, those growing the crop must agree to plant only allotted amounts. In some cases, this policy has paid farmers to keep fields idle or to destroy crops. Additional components include a program for dairy farming and a peanut support program. The acreage allocations and establishment of target prices have involved farmers and farm groups closely for decades. Many of the detailed decisions have been made at the county level, in conjunction with local USDA offices. In part as a result of budget-cutting pressure in Washington, the farm subsidy programs have been targeted for cutbacks.

Drastic cutbacks in agricultural subsidies may make efforts to influence agricultural behavior more difficult. Even under current conditions, the agricultural sector is notoriously resistant to regulation. There is little precedent for regulation over agricultural practices or land, virtually none from the national level (aside from controversial wetlands efforts). National efforts aimed at altering agricultural behavior must probably rely on subsidies, research, information, and/or technical assistance. Establishing institutions for common discussion of water-agricultural issues might also be a possibility.

12.3.2 SUBNATIONAL REGULATION

Except for the cases examined in more detail later, local or sub-state regional regulation is a topic omitted in the following coverage; the subject is simply too vast. Here the treatment is directed to state-level regulation, with one exception: local regulatory responsibilities for drinking water quality enforcement.

Most of the states play the lead role in enforcing the surface water provisions of the CWA, including the issuance of permits in the National Permit Discharge and Elimination System (NPDES). States also establish water rights for both surface and groundwater, particularly contentious issues in the arid West. The states play a lead role in pesticide regulation via FIFRA's provisions, which stipulate national requirements but also create a cumbersome process for reauthorization or withdrawal. Many states have established their own regulatory mechanisms. States also have the lead role in remediating problems in drinking water, when difficulties develop. And, as explained above, states are the principal regulators of groundwater quality and any planning efforts on its behalf.

The state regulatory effort regarding pesticides is one significant activity. In recent years, a number of states have tightened their regulation of pesticides and fertilizers. The points of regulation include pesticide application, sale, labeling, and disposal. Policy

> *... increasingly focuses on increases in pesticide and fertilizer registration fees, a portion of which may go toward monitoring for the presence of agricultural chemicals in groundwater and the development of educational materials and voluntary of mandatory best management practices to curb excessive chemical use. The submission of analytical data on the fate of pesticides in groundwater may be required during the registration process.* (NCSL, 1994: 11)

Some recent state enactments also focus on 'chemigation (the application of agricultural chemicals with irrigation water) and the creation of centers for sustainable agriculture.' (NCSL, 1994: 11)

State policy regarding groundwater protection has been an active sector in recent years. According to the National Conference of State Legislatures, all 50 states enacted legislation on this subject during the 1988-1992 period (NCSL, 1994: 1). These ranged greatly in type and likely impact.

> *If one factors out underground storage tank bills and looks at legislation where states have discretionary authority, then three categories become prominent - pesticides (23 states and 20.5 percent of the bills), comprehensive groundwater protection (13/8) and wellhead protection (13/8). The pesticide bills focused on increased fees for product registration ...* (NCSL, 1994: 1)

In place at the state level now, in at least some jurisdictions, are policies on such issues as groundwater standards, classification systems, financial support for groundwater protection, data gathering, educational programs, state planning regarding land use and other issues, and additional regulatory measures.

Groundwater quality standards illustrate the range of variation in regulatory policy across the states. Some states use narrative standards, some rely on quantitative measures, some employ both, and some states currently have no enforceable standards at all. Even numeric standards vary considerably. Certain standards may pertain to protection, others to remediation. The responses triggered by policy to violations of regulatory standards also range significantly (Benjamin and Belluck, 1994: 13-15; Environmental Law Institute, 1990; Morandi, 1992). Overall, state enforcement efforts vary widely, with clear strategies and comprehensive, well-supported monitoring and enforcement procedures in the minority.

Drinking water regulation is another sector of subnational influence. The enforcement of national SDWA requirements - including monitoring for compliance - is largely handled at state and local levels. The states generally oversee local drinking water systems' quality. The administrators of state drinking water programs are represented in Washington by the Association of State Drinking Water Administrators (ASDWA), formed in 1984.

Water supply agencies, typically local, have the primary responsibility for ensuring that supplies meet regulatory standards. Their efforts are reviewed by staff in state drinking water offices. The supply agencies in some parts of the country, particularly the West, are faced as well with water quantity challenges - or, to be more precise, being able to meet demand with water of sufficient quality. In many states in the West, this issue surpasses agricultural chemicals as a stimulus for water supplier-agriculture interaction.

12.4 Policy network: national overview

A huge number of suppliers, often referred to as water utilities, operate in the US At the broadest level, nearly 200,000 public supplies can be identified. Of these, more than 100,000 are transient non-community systems and more than 20,000 non-transient non-community systems. The arrangements serve more than 22 million people in places such as restaurants, hospitals, campgrounds, and other institutions. More than 90 percent of these systems are supplied from groundwater (Jehn, 1995: 35-37). Approximately 39 million people obtain their supplies from unregulated sources (like private wells or systems with less than 15 connections or 25 people), which are not counted in the total number of public water supply systems above. Very limited data are available on water quality for these unregulated sources. Again here, more than 90 percent of this total is reliant on groundwater. The remaining 58,000 public supply systems are so-called community water supplies, which serve roughly 240 million people. This core group of systems also contains many small ones: 88 percent serve less than 3300 people, 62 percent less than 500 people. The great majority of these smaller systems are also based on groundwater supplies. But most of the nation's people are in surface water systems, especially the larger ones. Surface water supplies approximately two-thirds of the population. Seventy-three million people are served by 193 larger surface water systems, and more than 105 million are served by the 1242 surface systems with a population base greater than 10,000 (AWWA, 1995: 3).

Surface water is exceedingly important for supply, and most of the nation's urban areas are particularly dependent on it. But groundwater is also significant, and it dominates as the source in a number of states. Groundwater is also the main source by far for the rural US Compliance problems in large systems (and surface water) are likely to affect more people, but problems tend to occur with much more frequency in the smaller (groundwater) systems. In federal fiscal year 1993, 26.5 million people were supplied from community water supplies in violation of an MCL for at least one contaminant. Of these, 62.3 percent were served by surface water systems, but 79 percent of the 'significant non-compliers' were very small systems with population bases of less than 500 (see Jehn, 1995: 36).

More than one half of the community systems are privately owned. However, the great majority of the national population receives water supply from publicly owned and operated systems. Once again, the pattern varies considerably by system size. In the small systems serving populations under 3300, two-thirds of the population is served by private suppliers. For those systems providing water to more than 10,000 people, however, approximately 87 percent of the population is served by government suppliers (see Haarmeyer, 1994). The modal arrangement in this category is a system owned and operated by a municipality. For public water utilities, the standard pattern has been for water to be supplied at subsidized rates. Private suppliers are subject to rate and service regulation by state public service commissions. Some public water agencies contract out the operations and management of their systems to private companies, thus creating another institutional variant. Most, nevertheless, remain publicly operated.

Private suppliers typically operate under a somewhat different grant of authority than do public utilities. In particular, the ability of private purveyors to use government authority to regulate land use in their watershed is limited to nonexistent. These, then, are especially dependent on either protection of the watershed (for instance, by purchasing land) or hoping the agricultural sector within their area does not create problems for them.

The SDWA legislation in 1986 enacted a number of provisions for monitoring, filtration, water quality, and regulatory compliance that have required many of the systems to increase substantially their capital and operating expenses. To a considerable extent, supplier efforts to deal with these obligations have been threefold: respond as required, pass increased costs on to consumers, and work through their national associations toward a reduction in regulatory requirements. Water suppliers are seen by some environmental groups as having a tendency to minimize quality concerns rather than as advocates for increased water purity.

The overall supplier array is a mass of discrete physical and organizational systems, organized largely as municipal bureaus and focussed primarily on their own needs and challenges regarding water sources, contaminants, technology, customers, and regulation. The atomistic setting of the supplier organizations is reflected, in turn, in the broader network structure in the water supply sector, as explained below.

Four other aspects of the supply sector's makeup should be mentioned. First, an important component of supply units' perspectives in the arid West is the dominance of water quantity issues. Finding enough water and constructing acceptable systems for water delivery have heavily outweighed concerns about quality. Indeed, developing infrastructure for supply and diverting water for domestic purposes from competing functions, like protecting natural habitats, has placed Western suppliers at odds with environmental advocates in a number of states. Second, some suppliers deal with the agricultural sector primarily as either a customer (water supply for irrigated agriculture) or a potential source of additional supply (via purchase of water rights). Again here, the issue is basically a Western matter. The intersectoral dependencies are not as straightforward in such cases as in other kinds of contexts. The result is that suppliers find their interests vis-à-vis agriculture linked in complex patterns.

Third, the threats to water quality stemming from agriculture include but are not limited to the issues of pesticides and fertilizers. For the most part, other agricultural issues are omitted from coverage here, but water supply-agricultural interactions are not necessarily limited to these chemicals. For surface water suppliers, in particular, sediment and pathogens may be the highest priorities. For this latter issue, *giardia* and recently *cryptosporidium* are often the most salient. And fourth, the mindset of those managing water supply organizations remains, by and large, engineering in orientation. Suppliers mostly see their role as identifying the purest sources possible, treating it with relevant processes, and transporting it to customers. Rare is the supplier involved directly in source protection by seeking behavioral alterations in sectors, like agriculture, that may pose threats to water quality. Suppliers see the job of controlling or influencing others' behavior as a matter for attention by regulators or agricultural agencies. With some exceptions, including some cases in this study, they are noticeable by their absence in interactions with the agricultural sector.

With few exceptions, these systems do not 'network' horizontally with counterparts elsewhere. They tend to operate independently, and most of their regular relationships are with the regulatory authorities at state or national levels. Their collective interests are represented by state associations (in many states) and organizations at the national level. Some, nonetheless, particularly in the West, have developed links with other suppliers, primarily for the purpose of reducing contingencies related to water quantity. Some large supply units in the West, like the Metropolitan Water District of Southern California (MWD), have also purchased water rights in recent years from agricultural users, including some located very far from the supplier's territory.

The regulatory units in the water supply sector are primarily state drinking water agencies in the state's environmental or public health bureaucracies. These too have associational representation in Washington (the ASDWA). In addition, there are other state-level units with influence over the quality of drinking water supplies, especially agencies focussing on pesticide registration and regulation. In Washington (and omitting the agriculture-oriented units, which are discussed below), the EPA is the obvious focus of involvement. Several offices within the agency have significant roles.

At the national level a congeries of organizations represent water suppliers and other associated organizations. These units are not well-integrated with either. They include the ASDWA, the American Water Works Association (AWWA), the Association of Metropolitan Water Agencies (AMWA), the Association of State and Interstate Water Pollution Control Administrators (ASWIPCA), the Association of State and Territorial Health Officials (ASTHO), the Ground Water Protection Council (GWPC), the National Association of Water Companies (NAWC), the National Rural Water Association (NRWA), and the National Water Resources Association (NWRA). The main supplier organizations are the AWWA, a large association of 55,000 members, including 3700 water supply utilities representing approximately three-quarters of the nation's water supplied; the AMWA, representing the interests of the 100 or so largest urban systems; and the NRWA, an association focusing on the systems at the other end of the size continuum. The agricultural sector of interest here consists of farmers, farm organizations - both commodity-specific and sectoral - industrial producers of agricultural chemicals (represented primarily by the National Agricultural Chemicals Association and the Fertilizer Foundation), and the government agencies involved in supporting agricultural activities. The lobbies for pesticides and fertilizer are perhaps the most vociferous opponents of tighter controls.

The number of farmers in the US has declined precipitously in recent years to approximately 2 million, although the political strength of the agricultural sector remains quite strong. Although the sector seeks to bolster popular support for itself by invoking the venerable and powerful image of the 'family farm' as the bedrock of the American culture, most agricultural production now is handled through large agribusiness. Producers of most crops are organized into commodity groups (wheat growers, soybean producers, hog growers, and so forth). Among those most active at the national level on key water policy legislative battles - SDWA reauthorization, Clean Water Act reauthorization, the Farm Bill - is the National Cattleman's Association. In addition, broad alliances of farmers continue to be important in the agricultural sector. Among the institutions significant at the national level on water-agriculture matters are the Association of Farm Cooperatives and the Farm Bureau. Such groups are well-organized and effective in lobbying against regulatory measures, suggesting sources other than agriculture as possible causes of water contamination, and seeking to demonstrate that the agricultural sector is economically hard pressed and should not be asked to shoulder more of the burden for water quality.

At the national level, the USDA is the major government agency providing support to the agricultural sector. The Department is organized into a number of subunits, some of which administer the commodity support programs via a broad network of field offices, others of which offer additional kinds of assistance. Among the bureaus within the USDA that play key roles, or potentially key roles, on water-agriculture issues are the Natural Resources Conservation Service (NRCS) and the Extension Service. The NRCS provides field assistance for conservation practices; and the Extension Service, with staff in every county in the country, offer technical assistance of all sorts to farmers. Both have had reputations suggesting great support for agricultural production goals. However, in the last few years these reputations have been changing. In the case of the NRCS in particular, newer staff and altered leadership has meant much more support for environmental goals, including water quality. In the Extension Service a newer generation of technical experts has begun to change the organizational culture here as well.

As a broad generalization, and despite the changing character of the field units referenced above, it can be said that the agricultural sector remains rather tightly integrated and normatively coherent. The 'network' in the agricultural sector is close to the notion of a policy community as sketched in the scholarly literature. This picture is quite different from that found in the water sector, as explained above. Further, the sector remains politically strong and largely resistant to direct regulation by other sectors.

An important point to make about the US water-agriculture linkages is that the extent of interdependence varies greatly across the nation. US water suppliers in a number of locales have been able to insulate their watersheds from agricultural and other threats.

This is possible in part because of lower population densities and less intensive agricultural activity in many cases. Also, policy in some states grants water suppliers land use regulatory control, which either allows for restricted uses or provides incentive for co-operative efforts (see the New York City case, below). And in certain regions, particularly in the Pacific Northwest, state law and watershed management by local authorities has greatly reduced agricultural activity within vulnerable areas. As a result, the interlinkages and dependencies have not developed as fully as in some other contexts. There are many areas, however, in which the link between water supply and agriculture is real and important. These include both surface and groundwater settings and are the focus of most of the case analysis below.

The contrast between the water supply and agriculture sectors is striking. The former is dispersed and atomized to the point that the national associations for supply are mostly inactive on the issue of agricultural chemicals. While the AWWA has recently taken some steps to become more involved in source water protection - including a June 1995 conference on the subject convened at the Association's headquarters, a national teleconference on the subject later that summer, and current work by the their Research Foundation to develop computer software with information on the subject - for the most part the water sector at the national level has been passive on this issue. Usually, when water organizations have taken this issue seriously, they have done so as individual supply units rather than as part of a larger water network.

Other peripheral participants include a couple of environmental organizations that tend to devote attention to drinking water quality and contamination from agricultural chemicals. The main actors here are the Natural Resources Defense Council, which is involved in policy advocacy, and the Environmental Working Group, an organization that conducts advocacy research on agricultural chemicals in drinking water. The latter efforts have resulted in studies with alarmist titles ('Tap Water Blues,' 'Herbicides by the Glass') that have attracted some national attention. These groups press for more stringent regulation, criticize the performance of water suppliers, and challenge farmers and chemical manufacturers to reduce use and develop more benign alternatives.

In most parts of the country, and certainly for most of the water suppliers, there are virtually no contacts between suppliers and the agricultural sector. Even where the issue is recognized as important, parties other than suppliers have been considerably more active. State agencies have developed source water protection management proposals, and many states have now developed wellhead protection strategies accepted by EPA. These typically involve approaches aimed at identifying vulnerable areas and significant threats, and finding ways to reduce risks. Land use controls are based primarily at the local level in the US, and in some regions these have been used to control agricultural practices. For the most part, however, agricultural behavior is not directly regulated by any level of government. Instead, efforts at the level of the individual farm have emphasized education, training, technical assistance, and support for improving farm practices. The main implementation arms for these kinds of programs have been agencies of the US Department of Agriculture, especially the Extension Service and the NRCS, and counterpart units affiliated with state governments' agricultural departments.

Water supplier agencies have been directly involved in some concerted efforts to improve source protection. An EPA review of ten innovative wellhead protection programs around the country indicated that certain suppliers are involved in certain efforts, including nonregulatory approaches (see EPA, 1992a). And an ambitious survey of hundreds of the larger surface water utilities showed that some of these had developed extensive efforts at watershed protection, including links with the agricultural community (see AWWA Research Foundation, n.d., but approximately 1990).

The issues of scale, complexity, and variety in the US experience compel a closer examination of several cases. The following four instances illustrate some of the more innovative efforts visible currently. In their successes and limitations, they provide an empirical basis for the analysis of control capacity in the following chapter.

12.5 Illustrative cases

Each of the following cases provides a more detailed examination of the water supply-agricultural interaction regarding agricultural-chemical threats. The cases are identified by state, and some coverage of the specific state's policy and institutional context is included. Certain examples concentrate more narrowly on locales within the states.

The cases are New York (with a focus on New York City area), California (with attention to special circumstances like the Bay Delta), Iowa (where threats from agricultural chemicals have prompted a nonregulatory state response), and Wisconsin (a state much like Iowa but with an aggressive regulatory policy).

It must be emphasized that these cases cannot be considered a representative sample of American states or locales with regard to water policy or threats from agricultural chemicals. Indeed, the cases are highly unrepresentative. The vast majority of drinking water supply organizations have stayed on the sidelines in the efforts to effect behavioral change in the agricultural sector. In part this passivity is a product of their small scale and relative inexperience. In part, it derives from the agricultural sector's normative and cognitive coherence, its political strength, and its resistance to intrusion or attempts at behavioral change initiated from outside. In part, as well, it derives from the water supply units' perspective on their own role, a viewpoint that remains for the most part engineering in approach, and thus far unaffected with sufficient turbulence to force a change.

The modal water supply sector-agricultural interaction pattern continues to be minimal or nonexistent. There are some signs of change on this score, however. The possibilities and limitations of a more active supply sector and an enhanced control capacity can be seen more clearly in light of some of the following cases, which document some of the dynamics currently visible.

12.5.1 NEW YORK

New York is the third most populous state in the US and home to the nation's largest city. The state has traditionally also been known as an innovative setting for policy. New York's environmental policies are generally considered among the most stringent in the nation, and regulation has been an important tool.

Despite the state's urban character, agriculture is an important industry. State residents rely on both surface and groundwater supplies. The majority live in areas served by surface suppliers. Six million receive their drinking water from groundwater suppliers. Surface water pollutants include industrial sources and diverse nonpoint sources, including agriculture. Groundwater is threatened primarily by organic solvents and petroleum products, although nitrates (including from fertilizers) and pesticides also play a role (Benjamin and Belluck, 1994: 446). The most pressing groundwater pesticide challenge is on Long Island, where the use of aldicarb in potato farming continues to contaminate groundwater supplies. Long Island was one of the sites where groundwater contamination from pesticides was first clearly documented in the US.

The primary agency with environmental protection responsibilities in New York is the Department of Environmental Conservation (DEC), which administers the pollution control regulations of the state for surface water and also handles groundwater policy. The New York State Department of Health (DOH) is also involved in groundwater policy, sets state drinking water standards, and enforces both state and national standards in this field.

Water supply in New York is handled predominantly by local public authorities. Agricultural chemicals pose difficulties for surface and groundwater in several regions of the state, but no statewide policy encourages the water supply and agricultural sectors to co-ordinate. Nevertheless a statewide initiative designed to draw from the New York City co-ordination experience (see below) is currently being considered; its acronym is REAP (Responsible Environmental Agricultural Planning). Some state (and federal) programs offered via agricultural agencies aim to educate farmers and encourage best management practices to reduce nonpoint source threats, but there has been no large-scale statewide effort at behavioral change. As in most of the country, the drinking water supply agencies are only sporadically and poorly networked.

However, one important part of the New York picture offers an interesting effort to link water supply and agricultural interests in a major program aimed at co-operative, nonregulatory solutions to water quality problems from agricultural chemicals. This instance is the program developed for the City of New York, in particular through its Bureau of Water Supply, to induce changes in agricultural practices with the goal of protecting the City's water quality. This program is currently attracting attention from several parts of the country.

The City's water supply pattern is one of the largest surface systems in the world. Parts of the watershed extend far from the City (approximately 200 kilometers). This one system serves the water needs of the eight million City residents, one million residents in upstate counties, and millions of commuters and tourists in the City (see Beckhardt, 1995: 1). Approximately 1.4 billion gallons of water are transported to the

City daily. New York City's watershed area began extending beyond the immediate urban environs long ago. In 1906, state law granted the City the right to oversee and regulate upstate watersheds (see Coombe, 1994: 2), a regulatory instrument that relates directly to the current nonregulatory effort. As the City expanded, it used its legal power of eminent domain to displace farms and small communities in the watershed region. The energetic exercise of the eminent domain power, combined with the City's reneging on promises to provide additional help and to maintain the new infrastructure need to support these development efforts, created much bitterness and distrust on the part of displaced residents and farmers upstate.

The federal SDWA amendments of 1986 generated a new set of drinking water regulations nationwide that were issued in 1989. These imposed a requirement that most local surface water systems begin to filter their water. For New York, accepting this requirement would mean a huge capital outlay and operating expenses - approximately $5-8 billion and $200-500 million per year, respectively. The Surface Water Treatment rule did include the possibility of a local system's avoiding the filtration requirement. The rule set specific standards for a variety of contaminants (Coombe, 1994: 2).

The City's response was to seek to avoid filtration by developing a program based primarily on regulation and land acquisition. The proposed changes included very stringent land use regulations that would have imposed significant costs on many within the watershed area, particularly farmers. (Most of the agriculture in the region is dairy and cattle farming on relatively small properties.) The upstate counties were angered by the proposals, and their politicians denounced the City's efforts, claiming that the regulations would put agriculture out of business in the watershed. Some agricultural leaders turned to the state agency most concerned with their interests, the Department of Agriculture and Markets. That unit's deputy commissioner for environmental affairs proposed a forum to create dialogue between the city and the farmers.

A series of meetings began with anger and distrust on the part of the rural participants but gradually resulted in progress toward a co-operative approach and also generated in a formal networking of the agricultural interests involved. Part of the agenda of the meetings was educational, 'to show farmers the links between their fields and herds and the drinking water for 9 million people.'[5] Agricultural experts from Cornell University, the state's land-grant university, assisted in these aspects of the process. Some agricultural interests argued that low density land use, including the agricultural status quo, actually served the City's interests by limiting development.

An option that held the prospects of avoiding filtration requirements was for the City to develop and implement a comprehensive watershed management program. Some form of action was needed, and the regulatory 'gun' at the City's head forced it to deal - somehow - with the agricultural issue. The farmers too had large incentives to find a solution other than the City's original regulatory approach. An Ad Hoc Task Force was convened by State Department of Agriculture and Markets and represented both the City's DEP and the agricultural community upstate. The objective of the Task Force was

---

5 *New York Times*, August 13, 1995, p 9.

to recommend regulations and/or programs that could protect both the water supply and the agricultural sector. More specifically:

> *The Task Force had three goals: to improve mutual understanding of the laws and public policies that shape the City's watershed program, to review and advance the scientific basis of farm planning conceived to protect water quality, and most importantly, the farmers['] business interests, and to explore ways in which the City may work in partnership with farms and* the network of agricultural support institutions to encourage a sustainable farm economy in the watersheds, yet achieve the City's water quality objectives. (Coombe, 1994: 3; emphasis added)

The Task Force eventually proposed the withdrawal of the proposed regulations regarding agriculture and recommended, instead, 'a voluntary, locally developed and administered program of best management practices.' This effort proceeded with substantial funding and the regulatory default option as a likely alternative, should the nonregulatory approach not work effectively. In late 1991 the City applied for filtration avoidance, and in early 1993 the EPA granted provisional acceptance for a period of a few years, provided the City would meet scores of conditions established by the Agency. These covered many issues. One set had to do with significant changes in agricultural practices, to be worked out co-operatively during the refinement and implementation of the plan. The Determination is effective until a further Determination is issued or until December 15, 1996. EPA ultimately has to pass judgment on the City's overall watershed protection effort, one component of which is the agricultural effort.

The lead actor and only water supply agency on the 'clean water' side of these developments is the City's DEP. From the agricultural perspective, several actors were partially integrated prior to this effort. These units include the State Department of Agriculture and Markets, the state's Soil and Water Conservation Committee (an umbrella group for agriculture), the county districts for the Extension Service, and the Farm Bureau. Extensive co-ordination and further networking developed as a result of this threat to sector autonomy. One resulting entity emerged from this organizing effort in 1993, the New York City Watershed Agricultural Council, Inc. (WAC). The purpose under-girding the creation of this institution was 'to provide a forum for farm industry input and leadership in the Watershed Agricultural Program,' the portion of the overall plan devoted to threats from the agricultural sector (Beckhardt, 1995: 1). This entity consists of a number of appointed agricultural sector representatives (farmers), along with *ex officio* membership from federal, state, and local government agencies. The only *ex officio* representation with a vote in Council decisions is the City's water bureau. The Council is chaired by a farmer-politician from the watershed region, and the representatives of the agricultural interests in the watershed dominate the Council in numbers.

Aside from the initial educational efforts developed at the time of the Task Force's work, the co-operative effort between water and agricultural sectors has encompassed a set of activities of substantial scale. Most significantly, the Watershed Agricultural Program commits the WAC to achieving explicit participation in the Program by farmers within the watershed area at a level of at least 85 percent of total farmers by 1997. Should this target not be achieved, the City's DEP can then revert to regulatory instruments. (The agreement is not explicit, however, regarding specific goals or limits of agricultural chemicals, a point of concern among some environmental groups.)

'In addition, the Watershed Agricultural Council ... has assumed administrative and operational responsibility for the Watershed Agricultural Program' (Beckhardt, 1995: 2). In effect, direct control is in the hands of the agricultural community, with regular communication with the other parties involved, particularly with the DEP.

The approach taken in the Program relies upon substantial subsidies for farm improvements aimed at reducing pollution by agricultural chemicals to surface waters in the region. The subsidies are aimed precisely at the kinds of improvements likely to be most effective in the circumstances faced on a given farm, and the level of technical assistance and support from experts on these issues is considerable. Current WAC policy has established $75,000 as the upper limit of assistance per farm, with the City's budget covering 100 percent of the costs.

The Program has been organized into two phases. Phase I focused on County Project Teams in the watershed counties developing individually-tailored plans for ten farms, in conjunction with the Technical Support Group (experts from the City DEP, the US Natural Resources Conservation Service, the state Department of Environmental Conservation, the state Department of Agriculture and Markets, and especially a wide range of specialists from Cornell University). The result in each case was a Whole Farm Plan, which covers specific farming practices that can be developed or modified to cover threats to the water supply from animal pathogens, sediment, as well as nutrients, pesticides, and 'other substances with direct or indirect human health effects' (Watershed Agricultural Program, 1994: 4).

Phase II is designed to extend Whole Farm Planning to the entire agricultural sector within the watershed. Individual farmers agreeing to participate in the program obtain detailed attention from the technical experts to the circumstances on their land, a list of specific changes that could minimize threats to the water supply, and complete financial coverage - up to the WAC-established maximum - along with continuing technical assistance in executing them. The City's DEP has committed $35.2 million, to be administered by the WAC, to carry out the effort during 1995-99. Much of the funding is devoted to executing the farm plans. The WAC has decided to reserve ten percent of its budget to advocacy research on issues of concern regarding agricultural threats to water supplies.

There are approximately 400 farms in the watershed. As of July 1995, 21 Whole Farm Plans had been developed and approved by the WAC; 13 have been implemented. Many more are in progress, and the WAC is currently ahead of schedule for getting farmers committed to the effort. Almost 200 farms had signed on to the initiative as of summer 1995.

Three features of the effort deserve additional comment. First, recruitment of the farmers into the program is being handled exclusively by leaders in the farm community. During the last year, these individuals have held a series of 'kitchen meetings' in homes of participants. At these sessions, refreshments are served while participating farmers extol the virtues of the program. Water supply officials are nowhere to be seen.

Organizers report a greater than 90 percent participation rate among those farmers attending the meetings.

Second, the WAC has institutionalized its efforts in a variety of ways and has begun to publicize its activities and agenda in the state and nationally. The Council now occupies a building in the watershed and employs a couple of staff, including an executive director. One goal, of course, is to reduce pollution problems. Another, more salient to the members, is to assist the farms in surviving and becoming more profitable. (Whole Farm Plans are scrutinized not only for their environmental soundness but also for their business impacts on the farm. Only plans deemed by the Council to be at least neutral economically toward the farmer, if not positive, are approved.) A third goal is to make the WAC and its approach sufficiently indispensable to a variety of other actors that funding to the Council will continue past this five-year period.

Third, the farmers on the WAC have recognized that the ultimate success of the Plans depends on implementation and continuing use of approved agricultural practices. Given the nature of the nonpoint source challenge, monitoring is difficult. However, the WAC has devoted attention to the question and developed plans to involve committees of farmers in regular revisits to participating farms, revisions and refinements of Plans, and at least informal sanctioning of farmers who ignore their obligation toward diligent execution.

Overall, then, both the water supplier and the agricultural sector, each faced with the threat of onerous regulation, are engaged in intensive efforts to use co-operative ties to deal with drinking water quality as well as the needs of the farmers. The approach taken emphasizes supplier-subsidized alteration of agricultural practices - with most control, and all detailed interaction and monitoring at the level of the farm, handled by trusted representatives of the agricultural sector and related institutions of technical experts. The farmers are seeking to use the threat to their autonomy to enhance their own ability to extract resources from the water sector and improve their business prospects.

12.5.2 CALIFORNIA

California, the nation's most populous state, draws equally from surface and groundwater sources. Supply for arid regions has been a highly salient issue ever since significant population growth began in the 1800s. Further, more than 90 percent of the water is used for irrigation. Public supply for domestic uses constitutes a tiny fraction of the total. To make matters more complex, approximately 70 percent of the precipitation in California falls in the northern third of the state, while the population is concentrated south of that - especially in the Los Angeles-San Diego region, hundreds of miles away. This disjunction has meant water battles between the state's important agricultural industry and large urban population over issues of supply (with agriculture in possession of legal rights to large quantities of water, thanks to prior appropriation doctrine), and the construction of expensive and controversial aqueducts, dams, pipelines, reservoirs, and other components of water storage and conveyance systems covering large territories. These issues continue to be fought today, and they are rendered more complicated in political terms by the emergence of strong environmental and conservation advocates, who argue against both the economic interests of agriculture and the drinking water supply claims of the state's urban centers. The fact that some of the nation's most explosive population and economic growth developed in California has exacerbated the challenge, as has the creation of key agricultural production in the California Central Valley, a region unable to support its tremendous crop output without massive irrigation. Water battles in California are complex and divisive in the best of times; they are made even more difficult during drought periods, which occur periodically.

Water is now imported via the Colorado River Aqueduct to help meet needs, especially for the Metropolitan Water District (MWD) of Southern California, a wholesaler. Within the state, north-south redistribution has engulfed the state in long-standing controversy and litigation. The most difficult issues center around the crucial San Francisco Bay Delta, through which water must travel on its way to Southern California. This region constitutes reclaimed land and has been used in part for farming; it is also an important wildlife and recreational site. Development and water redistribution have reduced the Delta's water quality, encouraged saltwater intrusion, and threatened some species of fish. A proposal for a Peripheral Canal that could be built to deal with some of these impacts while also assisting in water redistribution served as a focal point for water disputes in the state for more than 20 years.

In the midst of these large struggles, water quality has gradually emerged as an important issue, with quantity and quality issues interacting with increasing frequency. Along with other quality concerns, threats from fertilizers and pesticides have become significant issues, at least on a sporadic basis.

For surface waters in the state, there are currently no serious threats to drinking water quality stemming from fertilizers and pesticides. (One significant contamination a few years ago, a threat involving rice herbicides in the Sacramento River, was resolved successfully with a combination of regulatory pressure, technical assistance, and the partial participation of the water supply unit in the city of Sacramento.) Surface water systems receiving agricultural runoff do encounter some difficulties with trihalomethanes. Groundwater, which supplies roughly 45 percent of California consumers, faces a more complex situation. Nitrates remain a significant concern. In the 1970s and 1980s many wells were found to exceed the MCL, and the state forced suppliers to either treat the water or close wells for use. Monitoring shows continuing contamination, and many wells currently being used also have to be monitored, since their nitrate concentrations are significant, albeit below standard. However, education of the agricultural sector by the state's Department of Food and Agriculture seems to have succeeded in reducing nitrate use in sensitive regions. Currently, the Drinking Water and Environmental Management Division of the California Department of Health Services assesses the nitrate problem in groundwater as important, although surpassed by coliform levels in terms of urgency.

Pesticides in groundwater have also been an issue of concern, although primarily in the past. Use of the pesticide DBCP produced widespread contamination of groundwater in the 1970s. The state's drinking water office banned its use and ordered the continued monitoring of this and several other pesticides that had been detected. Fortunately, increased contamination has generally not materialized. The state required a number of groundwater systems, primarily in California's Central Valley, to treat their supplies to deal with the DBCP. Several of these, in turn, filed lawsuits against the pesticide manufacturer. In most cases these actions have resulted in settlements in the tens of millions of dollars.

As a consequence of these perceptions of problem severity, there are currently few nonregulatory initiatives now underway to deal with the threat from agricultural chemical contamination of water supplies in the state. Selected wellhead protection programs in certain local areas are the exception to this generalization, as is an effort underway by the State Water Resources Control Board (see below).

California has been known as a relatively strict regulatory setting by US standards. The state established pesticide limits in advance of and in many ways more stringent than federal requirements for many years. In recent years, the legislature has been less active in tightening regulations. Still, state standards for a number of pesticides are more stringent than federal. There are no state regulations regarding the use of nitrates and fertilizers. Both agriculture and drinking water suppliers themselves are accused of threatening environmentally fragile water systems, such as the Bay Delta. A consequence in political terms is a set of interests and coalitions somewhat different from those typically found in the East. Here suppliers are often considered to be tied to development, urbanization, infrastructure construction, and the destruction of important ecological habitats.

Some state policies have had an impact on the water quality-agriculture relationship over time. In 1984, for instance, a requirement placed on groundwater systems, particularly those with 200 connections or more (California's definition of large systems) stipulated that these monitor for potential pesticide contaminants. From the data the state drinking water office developed site-specific monitoring requirements for a number of pesticides. In addition, legislation enacted in the mid-1980s authorizes state regulators to undertake a series of actions if pesticides are detected in regions of the state. These actions range from deregistering the pesticide to setting up Best Management Practices (BMPs) where the pesticide has yet to be detected to banning the use of the pesticide within a certain distance from wellheads. This policy seems to have stemmed some of the worst potential problems. The state also imposes a fee on pesticide sales.

Institutional arrangements for water in California are extraordinarily complex. In addition to the 5000 community and municipal water suppliers (not to mention the 4000 to 4500 non-community water systems), a few hundred of which are for-profit water companies, the state is home to many mutual water companies, plus 22 kinds of public agencies involved in water supply. (Mutual water companies overlap in some cases with certain varieties of public agencies.) Many of the mutual companies, not-for-profit entities, supply groundwater. The suppliers range greatly in scale, from tiny suppliers of a few users to the MWD, a special district supplying water needs to approximately 20 million Californians.

Water suppliers in California are organized into a number of associations, based in part on special geographical or user focus. Among these are the California Agricultural Water Coalition, the California Urban Water Agency (CUWA), the Association of Water Districts of Southern California, and the Northern California Water Association. (This last group represents 51 water districts that are entirely agricultural suppliers.) The largest and most well known of the industry groups is the Association of California Water Agencies, or ACWA (note the pun), which represents both agricultural and municipal suppliers. With approximately 420 members, it is the oldest and largest of the associations acting on behalf of the state's water industry; ACWA members deliver 90 percent of all water used in the state.

Some of the California supplier associations and even some of the large individual suppliers are active nationally as well as within the state in efforts to influence policy. For instance, ACWA, CUWA, and the MWD all lobby on national policy. ACWA and MWD maintain full-time, staffed Washington offices to exert influence on federal policy (regulations on drinking water, Clean Water Act reauthorization, and such). Other individual suppliers contract for assistance in Washington.

Helping to bring some policy and regulatory order to this sprawling industry are a congeries of policies and state agencies. Central is the tendentious issue of supply. Water scarcity and water rights have driven much that has happened in California, in terms of economic development, settlement patterns, and politics. Early, huge public works projects, many supported by the federal government, constituted massive subsidies for the state's agricultural community. Federal water projects have now ended, but their legacy remains. And state projects remain on the agenda.

The California Department of Water Resources has responsibility for the State Water Project, the massive effort to develop a system to deliver water to Southern California. This effort has been highly controversial and remains incomplete after more than 20 years of turbulent decision making. The state Water Resources Control Board is responsible for issuing water rights and placing conditions on water use, some of which relate to water quality. The State Board determines water rights and general water quality policy; regional boards handle NPDES (point source) discharge permitting and aim to protect their watersheds from nonpoint source problems. The Board has been trying to develop BMPs for agriculture in the relevant locations. Currently a number of approved practices are being publicized by the State Board, after development through a system of technical committees comprised of major stakeholders on the relevant issues. In general, the State Board's policy is to encourage adoption of voluntary BMPs; to mandate them where necessary; and to impose waste discharge limits in the most pressing cases.

The California Department of Pesticide Regulation is the unit focussed on detection of pesticides in groundwater and determining allowable application limits of pesticides in certain regions. The agency is charged with informing individual farmers of their limitations. No one agency regulates fertilizers in the state. However, the Department of Food and Agriculture licenses fertilizers and administers a Fertilizer Research and Education Program, which is funded through a tax voluntarily paid by the fertilizer industry. Yet another agency deals with drinking water standards: the Division of

Drinking Water and Environmental Management in the Department of Health Services. This unit regulates for compliance with the SDWA requirements and California standards.

Some of the special features of water in California create the potential for water supplier-agriculture interaction on issues of quality. The presence of agriculture as the largest water customer, for instance, means that many suppliers have regular ties with the farming community. These could be used to exert leverage to control agricultural chemicals. One constraining factor here, nonetheless, is the fact that the larger suppliers are mostly segregated as to urban or agricultural focus. (An exception is in the urbanizing Central Valley, where the strong agricultural base is being diversified by urbanization.) Therefore, individual suppliers tend not to be a forum for intersectoral exchange. The same can be said for umbrella organizations. Even for groups like ACWA, there is little discussion of such issues. Where diversification in customer base does exist, the ties have mostly gone unused.

Why? Several reasons can be advanced. The quality of the state's water has been good until fairly recently. Also, the issue of quantity and water rights has long dominated water policy making. Urban suppliers have not been eager to press the agricultural sector for changed practices, since their foremost problem has been to acquire supply *per se*. Exerting pressure for improved quantity runs the danger of losing out to others in the competition for supply. The typical pattern has been to secure supply and then treat to standards - thus internalizing the quality issue. A third reason is the fact that the state has seen several consecutive terms of Republican governors who have been supportive of the state's agriculture industry and have not dealt aggressively with controlling the industry's practices. Finally, as in other states, managers of supply organizations have tended to consider treatment as the preferred option. Among the larger suppliers in California, this pattern is slowly but steadily changing.

Still, despite the relative quiescence between the sectors on issues of fertilizers and pesticides, some developments suggest that drinking water-agricultural ties may be in the early stages of development. Two can be discussed briefly: water suppliers' activity in encouraging national legislation to enhance voluntary efforts; and possibilities in the Bay-Delta agreement.

California water suppliers press for national policy to encourage watershed protection. Some California suppliers have become active in efforts to shift national policy. Protection from nonpoint source problems seems increasingly significant to some of the state's largest suppliers, and they have sought support from national legislation to subsidize their own source protection efforts. Such important actors as MWD, ACWA, and CUWA lobbied for provisions to be included in the US SDWA to use a portion (ten percent) of the newly-created revolving loan program for water treatment to support 'voluntary, incentive-based partnerships' between suppliers and those in their watershed whose actions pose potential nonpoint source problems. The financial resources could be used to subsidize BMPs or to provide incentives to change behavior. In essence, such a program offers a national source of support for some of the kinds of activities being undertaken now in the New York City watershed. This proposal continues under discussion.

The Bay-Delta Accord and the institutionalization of a new forum for intersectoral negotiation. It would be difficult to exaggerate the complexity or importance of the long-running dispute centered on the San Francisco Bay-Delta estuary at the confluence of the Sacramento, San Joaquin and a number of smaller rivers. The Delta is the largest wetland habitat in the Western US, provides approximately 60 percent of the fresh water used in the state, and is the source of irrigation water for 45 percent of the nation's fruits and vegetables. The Delta is an ecologically-sensitive setting which is home to a number of endangered species. Agriculture and urban water users have drawn fresh water from the Delta at rates that have increased salinity and threatened wildlife and fish. A virtual stalemate among the interests involved persisted over two decades and was broken, at least in principle, only in December 1994. One result may be the development of an institutional setting for urban water and agricultural interests to work co-operatively on some important issues.

In 1982, after years of discussion and recommendations, the state's voters sent a proposal for the so-called Peripheral Canal (which would have passed fresh water around part of the Delta on its way to users to the South) down to defeat. A variety of alternatives were subsequently considered. The US EPA pressed California to tighten water quality standards in the Delta, especially because of threats to fish. No state-endorsed proposal commanded general support, however. The federal government threatened to impose its own standards. EPA's requirements would have significantly reduced the quantities of fresh water that urban and agricultural users would be able to draw from the Delta. These interests, working through a coalition of CUWA and major agricultural users, suggested modifications in EPA's plan that were aimed at preserving the quality of the Delta habitat while also allowing increased flow. The prospect of EPA control, coupled perhaps with Congressional elections suggesting a changed regulatory climate at the national level, triggered an agreement among federal and state agencies, key water users (urban and agricultural), and important groups like the Environmental Defense Fund. The Accord, signed with much fanfare in December 1994, creates the beginnings of potential long-term agreement. The Accord establishes a set of understandings among most of the stakeholders for a three-year period while the parties work out long-term solutions.

The agreement commits the federal government to be more flexible in its enforcement of the Endangered Species Act and provide water from its own projects at specified levels during drought periods. Levels of fresh water use from the Delta are specified by season and inflow conditions. Users, especially urban users, are granted much more certainty of supply. A number of standards to protect the habitat are put into place. And three 'Categories' of action are specified, to be executed by a joint federal-state structure, CALFED, which also involves important stakeholders. Category III is of special interest for present purposes. This set of activities is focused on non-flow-related factors. The interest in such matters was stimulated by a coalition of water suppliers called CUWA-Ag (the urban water association and representatives from some of the largest agricultural suppliers) who suggested that water exports from the Delta flows might be increased if certain non-flow activities were initiated to increase the quality of Delta conditions. Category III projects are to deal with several issues, including agricultural chemicals. The projects are to be selected by the stakeholders. This Category is slated for $60 million of spending per year. For the initial three years, MWD has committed $10 million per annum. CALFED, water suppliers, and environmentalists are to develop the implementation plan.

This Category, part of the most significant water accord in the state for many years, has the potential to begin to integrate water supply and agricultural practices in ways that have not been seen in California. The plan includes financial goals, nonpoint source/agricultural targets, and most of the state's significant water and agricultural stakeholders or their networked representatives. The CALFED institution, plus the highly visible forum for continuing discussion (and the potential of federal imposition of regulatory standards) suggests that the Accord may help to initiate important cross-sectoral negotiations.

12.5.3 IOWA

In Iowa, in the upper midwest, agriculture is the leading industry and the most important employer. The midwest 'Corn Belt' states are key sites of grain production, hog farming, and other forms of agriculture. The state has considerable water resources, both surface and ground, although these resources are unevenly distributed. The northeast half of the state is groundwater-plentiful, with high quality aquifers. Streams are also abundant here, and groundwater recharge is substantial. The southwest half has very poor and highly mineralized groundwater, and surface water sources are also not substantial. Most of the economic and agricultural strength of the state is in the former portion. Evidence in recent years has shown some deterioration of water resources, and agricultural chemicals - both nitrates and pesticides - have been main sources of contamination. Some 80 percent of Iowa's population depends on groundwater for drinking, so groundwater contamination is a particular concern; but problems from agricultural chemicals have affected surface water systems as well.

Groundwater used in community systems in Iowa is relatively deep, so most nitrate and pesticide contamination is not yet demonstrable. However, important recharge areas in the state are not well monitored for contamination from surface supplies. Once contamination is identified there, substantial, long-term problems can be expected. In private farm wells and other individual groundwater sources, the state has found some contamination, including in recharge areas. To date little action has been taken to deal directly with these findings.

Several years ago, considerable effort was made at the state level to develop a policy that might recognize the importance of groundwater protection from many threats, particularly fertilizers and pesticides. The result was the 1987 Iowa Groundwater Protection Act, a fully nonregulatory approach. Given the importance of agriculture to the state's economy and politics, and the antipathy to regulation in the state, this strategy was seen as the most feasible option. Within the US, Iowa provides what may be the best example of an almost purely nonregulatory, but not passive, method. Indeed, the Iowa approach has received substantial attention and been analyzed in some depth elsewhere (see John, 1994).

The Iowa policy stipulates that groundwater is essential and that contamination will be remedied. There are no specific regulatory standards for groundwater per se, and the state does not rely on a groundwater classification system. The Department of Agriculture handles the usual pesticide regulatory tasks, and the main prevention efforts regarding groundwater are under the jurisdiction of the Department of Natural Resources. These are basically aimed at research, education, and demonstration programs; and these efforts in connection with agriculture are funded by a fee placed on fertilizer and pesticide sales. A potentially important component of the Iowa approach has been the establishment of the Aldo Leopold Center for Sustainable Agriculture at the Iowa State University. The Leopold Center, funded on the basis of these fees, is dedicated to supporting research and outreach to advance sustainability. A major transmission route for the information developed through the Center is the state's Extension Service. A variety of interdisciplinary research efforts and education programs have been sponsored.

The main agricultural-chemical targets of concern at the time of passage of the 1987 Act were atrazine and nitrates. Atrazine and fertilizer use have declined dramatically, recently in the case of fertilizer. These developments are due principally to market forces rather than efforts to encourage more sustainable practices. Despite the substantial programs undertaken through the Leopold Center and otherwise, it is impossible to say whether there have been any measurable water quality improvements as a consequence. Despite several years of effort and consensus concerning the wisdom of research and education, no one knows the impact.

One additional point can be made about the Iowa experience: water suppliers have not been involved in the state-established policy or implementation of its programs. While Iowa's approach has received high marks from some analysts for its involvement of a wide array of stakeholders, water suppliers have not been among these. The case illustrates, therefore, the general point made earlier regarding the relative inactivity of US suppliers.

One additional Iowa-related element, nonetheless, should be presented. The state's largest water supplier, the Des Moines Water Works, has made substantial efforts on its own (without incentive from state or national government) to deal with issues regarding agricultural chemicals. The manager of the agency is responsible for the organization's progressive efforts on a number of issues, including agricultural ones. Des Moines is a surface water system supplied by the Raccoon and Des Moines Rivers. Significant challenges from agricultural chemicals have been experienced for several years. The major pesticide threat has been atrazine, and the immediate mode of response has been to use a technical solution (activated carbon). A more serious difficulty has been nitrates. During peak seasons, the MCL has been exceeded - in 1995, by 50 percent. The Water Works has installed an expensive ion exchange process to deal with the short-term issue of meeting the standard. On both pesticides and nitrates, however, the organization aims to emphasize watershed management.

The strategy adopted in Des Moines has been a conscious effort to develop links with agriculture, demonstrate the salience of the issue and who must take action to solve it, show reasonableness and empathy, and develop co-operative efforts. The initiative has included such actions as giving speeches (including to agricultural groups), sharing data, publicizing research, maintaining links with state and national legislatures to remind them of the problem, negotiating on state and national legislative proposals, and building coalitions bridging water and agriculture. On this last-mentioned front, the Raccoon River Project is the most significant effort. The Water Works has helped to organize several of the state's principal commodity groups to sponsor and manage demonstration projects on numerous potential control strategies. The Water Works was instrumental in developing grant support from a private foundation to assist the efforts for a three-year period. The principal objective has been to encourage some of the main agricultural actors to consider themselves full participants in the issues. Ultimately, the demonstration projects themselves could also serve to alter behavior over a wide set of agricultural practices as well. The effort began well but has slowed. One reason appears to be that changes in the composition of Congress have signalled to agricultural stakeholders that they may face less regulatory pressure at the national level. The case illustrates once again the links between regulatory policy and the success of nonregulatory initiatives.

The experience of the Des Moines supplier suggests the possibilities and limitations in the Iowa experience. Strategic approaches emphasizing nonregulatory and collaborative efforts are possible, but they rely on unusual and highly committed management, as well as some baseline of regulation.

12.5.4 WISCONSIN

Wisconsin, another state in the upper midwest, is also a heavily agricultural region and is predominantly groundwater dependent. Among the major agricultural products are corn, soybeans, alfalfa, and potatoes. However, unlike for Iowa, Wisconsin's approach is heavily regulatory and involves considerable enforcement by the state's agriculture department.

In 1980, aldicarb contamination was found in groundwater in the state's Central Sands potato region. This finding, along with concerns about other threats to groundwater quality, stimulated passage of a major regulatory effort. Aggressive action by Wisconsin, and reliance on strong regulation, are consistent with the state's progressive reputation. Groundwater sources for public supplies in Wisconsin are generally located very deep below the surface. Although some community suppliers have initiated protection efforts (see below), and despite low-level contamination experienced in some public supplies (particularly with regard to atrazine), the main state regulatory initiative is directed at private wells. These, numbering 750,000, supply a sizable minority of Wisconsin's population with drinking water.

A core of the Wisconsin system is a data base of monitoring results for 16,000 private wells. These data are updated and shared between the main state agencies involved in groundwater quality: the Department of Agriculture, Trade and Consumer Protection (DATCP), the Department of Natural Resources (DNR), and the Department of Health and Social Services (DHSS). The diverse tasks of these agencies on groundwater quality issues are guided to some degree by an advisory Groundwater Coordinating Council. A much smaller set of monitoring wells placed immediately down grade from agricultural fields, by agreement with individual farmers, are examined periodically. (The agreements grant permission to monitor, in return for precluding regulatory action against the farmer.) These efforts help to provide the data Wisconsin needs to determine the early development of groundwater pesticide problems and undertake preventive actions. The Groundwater Unit of the DATCP conducts this monitoring and has the most active role in regulating groundwater quality. Costs of the Wisconsin monitoring program are primarily supported by a fee structure based on the registration of agricultural chemicals in the state.

Currently the largest pesticide problems stem from atrazine contamination, particularly in the southern part of the state, a corn-growing region. Alachlor is also a threat in some locales. Regulatory restrictions regarding atrazine are now in place in many regions of the state. Ironically, aldicarb has returned to use in the state on soybean fields. Contamination from this chemical was a principal stimulant for the state's Groundwater Law, but after a period of prohibition regulations now permit the use of this chemical.

Monitoring supplies the data used to support the Wisconsin groundwater program, but the core of the effort is a set of standard setting and enforcement processes. DHSS recommends health-based pesticide standards to DNR. These quantitative concentrations are provided at two levels for each regulated chemical. The Enforcement Standard, or ES, is set in accord with public-health and public-welfare considerations, as outlined in law. Strict regulations including use prohibitions can be enforced if these levels are reached. A lower concentration, the Preventive Action Limit (PAL), is a fraction (between ten and 50 percent, depending on the contaminant) of the ES. The PAL must be used in development of management practices like pesticide use rules, and the lower level is also used as a trigger for certain remedial steps. Actions triggered by the PAL must be 'technically and economically feasible'; this limitation does not apply for the ES.

No predetermined set of actions is taken is undertaken if PALs are reached for a given chemical. Instead, the DATCP assesses the situation and determines what steps should be taken to maintain concentrations at the PAL or lowest feasible level. Even prohibitions are allowed, provided certain statutory requirements are met. The general idea is that the PAL is to generate preventive action to preclude groundwater concentrations reaching ES levels. Currently, an Atrazine Rule adopted by the state prohibits use of the pesticide in many specific parts of the state (depending on monitoring data) and limits application amounts elsewhere.

Enforcement of the Wisconsin policy is aided by the fact that the DATCP, which establishes the rules regarding use restrictions, is known and respected in the agricultural community. DATCP has had the rapport, contacts, and good will sufficient to induce compliance. The agriculture department has even had some success in developing assistance from agricultural chemical companies. Once restrictions are enacted, information on regulatory measures is transmitted to agricultural users through the university extension districts, conservation districts, agricultural chemical suppliers, and the NRCS.

ES and PAL levels have also been set for fertilizers, including nitrates. However, DATCP has not been granted clear enforcement authority to allow the agency to restrict or prohibit use, so the two-tiered system is relatively ineffective for these chemicals. An effort to modify state policy on this issue was narrowly defeated recently, in part because of agricultural opposition. Another source of resistance is the fact that nitrates derive from multiple kinds of sources, and background levels also complicate the picture.

One additional component of the Wisconsin approach can be discussed briefly. A recent state policy enactment now stipulates that any new municipal well approval requires that a wellhead protection program first be enacted. Since the requirement is mandatory for well authorization, the component is essentially regulatory. However, in the shadow of this regulatory instrument, the locality has considerable discretion as to how the protection program will be developed and structured.

The Wisconsin Rural Water Association has been very active in assisting in the development of municipal protection programs, and DNR's Bureau of Water Supply has worked closely with the Association in supporting local efforts. The main problem in practice has been that narrow political-jurisdictional boundaries mean that municipalities depend on co-operation in land use and property rights decisions on the part of local units outside their own territory. These planning efforts can be difficult, and some localities experience difficulty in protecting the full wellhead area. In several cases, however, municipalities have worked with surrounding jurisdictions to reach agreement. In the most obvious successes, a principal feature of the local situations has been the presence of contamination. This experience supports the generalization that regulation in combination with high problem salience are important contributors to co-operative efforts.

12.6 Conclusions

A combination of lower problem severity, less stringent pesticide standards, substantial surface water supplies, and diffuse supplier organization has meant that efforts by the water supply sector to induce co-operation on the part of agriculture have been relatively limited in the US. There are exceptions, however, including the cases outlined above. And evidence suggests the likelihood of greater cross-sectoral interaction in the future. The issues of control capacity are, accordingly, examined in the following chapter.

References

American Water Works Association [AWWA] (1995) Comments on the Notice of Initiation of Special Review for Atrazine, Syanizine, and Simazine. Published in the November 23, 1994 *Federal Register* (Vol. 59, No. 225, pg. 60412), unpublished document submitted to EPA.

AWWA Research Foundation. (n.d., but approximately 1990) *Effective Watershed Management for Surface Water Supplies*. Prepared by Richard W. Robbins, Joseph L. Glicker, Douglas M. Bloem, and Bruce M. Niss. Denver Colorado: AWWA.

Beckhardt, Laurence (1995) *New York City Watershed Agricultural Program Overview: Water Quality Protection through Public-Private Partnership between New York City and the Watershed Agricultural Council.* New York: New York City Watershed Agricultural Council, Inc. Unpublished paper.

Benjamin, Sally, and David Belluck (1994) *State Groundwater Regulation: Guide to Laws, Standards, and Risk Assessment.* Washington, D.C.: Bureau of National Affairs, Inc.

Bosso, Christopher J. (1987) *Pesticides and Politics: The Life Cycle of a Public Issue*. Pittsburgh: University of Pittsburgh Press.

Coombe, Richard I. (1994) *Watershed Protection: A Better Way.* Unpublished paper.

Environmental Law Institute (1990) *Survey and Analysis of State Groundwater Programs - Policies, Authorities, and Management.* Washington, D.C.: Environmental Law Institute.

Environmental Protection Agency, U.S. [EPA], Office of Pesticide Programs (1988) *Pesticides in Ground Water Data Base: 1988 Interim Report.* Washington, D.C.: U.S. Environmental Protection Agency.

Environmental Protection Agency, U.S. [EPA], Office of Water, Office of Pesticides and Toxic Substances (1990) *National Pesticide Survey*. Washington, D.C.: U.S. Environmental Protection Agency.

Environmental Protection Agency, U.S. [EPA], Office of Water (1992a) *Case Studies in Wellhead Protection: Ten Examples of Innovative Wellhead Protection Programs*. Washington, D.C.: U.S. Environmental Protection Agency. (EPA 813-R-92-002).

Environmental Protection Agency, U.S. [EPA], Office of the Administrator (1992b) *Final Comprehensive State Ground Water Protection Program Guidance*. Washington, D.C.: U.S. Environmental Protection Agency. (EPA 100-R-93-001).

Environmental Protection Agency, U.S. [EPA] (1992c) *Water Quality Inventory: 1009 Report to Congress*. Washington, D.C.: U.S. Environmental Protection Agency. (EPA 503/9-92/006).

Environmental Protection Agency, U.S. [EPA], Office of Water(1994) *National Water Quality Inventory: 1992 Report to Congress*. Washington, D.C.: U.S. Environmental Protection Agency. (EPA 841-R-94-001).

Haarmeyer, David (1994) Privatizing Infrastructure: Options for Municipal Systems. *Journal of the American Water Works Association* 86, 3 (March): 43-55.

Heilman, John G., Gerald W. Johnson, John C. Morris, and Laurence J. O'Toole, Jr (1994) Water Policy Networks in the United States. *Environmental Politics* 3, 4 (Winter): 80-109.

Jehn, Paul (1995) *The National Ground Water Status Report*. Oklahoma City (Oklahoma): The Interstate Ground Water Management Network.

John, DeWitt (1994) *Civic Environmentalism: Alternatives to Regulation in States and Communities*. Washington, D.C.: CQ Press.

Knopman, Debra S., and Richard A. Smith (1993) 20 Years of the Clean Water Act. *Environment* 35, 1: 16-20, 34.

Morandi, Larry (1992) *An Outside Perspective on Iowa's 1987 Groundwater Protection Act*. Denver, Colo.: National Conference of State Legislatures.

National Conference of State Legislatures [NCSL] (1994) *Groundwater Protection Legislation: Survey of State Action, 1988-1992*. By Larry Morandi. Denver, Colo.: National Conference of State Legislatures.

Nielsen, Elizabeth G., and Linda K. Lee (1987) *The Magnitude and Costs of Groundwater Contamination from Agricultural Chemicals*. Washington, D.C.: U.S. Department of Agriculture, Economic Research Service (Agricultural Economics Report No. 576).

Office of Technology Assessment, U.S. Congress [OTA] (1995) *Environmental Policy Tools*. Washington, D.C.: U.S. Congress, Office of Technology Assessment.

Pease, William S., David Albright, Claire DeRoos, Laura Gottsman, Amy D. Kyle, Rachel Morello-Forsch, and James C. Robinson (1995) *Pesticide Contamination of Groundwater in California*. California Policy Seminar 7, 8 (May). Berkeley: University of California and State Government.

Reichelderfer, Katherine, and Maureen Kuwano Hinkle (1989) The Evolution of Pesticide Policy: Environmental Interests and Agriculture. In Carol S. Kramer, ed., *The Political Economy of U.S. Agriculture: Challenges for the 1990s*, pp: 147-73.Washington, D.C.: National Center for Food and Agricultural Policy, Resources for the Future, Inc.

Watershed Agricultural Program [New York] (1994) *First Progress Report of the New York City Watershed Agricultural Program*. New York City Watershed Agricultural Council, Inc.

13. CONTROL CAPACITY - UNITED STATES

LAURENCE J. O'TOOLE, JR.
Department of Political Science, The University of Georgia, Athens, USA

13.1 Introduction

The foregoing coverage treats the policy context for water-agriculture questions, provides a general characterization of the institutional setting, and explores some cases in more detail. The cases are chosen to highlight promising, successful, and/or innovative efforts. The preceding chapter provides implicit evidence regarding control capacity of the water supply sector in the US In the following pages, the focus on control capacity is made explicit.

In the next section, the US water supply sector is assessed for the degree of control capacity available. This analysis is organized via the types of resources with potential to provide capacity: natural resources, organization, financial resources, information, confidence, authority, and time. The section following then examines the prospects for increasing the control capacity in the hands of the water supply sector. The last section offers some general conclusions.

One preliminary comment can be offered. Any effort to document and analyze, let alone enhance, control capacity is bound to be somewhat contentious. How to measure control is not a subject of general agreement; and assessing the *capacity* for control is considerably more problematic. Here the indicators of control capacity are: the successfully demonstrated use of one or more resources by actors in the water supply sector to influence behavior in the agricultural sector in ways protective of water resources (as seen by the principal actors involved in the two sectors), as well as the likely ability of the water supply sector to stimulate such change without infusion of substantial additional resources (as judged by the principal actors). In some instances, the principal actors are not in full agreement, and the evidence on control capacity is therefore more murky. Greater amounts of the desired behavioral change in agriculture, visible or possible, are treated as requiring greater amounts of control capacity.

Geerten J.I. Schrama (ed.), Drinking Water Supply and Agricultural Pollution, 341-364.

13.2 Actual control capacity and its application

As is obvious from the preceding chapter, the full water supply sector in the United States is a huge conglomeration of actors, including hundreds of thousands of organizations. For purposes of the discussion here, the water supply sector consists of the drinking water suppliers themselves, the principal associations of suppliers at state and national levels, and the public health and environmental regulatory agencies at relevant levels of government. These actors range greatly in scale, sophistication, and perspective. Still, generalizations can be offered regarding the capacity of the sector and its principal sets of actors, by virtue of the resources available vis-à-vis the agricultural sector and its network. Certain exceptions, possibilities, or tendencies highlighted by the selected cases mentioned earlier are included as appropriate.

In the United States, the water supply sector can be seen as an exceedingly loose, amorphous, marginally coherent set of actors. Many of these have been largely unintegrated into any larger network, except for necessary regulatory oversight. The overwhelming majority of actors in the sector have been inactive in seeking influence on the agricultural sector. A segment of the water supply industry in certain regions of the country, particularly in the West, focuses primarily on agricultural rather than urban supply and thus can be viewed as belonging to the other sector (or both). Some larger, financially-advantaged suppliers, and suppliers led by more active management are exceptions to this generalization. Even government regulators have not been energetic activists in dealing directly with the agricultural sector. Indeed, many state regulators, along with some of the associations of suppliers, have viewed one of the principal challenges as reducing the nationally-established regulatory responsibilities imposed on the water sector, rather than engaging in direct efforts with the farming community.

The term 'community' is used advisedly. As in other countries, the US agriculture sector has long been politically strong, well organized, normatively integrated, and tightly linked. With high levels of mutual commitment and interrelatedness, the sector comprises a web of networks which, taken together, are close to the standard meaning of policy community in the network literature. American agriculture is under challenge from economic forces and political pressure to trim subsidies. Any assessment of the relative influence and network strength of the two sectors, nonetheless, would have to conclude that agriculture has been dominant. The water supply sector has not made many efforts to alter agricultural behavior, and a number of the efforts that have been made have been successfully resisted by farming interests. The agricultural sector retains strong political support, powerful patrons, and high levels of integration. In addition, almost all the policy instruments directed thus far at agricultural practices have been based on incentives, information, and technical assistance. US agriculture is strongly resistant to regulation and has seldom had to deal seriously with regulatory requirements.

13.2.1 NATURAL RESOURCES

To consider natural resources as assets potentially under the control of water suppliers, or others for that matter, distorts somewhat the ordinary meanings of resources and control capacity. Here, the natural resources are considered in two related senses: the 'objective' quality of the water resources, which imparts greater or lesser urgency to the cross-sectoral control question; and the regulatory standards regarding drinking water quality, since these can confer a greater or lesser salience to the issue of natural resource protection.

Sources of drinking water in the United States are, generally speaking, of high quality. The US probably also enjoys fairly high quality with respect to most pesticides, although here the issue is complicated by gaps in data and also differences in regulatory standards. Certainly, surface drinking water supplies, which serve the majority of the population, are largely of acceptable quality. Problems tend to be sporadic and are seen as addressable via technical rather than behavioral means. There are exceptions to this generalization (see, for instance, the New York City and Des Moines, Iowa instances), and more can be expected in the future. But the quality of surface water resources as regards fertilizers and pesticides has not created broad concern. Certain environmental organizations have sought to document threats and thus increase the salience of the issue. These efforts have become more widely publicized recently and may ultimately provide significant leverage in dealing with the agricultural sector. The current situation, nonetheless, can be characterized as one of generally good quality surface resources.

Summarizing the status of groundwater quality is more difficult. The main problems include much less reliable and complete data and also great variation within the country on the extent of reliance for drinking water purposes. Some states and regions face serious present and imminent groundwater problems stemming from agriculture; others are suspected of being under threat, but little agreement on the issue is possible given the absence of data; and still others have yet to face the issue. Municipal water suppliers are often able to rely on deep wells to avoid facing the issue. In general, there is no national sense of urgency; the EPA has proceeded to try to stimulate states to develop groundwater protection strategies, but without clear, legislatively-authorized national effort the state initiatives are likely to continue to develop unevenly. These generalizations encounter exceptions in states with either a history of some groundwater contamination or a high level of support for the public health or environmental issues at stake.

In general, the supply sector has not encountered serious and sustained quality threats stemming from fertilizers and pesticides. Most threats from agricultural chemicals are able to be dealt with, at least for the moment, through technical responses; and evidence on trends is generally not alarming.

In addition, US drinking water regulations effectively reinforce the relative lack of urgency in the perceived state of the country's water supply. Nitrate standards are marginally more stringent than in the European Union; however, in combination with the state of the resource, most water suppliers have not been forced to respond. Those that have, including the city of Des Moines, have been able to do so by applying technology.

Pesticide standards are considerably more lenient than those prevailing in the EU, in two respects that matter for control capacity. First, most of the standards for individual pesticides are less stringent. And second, the lack of an overall pesticide concentration limit means that when regulators do impose severe restrictions, the agriculture sector can shift to another product. Substitutability creates a mechanism to ease the regulatory pressure. Some critics challenge the wisdom of this approach to pesticide regulation, but the absence of an overall ceiling continues to influence behavior. Taken together, these two characteristics of the policy setting mean that salience for the pesticides issue is lower, options for the agriculture sector more accessible, and therefore that the water supply sector is provided relatively little control capacity via regulatory requirements in force. Enhancements in control capacity will likely await either convincing new evidence of serious water quality threats or improvements in some of the other resources contributing to control capacity in the water supply sector.

The most interesting exceptions to these assessments can be found in circumstances in which some regulatory standard or directive has forced a supplier to respond by contacting the agricultural sector. The surface water filtration requirement in New York is a clear instance. The threat of national imposition in the case of Bay-Delta in California triggered similar behavior, although the latter case is complicated by non-drinking water environmental issues at stake (species protection). Several other examples can be found in the cases covered previously. Without the external, particularly regulatory, 'stick behind the door' (as it were), it is likely that few supplier-initiated efforts would have occurred.

13.2.2 ORGANIZATIONAL CAPACITY

Some features of sectoral organization have been reviewed in this chapter already. Here discussion focuses on the sector's organizational ability to mobilize people and other resources to effect changes in agricultural behavior.

In the United States, the water supply sector is almost impossibly large, dispersed, and loose. Organizing the millions of private systems, many at the household level, would require solution to a virtually unsolvable collective action problem - given the impossibility of imposing and enforcing their contributions to protecting common-pool resources. Most of the community systems themselves are small, undercapitalized, and ill-equipped, with staff lacking advanced training. Medium-sized suppliers may have some organizational capacity but are effectively precluded from exercising it by political or constituent constraints: suppliers in largely-agricultural areas find it very risky to point the finger at farming as a problem that needs addressing. The nation's largest suppliers include exceptions to these generalizations. Since most of the largest supply authorities are heavily dependent on surface water, many of the large suppliers have a somewhat easier time dealing with agricultural threats than do groundwater-based suppliers facing challenges from fertilizers or pesticides. Some of the largest suppliers, however, have the organizational capacity to initiate or contribute to efforts to alter agricultural behavior.

Of those who have organizational capacity, in the sense of structure, personnel, and funding (and this last item is treated separately below), only a relative handful have become active. Leaving aside the issues discussed above regarding natural resources, one can offer another reason: self-imposed capacity limits. The great majority of US water suppliers simply do not see it as their job to contact the agricultural sector and become involved with efforts to initiate behavioral change. Even for suppliers possessing significant organizational capacity and documented water quality problems stemming from agriculture, most have simply not considered initiating cross-sectoral contact.

The reasons appear to be several. One, an issue of *relative* organizational capacity, is that most suppliers see organized agriculture as significantly stronger than they are in political and organizational attributes. A second is the background and orientation of the suppliers' management. Water suppliers tend to be led by engineers and technical specialists. This point remains true today but is changing primarily in the largest supply authorities (for instance, MWD). The notions of undertaking efforts toward behavioral change, entering contested social terrain, and seeking alternatives to technical options are unconsidered possibilities for most in the sector. Some of the cases examined for this study, particularly in New York City, California, and Des Moines, are clear exceptions. They serve as evidence of shifts occurring. However, the much more common pattern is noninvolvement, in part because of what is being called here a self-imposed limitation in capacity. A third, related and reinforcing, reason is the availability of adequate and reliable technology. Fourth, the water supply sector is constrained by financial considerations that mitigate against untested alternatives. This last point is discussed below.

Other parts of the water supply sector may also be considered. No single industry association has the capacity to speak for suppliers generally. The American Water Works Association is the closest institution to an 'umbrella' for the industry. The AWWA has organizational capacity to be influential at the national level and with agricultural interests, for instance on legislative proposals. That capacity has recently become visible with AWWA's involvement in legislative efforts connecting drinking water concerns with agriculture. The AWWA, nonetheless, can only be effective where there is substantial support among the membership. The Association of Metropolitan Water Agencies also has capacity sufficient to become involved in national discussions, particularly regarding issues of concern to the largest suppliers. They too have begun to get involved in the issues. And the National Rural Water Association, while perhaps less influential in the legislative arena, has the field presence and organizational capacity to provide technical assistance to smaller suppliers.

One additional aspect of organizational capacity should be mentioned. Suppliers in the Western part of the US are even more fragmented than elsewhere. The West is a region in which suppliers do not even populate a loose 'issue network'. Some suppliers have close alliances with the agricultural sector and see their common interests challenged by the demand for urban water. In such settings (and leaving aside the further complications of struggles with conservationists), the water sector itself is polarized. The sector is a very different unit than the network. And umbrella organizations, like

ACWA, seem to offer little potential for integration. Here the overall organizational capacity of drinking water suppliers can be weaker still than elsewhere.

Additional limitations constrain the capacity of many regulators. EPA, never one of the strongest federal agencies, is in a particularly vulnerable position currently; proposed budget cuts could weaken the agency considerably more. To some extent, EPA's renewed interest in nonregulatory approaches stems from the vulnerable situation in which environmental programs and interests find themselves now. Many state-level drinking water regulators enforce standards but generally do not induce cross-sectoral efforts. The same can be said regarding environmental regulators. Some of these, however, are alert to the benefits of watershed and wellhead protection, and a number of states have regulatory authorities with sufficient interest and organizational capacity to encourage innovations. Wisconsin is an instance; the state drinking water (natural resources) agency is informed, supportive, and effective in encouraging nonregulatory understandings across the sectors.

Omitted thus far are two additional institutional loci through which significant organizational resources could be leveraged: progressive units within the agricultural sector itself, and what might be called 'bridging organizations': nascent structures located at the boundary between the sectors and involving participation by both groups of stakeholders. These units may be the best organizational option for efforts at indirect regulation or steering of the agricultural sector (see below).

Institutions in the agricultural sector have not been the primary focus of analysis in this project. It has become clear, nevertheless, that some of these possess substantial organizational resources - as well as finances and authority (see below) - and could serve as an indirect channel for enhancing capacity of the water supply authorities. Field units of the US Department of Agriculture (Natural Resources Conservation Service, Cooperative Extension Service, and state land-grant universities) and state agriculture departments are the most obvious cases. These organizations have shifted measurably toward more sustainable approaches in recent years, at different paces in different locales and regions. They are primarily nonregulatory and, perhaps most importantly, are known and trusted by the agricultural community. They also frequently possess skilled, capable personnel; extensive field apparatuses; knowledge of farming constraints and needs in diverse settings; and political strength. In the New York City watershed and elsewhere, they may be the most potent organizational resource available for water quality protection. EPA and some state environmental agencies have begun to develop cooperative efforts with certain organizations in the agricultural sector; these connections have proven productive and are likely to be expanded in the future.

Organizations bridging the sectors are also important in some instances. Note, as examples: the New York City Watershed Agricultural Council, with water supplier and environmental representation in a primarily agricultural apparatus; the administration of pilot projects between the Des Moines water supply unit and a series of commodity groups; the decision-making forum for Category III projects in the Bay-Delta agreement; the Aldo Leopold Center for Sustainable Agriculture in Iowa; and the brokering efforts between water and agriculture centered at the AWWA. In each of these cases, both constellations of interests are explicitly incorporated into a structure designed in part to accommodate concerns of both groups. In none of these cases can it be said that the bridging organizations themselves are substantial, independent forces capable of exerting control over either sector. In some of the cases, there is as yet relatively little evidence of successful accommodation. However, some of the structures are correctly viewed as important, potentially influential, and promising channels of change.

13.2.3 FINANCIAL RESOURCES

The general state of the sector's financial resources must be characterized as limited. The great majority of suppliers face significant financing constraints. Water pricing has been subsidized for years but this support has become less generous. And SDWA implementation has caused local water prices to rise substantially faster than the cost of living. Authorities are under pressure to hold down prices, and most have little slack for undertaking new ventures. A few of the largest suppliers possess considerably more favorable financial resources, and these tend to be the units that have become actively involved in efforts to influence behavior in the agricultural sector. The cases of MWD and the California Urban Water Authorities, New York City, and Des Moines illustrate this point.

Given the budget battles in Washington, the likelihood of substantial alterations in this general picture in the near term is small. A policy initiative via the 1996 SDWA re-authorization - to make a modest level of funds available to water suppliers interested in source protection - enhances capacity. Also, recently enacted changes to the SDWA have reduced some ancillary regulatory responsibilities for suppliers and therefore might release some financing for agriculture-related issues. Beyond these possibilities, it is unlikely that significant improvements in suppliers' financial resources will soon be achieved.

Two additional points regarding financial resources in the sector are worth noting. First, drinking water supply is the most capital-intensive industry in the country when considered in terms of the revenue received. The investment needed to deliver a reliable supply of high-quality water means that officials responsible for management undertake substantial long-term risks when they commit their capital for the future. This constraint strongly encourages a conservative financial perspective. This institutional pressure toward conservative financing decisions tends to reinforce the engineering world view. Water suppliers are more likely to opt for approaches that have calculable, predictable consequences for some years into the future; and these options are most likely to be technical. It can be expected, therefore, that more innovative initiatives to deal with behavioral change in agriculture will require: demonstrated, unequivocal successes leading to a slow diffusion of these practices to others; subsidization of such efforts by state or national governments; and/or a substantially heightened problem salience.

And second, it is not so much the financial resources of the water supply sector, in some absolute sense, that is the central issue; but rather the *balance* of financial resources and incentives - when compared to opposing perspectives and, especially, when options are being considered by decision makers in the agricultural sector. In both the New York City and the Bay-Delta cases, innovative steps toward cooperation have been taken. In both cases, the short-term cost of the cooperative venture was significant enough so that the effort would probably not have been undertaken, had the alternatives not been even more costly. These examples show the importance not so much of the sheer financial resources available to the sector, but of the relative costs and benefits, for all parties including water suppliers, of the options available. Cooperative efforts seem to develop when the alternatives (often including regulatory default options) are viewed as more costly.

13.2.4 INFORMATIONAL RESOURCES

Water suppliers are required to monitor and test on a regular basis for the variety of threats to water quality that have been incorporated into policy. Therefore, most possess sufficient current information of this sort. Information regarding the rate of degradation (and options for pollution prevention), however, is much less complete and reliable. This point is especially pertinent regarding groundwater. A variety of steps are underway in many states to encourage or require source protection efforts, a part of which include some assessment of current conditions and principal risks. However, these efforts have been unevenly developed, and they are limited by the available data on groundwater degradation. The imponderables associated with groundwater contamination - variations across chemicals, soil types, and aquifers - and the lag time between pollution and contamination mean that any current assessment reflects a retrospective snapshot. Changing agricultural behavior today may or may not have an impact on water supplies for decades into the future.

The issue is further complicated by the finding, in states like Wisconsin, that the prevailing models often used to predict impacts of agricultural chemicals are seriously flawed. This point, in combination with the political challenges involved, means that suppliers and other authorities have a very difficult time initiating preventive measures. The Wisconsin two-tiered effort is an exception. And Wisconsin authorities note that the actions they initiate in response to threshold warnings are necessarily ad hoc rather than the product of a comprehensive strategy. Scientific information regarding how to interpret these kinds of data is simply lacking, and the consequence is to limit the capacity of suppliers and other authorities to act with confidence.

Another information resource potentially in the hands of water suppliers is knowledge about how the agricultural sector can reduce or prevent deterioration in water supplies. On this subject, resources are spread rather unevenly. Many small suppliers know little about their problems and the kinds of changes that would help. Larger suppliers are generally knowledgeable regarding their principal pollution threats, as are regulatory officials. Most of these suppliers, nonetheless, are considerably more knowledgeable regarding technical options than they are with respect to behavioral change in agriculture.

However, this generalization encounters exceptions. In some of the more innovative water supply organizations, management has begun to seek out agricultural audiences. In

Des Moines, for instance, the general manager frequently communicates with farming groups to convey a sense of the importance of the issue, empathy with the farmers' situation, and broad ideas regarding what changes the agricultural sector might make. Similar efforts are underway by regulators and larger suppliers in states like Wisconsin and New York. Still, neither regulators nor water suppliers are agricultural experts, nor are their suggestions necessarily trusted and easily accepted by farmers. Some suppliers and bridging organizations have more success by conveying information indirectly: subsidizing research on reduced fertilizer application, for instance, or assisting in efforts to obtain foundation funding for demonstration projects organized by commodity groups. Both efforts are currently underway in Iowa. Similarly, in New York, the supplier is funding the development of research, Whole Farm Plans, and the implementation of BMPs. But all the direct effort is executed through agriculture-based organizations.

In the US, the development and communication of information useful in altering agricultural practices is primarily in the hands of the agricultural sector and has been distributed primarily via informational and learning tools, like assistance with BMPs. Further, it is most likely to be treated seriously if provided by organizations within that sector, with suppliers and direct regulators in the background.

13.2.5 CONFIDENCE

The issue of mutual confidence between the water supply and agricultural sectors - and especially the question of how much trust and confidence the former possesses as a resource that can be used to induce agricultural change - can be addressed generally. In the great majority of instances, there is little interaction between the two sectors, and therefore little basis for confidence and trust. Some of the case material included in the preceding chapter documents the most prominent efforts occurring currently. Most of the programs and policies underway across the country to stimulate source protection and organize stakeholders to reduce nonpoint pollution, including from agriculture, have not involved water suppliers (EPA, 1994).

In a number of instances, agricultural interests have begun to attend to the issues. It is likely that the attention of the agricultural sector will be directed increasingly at fertilizer and pesticide problems, although the difficulties of identifying precise sources of pollution will continue to pose monitoring difficulties and encourage some evasion of responsibility. However, the residuum of confidence and trust necessary to attract attention to the issue in the agricultural community and to change behavior can be expected to be difficult for water supply organizations to develop. (The cases sketched in the preceding chapter indicate some exceptions to this rule.)

This circumstance is understood clearly by national and state regulators. EPA, for instance, continues to battle a legacy of distrust - one stimulated in part by its largely regulatory mandate, in part by farmers' lack of knowledge of the agency and its approach, and in part by unhappy experiences some farmers have had dealing with EPA on issues of wetlands protection (the one national environmental-regulatory initiative with a direct impact on agriculture). Experiences of farmers in dealing with some state environmental units have been roughly similar. Given farmers' antipathy to regulation, political strength vis-à-vis environmental actors, and potent organization, there is little likelihood that environmental groups generally, let alone water suppliers in particular, can build substantial confidence nationwide. While the backlash against regulation may encourage some efforts to build confidence through nonregulatory efforts, the context also discourages the agricultural sector from concluding that it will have to respond seriously to pressure from environmental agencies. Therefore, regulatory units concerned about agricultural threats are making efforts to strengthen their role by working through organizations in the agricultural sector that already have built substantial confidence among farmers. Even in settings with strong regulatory efforts, like Wisconsin, one of the keys has been the active involvement of the state's agriculture department as the primary regulator for groundwater.

13.2.6 AUTHORITY

Authority, or the ability of water suppliers to induce willing compliance on the part of agricultural actors, can be considered in both formal and informal terms. In a formal sense, water suppliers are in a weak position to compel cooperation on the part of the agricultural sector. They have no standard-setting authority and, in most cases, little or no land-use control. There are some exceptions. Some municipal water bureaus, or the cities of which they are a part, either own land in the watershed or possess some form of controls. In these cases, the suppliers possess decisive authority. New York City is an example, and the authority the City possesses on its large watershed upstate guarantees serious attention to problems from agricultural chemicals. But this form of authority is relatively unusual. And even in the circumstances most 'favorable' for water suppliers, political feasibility may place sharp limitations on the exercise of formal authority (see AWWA Research Foundation, n.d.).

It should be noted, as well, that many water suppliers are not anxious to obtain more authority in this sphere. Enhanced authority means enlarged responsibility, and suppliers generally prefer narrower, more technocratic definitions of their jurisdictions.

One possibility little used thus far in the US is *partial* land-use authority. An option available is the issuance of 'conservation easements' or similar authorizations. This mechanism would allow the granting of restricted property rights - for residence, possibly some non-intensive forms of agriculture - to current farmers, in exchange for freedom from threat of eminent domain or detailed regulation of agricultural practices. This kind of exchange, perhaps most feasible where public authorities currently possess significant land use authority, could grant something to both sides. Small farmers could have guarantees of access and certain uses of the land, while those concerned with water quality could ensure protection. Of the cases examined here, the New York City watershed offers the clearest opportunity. And the WAC is now exploring the option.

This subsection has focused on the limited formal authority of water suppliers. The same can be said for national environmental authorities, particularly in the EPA. Most of the EPA's non-point source efforts are relatively small, slightly funded, and of undemonstrated effectiveness thus far. Part of the problem derives from the nature of nonpoint source pollution per se. Part, nonetheless, follows from formal authority available to EPA. The statutes authorizing EPA's involvement in water quality have been largely geared toward point sources until recently, and the vast majority of funding has dealt with point sources, especially wastewater treatment. The most recent efforts to direct attention at nonpoint source threats, including from agriculture, have not thus far resulted in a rewriting of water law or regulation to deal comprehensively with the full range of threats to drinking water and other water uses.

Furthermore, the issue of groundwater quality faces even more substantial limitations. There is no authoritative national groundwater policy, no clear authorization for the EPA to develop regulations or a national strategy, and no prospect that the limited present authority will be enhanced soon. EPA has made efforts to create a sense of national energy and encouragement to the states on the subject. But national authority remains very limited.

State authority, and great state-to-state variation, is the most important point to be made regarding groundwater. Some states have done almost nothing to enhance protection and face few threats anyway. Other states, like Iowa, have taken some steps but may encounter serious challenges before long. Still others, like Wisconsin, have developed active policies. Most states have been producing wellhead protection plans. And states are also making progress in anticipating the impact of certain pesticides. Still, most state policy does not clearly address questions of prevention. In most of the state cases, as well, strong regulations have been eschewed in favor of other, nonregulatory options. It remains to be seen how effective these efforts will be in dealing with the developing challenge to groundwater quality.

Informal authority, or the actual ability of water suppliers to induce voluntary compliance with their needs on the part of the agricultural sector, was discussed in the coverage of confidence and trust, above.

13.2.7 TIME

Time as a resource in the hands of water suppliers is a difficult item to analyze. Only a few comments can be made. First, for groundwater protection time is one of the most problematic resources. The difficulty is handling groundwater contamination ex post facto. At this stage, the best that can be hoped for is minimization, and ultimately (perhaps over decades) remediation. All the basic features of the challenge, nevertheless, are shrouded in uncertainty. The constraints on time, therefore, can be considered inversely related to informational resources. The relative absence of clear information, especially regarding groundwater and its impending contamination, is what imposes time needs and challenges on water suppliers.

Prevention of contamination, rather than mere remediation, is a priority. Prevention works to reduce resource constraints on several fronts. Successful protection of water resources means less imposing organizational and financial demands, lower stress on additional natural resources, and less need to use authority to change behavior quickly and drastically. Prevention also reduces the demand for quick action.

Resource constraints on some of the other items already covered in this chapter mean that the time available for water suppliers to develop contacts with and seek forums for influencing the agricultural sector is relatively limited. Many other issues, including short-term compliance with regulatory standards, are deemed more important. In effect, then, the challenges deriving from fertilizers and pesticides often appear as a contest between short-term and long-term issues facing water suppliers. In the short term, even agricultural chemicals can typically be dealt with via technical means. In the long term, efforts at behavioral change may be more productive. The resource constraint of time, then, manifests itself as suppliers' experiencing difficulty finding sufficient time to develop programs to address long-term issues.

13.3 Prospects for increasing control capacity

The review of control capacity suggests that substantial changes in agricultural behavior, at least as induced by suppliers and their supporters, are likely to await enhancements in key resources available to the supply sector. Some of these kinds of increases are discussed shortly. In addition, the limited successes already experienced in certain states and locales in the US provide information regarding what options can be expected to be most feasible. Prospects for increasing sectoral control capacity, with an emphasis on requisites and options, are the subject of this section. Particular attention is given to alternatives that may have relevance beyond the US.

13.3.1 REGULATING AS 'SINE QUA NON'

If the US situation regarding control capacity teaches an unambiguous lesson, it is that stringent regulation is the basic requisite for any significant effort on the part of the water supply sector. Nationwide, the issue of agricultural chemicals has only become salient with the enactment of chemical-specific regulatory standards that carry significant sanctions and require action in some locales. And, at least to some extent, *national* direct regulation may be an important component of the regime necessary for effective indirectly regulatory or nonregulatory efforts. Where there are strong agricultural interests, state governments may be reluctant to impose forceful regulatory measures, even in the face of deteriorating water quality. The case of Iowa is instructive in this regard.

Newer efforts underway at the national level to support source protection have followed in the wake of regulatory experience and concern about increasing burdens. The case evidence also points in the same direction. In New York and California, for instance, the presence of regulation combined with the all-but-certain imposition of unattractive sanctions have been the stimuli for new nonregulatory collaboration. States emphasizing nonregulatory strategies without an effective regulatory base, like Iowa, have found that suppliers as a group have generally not become involved in efforts to modify agricultural practices. (Des Moines' activism is quite unrelated to state policy initiatives but is traceable, rather, to an unusually aggressive supplier management.) The importance of regulation is perhaps especially crucial for groundwater, since uncertainties about the impacts of current agricultural practices impose substantial barriers to intersectoral efforts.

This finding is consistent with a generalization reached in another recent study: DeWitt John has found that the states with the most active efforts to implement nonregulatory options tend to be those that are also national leaders in regulation (John, 1994). The conclusion may be especially pertinent in the current political and regulatory climate. The practical implication, furthermore, is to maintain and perhaps strengthen the regulatory base as a complement to the use of other policy tools, not as a replacement for them. Furthermore, regulation works best in circumstances where regulators do not signal a lack of interest in source protection. The New York City case, for instance, is complicated in practice by regulators' suspicions about the City's motivation for source protection (that is, filtration avoidance). In this instance, and for some other large urban suppliers, the national regulation on surface filtration endorses a preferred technique instead of offering alternatives for suppliers interested in source protection and pollution prevention.

13.3.2 DEVELOPING A NATIONAL GROUNDWATER POLICY

As documented above, there is no formal and enforceable groundwater policy for the United States. Several efforts to enact one in earlier years were unsuccessful, and the climate for doing so has grown more hostile in the current Congress. The EPA has clearly sought to move in this direction over several years, but the ability of an administration to develop and enforce a policy with insufficient support in the legislature is questionable. The EPA's groundwater 'strategy' has served to encourage state action, especially where there is support for such efforts already. More recently, the EPA's development of a proposed rule to require state planning to ward off threats to groundwater from selected pesticides can be viewed as a helpful effort, albeit one that also relies on states as prime actors and has few sanctions to encourage speedy action. But there is unlikely to be broad movement across the many states for which groundwater is an important drinking water source without a national policy to provide a backdrop and some mandated minima to support state efforts.

13.3.3 DEALING WITH SELF-IMPOSED CONSTRAINTS

As the preceding analysis indicates, one of the most significant limitations on the control capacity of the water supply sector is the sector's self-imposed, or self-adopted, conception of its role, capabilities, and responsibilities. With a few exceptions, US water suppliers do not see themselves as potentially major actors in influencing agricultural behavior, even when agricultural challenges are important for protecting and preserving water quality and a steady supply of the resource. The reasons for the limitation were detailed above. Some, but only some, are subject to influence in the short term.

A political strengthening of the sector must await an increase in salience for the issues involved, perhaps stimulated by serious degradation of the nation's water supply - hardly an attractive prospect. Similarly, the presence of feasible technical options and the conservatism encouraged by the capital-intensive nature of the industry are not directly manipulable by policy makers, nor would such influence be productive.

However, the backgrounds and role conceptions of those governing the sector - directly within water supply organizations and indirectly via regulatory overseers - can be influenced. In fact, as some of the case evidence presented above suggests, in the nation's larger supply organizations shifts have been occurring in the patterns of recruitment and designation of essential skills at management levels. The MWD of Southern California and the water supply Bureau of the City of New York are clear illustrations of the changes underway. Such shifts have not percolated down into the smaller supply organizations. However, the direction of change is clear. Further encouragement of these changes, by those in the sector as well as policy makers, is likely to assist in reducing the self-imposed limitations. As such changes develop, it can be expected that supply authorities will become somewhat more involved in such tasks as identifying source protection strategies, negotiating agreements with those involved in activities threatening water quality, communicating to groups such as farmers on the nature of the vulnerabilities to agricultural practices, and developing approaches aimed at behavioral change.

Training and informational campaigns, with support from the principal industry associations, could be useful means of raising the consciousness of supply managers regarding altered conceptions of their roles. Involvement of some of the more innovative suppliers might be especially influential in providing concrete illustrations of what kinds of efforts can be successfully mounted.

An additional way of encouraging this shift is for policy makers at all levels of government to begin to involve water suppliers in the initiatives underway to deal with threats from fertilizers and pesticides. To date, most of the source protection efforts undertaken - whether by government, NGOs, or some combination - have not involved water suppliers. The evidence suggests that the self-imposed limitation in role has a broader resonance. That is, most actors involved with water quality and agricultural pollution do not consider water suppliers to be centrally important in programs aimed at changing behavior. As suppliers become involved, and as their industry associations take a more active role in discussions of the issue, the perceptions of this broader set of actors can be expected to change gradually.

13.3.4 REMOVING POLICY INSTRUMENTS THAT DISCOURAGE PROBLEM-SOLVING BEHAVIOR

Prospects for enhancing control capacity of the water supply sector depend not only on reducing self-imposed limitations and directly addressing resource constraints. It can also be important to remove, or reduce the influence of, any policy instruments currently in force that have the effect of discouraging problem-solving behavior between the sectors. The most obvious instances are agricultural initiatives aimed narrowly at production goals. Additional examples are local practices of subsidizing clean water prices from general government budgets (thus encouraging inefficient consumption), and in some regions granting early customers - including heavy agricultural users - priority. These policies have encouraged waste and discouraged careful allocation among users.

13.3.5 ENHANCING SECTORAL ORGANIZATION AND SUPPORTING AGRICULTURAL ALLIES INDIRECTLY

The water supply sector is populated with an extraordinary number of organizations. For present purposes the focus is not so much on these individual units as on the structure of the sector as a whole. In particular, the coverage of network features, as developed earlier in this study, can be emphasized. The US water supply sector scores quite low on the dimensions of interrelatedness and commitment. Most suppliers interact at most sporadically with other suppliers, the major national organizations are dominated by the largest suppliers (with the exception of the National Rural Water Association), and the interests of individual suppliers are sufficiently diverse as to render tight, committed alliances improbable on many issues. This situation is unlikely to change drastically in the near future. Several implications for increasing control capacity can be sketched.

First, it is likely that the most effective water supply efforts in the near term will be made by a few individual supply water organizations, particularly some of the largest authorities. These efforts can be monitored, documented, and publicized by regulators, industry associations, and others. This activity should receive some priority, since it is likely to multiply the receptivity for innovative efforts among others in the sector. Information sharing from innovators to others in the industry is not an intensive form of organizing, but it has the potential to increase organizational and informational resources in the sector. The importance of the largest supply organizations suggests, in turn, the potential for the Association of Metropolitan Water Agencies to occupy a lead role in national discussions and the development of innovations.

Second, the structure of the sector nationwide suggests that most of the potentially productive links must be built on subnational levels. States, watersheds, and aquifers provide manageable jurisdictions for initiatives. The fragmented nature of the sector nationwide, however, portends relatively less success in mobilizing capacity at that level. Several of the cases included earlier illustrate the possibilities. Even in these instances, however, much of the effort is undertaken by individual suppliers rather than a network of them. State and watershed associations or consortia of suppliers can play a role in building network links at this level.

Third, despite the preceding point some potentially important albeit limited forms of network development are observable at the national level. These involve major associations, some of the largest suppliers, agricultural stakeholders, and others in bridging efforts. Thus far the discussions have revolved almost exclusively around issues raised in the context of national legislation. The extant links cannot be expected to serve purposes such as the development of ongoing programs of cooperation or the gradual incorporation of the full sector in the developing network pattern. Encouragement by policy makers and others of these continuing cross-sectoral network developments would be sensible.

Fourth, the water supply sector could provide encouragement and support for increased organizational capacity on the part of complementary parts of the agricultural sector, particularly environmentally aware agricultural agencies and technical assistance units. This indirect channel of influence may be one of the most significant available in the US context; the advantages of authority and commitment typically possessed by these organizations often render them the best options for inducing change. The water supply sector might be encouraged to develop more links to these units, convey information to them on a regular basis, indicate priorities from the perspective of water supply, express willingness to participate in data gathering and information sharing initiatives, and perhaps support some of their efforts financially. Especially promising are those efforts that can include opinion leaders within the agricultural community. (The New York City watershed instance is a clear illustration.) Potentially effective organization within the agricultural sector can facilitate water suppliers and others to use more indirect efforts, and this approach is likely to reduce resistance and enhance confidence.

Fifth, the industry associations might increase the sector's organizational capacity by undertaking more informational and demonstration programs on fertilizer and pesticide issues, with the objects of communicating results to their membership, serving in limited representational roles in cross-sectoral discussions, and aiding in the process of technical assistance with smaller supply organizations.

Finally, for the US case, it is important to bear in mind the regional differences in sectoral organization. The water supply sector in the West faces two significant organizational differences from the typical pattern in the East. One is that the sector contains both drinking water and agricultural suppliers. The other difference, less striking than the first, is that conservation and environmental protection interests are typically an important part of the stakeholder set, and these groups are likely to see suppliers as threats to environmental quality rather than supporters of it. The practical results are a fractionated water supply sector, more complex and multilateral negotiations, and less broad and uncritical support for water suppliers. The drinking water (sub)sector is in a relatively weaker bargaining position when dealing with agriculture. This relative disadvantage in terms of organizational resources may be partially compensated by a kind of natural resource advantage: water supplies in the West are somewhat more isolated from nonpoint source threats, including those from agriculture. The overall impact is nevertheless an organizational weakness of drinking water suppliers in arid regions.

The further implication, then, is that Western water organization requires significant strengthening. The task is difficult to accomplish, given the political strength and relative advantage in water rights accorded to the agricultural sector. A few concrete steps may be worth considering. One is for those leading umbrella water associations in the West to make a greater effort to identify a productive long-range *modus vivendi* between urban and agricultural suppliers. A second is for policy makers to consider modifying water rights where feasible to encourage conservation, require environmentally benign extraction practices, and integrate water law with the needs for and principles of sustainable development. A third strategy derives from the analysis of sectoral fragmentation just summarized. Urban and agricultural suppliers are competitors in a circumstance of heavy demand and large year-to-year fluctuations in supply. This mutual interdependence could be the stimulus for intra-sectoral communication, negotiation, and gradual development of institutions promoting (limited) cooperation.

13.3.6 ENHANCING FINANCIAL RESOURCES IN THE WATER SUPPLY SECTOR

The water supply sector in the US is tightly constrained by financial limits. Implementation of more stringent drinking water regulations has forced substantial price increases - and thereby consumer resistance, attention, and some increased salience for water quality issues. The largest suppliers, with substantial cash flows, possess sufficient slack to undertake initiatives to deal with fertilizer and pesticide issues; a few have done so. The great majority of suppliers see their situations as severely constricted. And yet financial resources are clearly needed if the sector is to deal more productively with agriculture: theoretical arguments regarding the relationship of policy instruments and network characteristics suggest that instruments based in positive inducements and non-coercive efforts may be necessary in dealing a sector structured as agriculture is (see Bressers, 1993). Water suppliers and their institutional/policy supporters need to consider education, training, promotion (of BMPs, for instance), and perhaps ultimately direct cross subsidies. These kinds of initiatives require financial resources; those involving direct or indirect subsidies are the most costly.

Four options for enhanced financing can be identified. The first is for suppliers as a group to accelerate their price increases and designate a portion of the increment for source protection efforts. This possibility is problematic, particularly in the wake of the large increases already experienced in recent years.

A second possibility is for industry associations to devote an increased portion of their budgets to source protection and the challenge of agricultural chemicals. Some movement in this direction has already taken place. For instance, the AWWA has become active in EPA reviews of selected pesticides regarding their registration status. And the National Rural Water Association, along with state-level parallel units, has devoted significant effort to its 'circuit riders' initiative, through which expert association staff travel among small communities to advise suppliers of their most serious water quality problems, identify sources of contamination, and develop options for source protection. For selected kinds of initiatives, like education, training, and cross-sectoral negotiations, water industry associations can provide undeniable economies of scale and some degree of authority. And such efforts are relatively low cost, at least as compared with certain technology-intensive programs.

Third, financial resources of the sector could be increased as a matter of national policy. (The state drinking water agencies are not currently in possession of substantial quantities of financial resources, unlike their counterparts on the wastewater side. State aid, therefore, is unlikely under current conditions.) The federal government could choose to devote more resources to protection by funding initiatives in the water supply sector aimed at altering agricultural (and other) practices. Modest versions of this idea are currently under discussion in Washington. One possibility is a proposed expansion of the so-called revolving loan program for wastewater facilities, an initiative aimed at capitalizing a set of state-level subsidized loan funds for infrastructure development. Some advocates for drinking water quality have proposed an extension of the revolving loan idea to drinking water infrastructure, with part of the assistance available for competitive application by water suppliers to be used for source protection. Other versions of the same general idea have been considered, and the issue is currently part of the negotiations underway on re-authorization of the SDWA. Some very modest additional support for nonpoint source projects might also become available under a re-authorized Clean Water Act. Indeed, a part of the policy debate now in Washington is devoted to seeing if the several partial national laws which affect water quality can be more closely integrated into a coherent strategy. Doing so would likely create certain efficiencies in control capacity for water suppliers and also enhance confidence and authority resources.

However, the political climate for such an effort - and in particular for substantial increases in funding - at the national level is not propitious. The budget impasses of 1995 and 1996 suggest continuing difficulties generally, and funding for environmental programs in particular is in substantial jeopardy. Over the longer term, some national funding (with considerable state and/or water supplier discretion as to use) and consolidation of national policies may be enacted.

A fourth financing option worth serious attention is a state-level choice: fees on fertilizers and/or pesticides. Some states currently use these as a source of revenue. The option can be considered by other jurisdictions and may be politically feasible in several additional states. A distinction should be made between two uses of fertilizer and pesticide fees: (1) as a disincentive for application, and thus as an instrument for reducing agricultural chemical threats to water quality; and (2) as a source of revenue for protection programs, including nonregulatory ones. The former purpose is likely to be much less feasible than the latter (and is also not aimed at an enhancement of financial resources as a route to increased control capacity). From the perspective of most farmers, chemical applications surely involve a cost, but not a large one. What is more, the perceived costs of under-application of chemicals are greater than the likely expense of heavy treatments. It would require very large fees to influence behavior appreciably. (A more efficient effort is likely to rely on education and information: clear, believable demonstrations that productivity is not sacrificed by lower applications of pesticides or fertilizers.) The political difficulty of enacting fee structures at a level sufficient to influence application rates significantly is great.

Fees as a source of revenue, on the other hand, have been shown to be a feasible policy option. The most important question has to do with the designated uses of the resulting revenue. A variety of defensible purposes could be identified. Some of these might enhance control capacity over agriculture while not providing direct support to water suppliers; other uses could channel financial resources to the water supply sector. An example of the former is the current policy in Iowa, with dedicated state assistance for a center for sustainable agriculture as the direct consequence. An alternative would be an initiative to channel revenue to the supply sector for their use in programs aimed at behavioral change in agriculture: training, demonstration projects, dissemination, subsidies for source protection, or other efforts.

A few additional points can be made regarding financial resources and control capacity. First, it should be remembered that budget cutbacks may also have an impact on agriculture, particularly with regard to trimming subsidies for production. To the extent that the agricultural sector becomes more market driven and less dependent on national programs of support, the results as regards control capacity in the water supply sector may be complicated. If farmers do not receive subsidies, their behavior may be even less subject to regulation and other forms of influence by governmental authorities; noncompliance involves fewer potential costs. Second, certain forms of *exchange* rather than increases in actual financial resources may have an efficiency effect, thus increasing the impact of a given quantity of resource inputs. Thus, the notion of 'pollution trading' within source protection areas, attempted in parts of California and under consideration within EPA, offers potential for freeing up some currently-allocated funding to be directed at serious agricultural contamination in the same watershed or wellhead. Practical impediments related to the mechanics of administering such a system and the need to limit the use of such an approach to contaminant-compatible trading settings suggest that this instrument is no panacea. However, it can have utility under the right circumstances. And third, some supplier financial resources may be able to be targeted more strategically as an inducement to encourage cooperation from the relevant elements of the agricultural sector. Note, for instance, the active discussion in New York of the possibility of the City's purchase for institutional use of agricultural products like milk from the watershed area, thereby cementing the sectors' interdependence in a mutually beneficial fashion.

And finally, direct subsidy initiatives from any source to assist behavioral change in agriculture may have to deal with an equity issue. In certain regions environmentally-committed farmers have already undertaken changes in their practices, some of which may have been costly, in response to calls for a cleaner water supply. New programs of financial assistance aimed at more recalcitrant actors may inadvertently penalize the earlier farmers by comparison. In the New York City watershed, the WAC is currently wrestling with such issues.

13.3.7 IMPROVING INFORMATION RESOURCES

Fertilizers and pesticides cause problems for many drinking water supplies. However, reliable data on the extent of the problem and its trend over time is very spotty. National support may be required for the kind of information gathering that is essential. The likelihood of satisfactory levels of assistance, nevertheless. is very low. Furthermore, as suggested earlier, there is an important role here for the water supply industry associations. Raising the issue regularly, reminding policy makers and the agriculture sector of the importance of the topic, and publicizing the intersectoral dependence in ways that all can understand can be significant parts of the informational basis for more detailed deliberations among both specialists and the broader, interested policy setting. This latter audience includes legislatures at both state and national levels.

The importance of this sort of assistance can be understood by considering some of the most well-known current theories of public policy. Work by scholars such as Ostrom (for instance Ostrom, 1990) and Sabatier (especially in his work with Jenkins-Smith, such as Sabatier and Jenkins-Smith, 1993) suggests that policy change of this order requires modifications in the cognitive structures of interested parties ('learning'), and not simply shifts in political strength. In the terms used in this study, control capacity is not merely an issue of leveraging opposed interests but also involves influencing perspectives of others (Sabatier's advocacy coalitions). The point can be put in terms of resources by noting that information can influence confidence or trust, thus an expansion of one resource can catalyze additional resource enhancements. Power in the usual sense surely matters as well, and the process of policy learning is unlikely to be direct, linear, and free from contention. But the evidence in other policy fields suggests that information matters in determining what happens, in influencing parties' capacity to exert some control over the actions of others.

It is clear from the empirical work undertaken for this investigation that there remains some degree of good-faith cognitive resistance or denial within the agricultural sector to the causal connection between agricultural practices and drinking water quality. Widespread behavioral change in the agricultural networks is unlikely absent clear and convincing evidence, reliable trend data, and channels for communicating the information in acceptable ways to those in the agricultural sector. It is also clear, moreover, that when such information is provided, behavioral change can be induced. The development and dissemination of BMPs does affect agricultural practice, albeit unevenly. More intensive, site-specific efforts when combined with other resources like financing can be particularly effective. The New York case demonstrates the potential for information - and subsidizes for altered practices - to effect significant behavioral change. Similarly, an emphasis, as in New York, on aiding behavioral change that can also be expected to increase profitability can be especially effective.

Furthermore, the experience in several states suggests the importance of the suppliers of agricultural chemicals for information dissemination. These parties typically enjoy the confidence of the farmers and are few enough in number so that essential information regarding water quality issues, application rates, and related matters can be transmitted to them for communication to farmers in the region. Some states make special efforts to educate these actors, require their certification, and monitor their performance. Such an approach has the potential to increase both informational and confidence resources.

13.3.8 BUILDING CONFIDENCE

Efforts to develop greater mutual trust between the water supply and agricultural sectors, in particular steps to enhance the confidence farmers and other agricultural stakeholder groups have in water suppliers, can be difficult but ultimately crucial components of any strategy for boosting control capacity. Changing behavior within the community-like agricultural sector is likely to be difficult, costly, politically contentious, and often unproductive if attempted primarily or exclusively via coercive and presumptively authoritative policy instruments (Bressers, 1993). And inducing behavioral change via financial incentives is more feasible in theory, but practical limitations on the financial resources available constrain the options. Under these circumstances, influencing the perspective of the agricultural sector - by providing persuasive information and also by enhancing mutual trust between the sectors - is likely to be a necessary element of any successful strategy. Certainly these notions are central in the minds of policy makers and some water suppliers in several states.

The subject of information resources was treated earlier, although in some respects it emerges here as well, in connection with the kinds of data that might enhance confidence and trust. Information resources and confidence cannot be considered independently of each other when it comes to finding ways of encouraging cooperation across sectors that are interdependent but partially at odds.

Those water suppliers most sensitive to this point have become involved in concerted efforts to assuage fears and develop trust on the part of actors in the agricultural sector. The efforts thus far have included such initiatives as managers of water supply institutions giving speeches and presentations to community agricultural groups, commodity organizations, and similar audiences; supply managers' frequent expression of empathy for farmers' constraints and intentions, combined with reasonable expositions regarding the challenges posed by agricultural chemicals to water supplies; support for agriculture-managed demonstration projects regarding BMPs and other innovations, like the commodity group-administered projects supported by the Des Moines, Iowa supplier; and subsidization of research and dissemination with substantial agricultural-sector participation (note the efforts along these lines in California and New York). It is likely that water sector projects alone, and supplier exhortation in and of itself, will not produce substantial increases in trust and confidence. Rather, clear evidence of suppliers' understanding of farming as a business enterprise, combined with a demonstration that suppliers are willing to treat the agricultural sector responsibly (and not as a mere target for manipulation), are important requisites of successful confidence building.

Bridging institutions, therefore, may be especially important in facilitating the confidence building that can be so important. Likewise, as suggested above, efforts to enhance intersectoral trust are likely to be assisted if suppliers develop regular ties and communication channels with potential allies in the agricultural sector; involve these latter actors when possible in direct contacts with farmers; and seek to build broader coalitions for simultaneously assisting and influencing agricultural behavior. Support from local offices of such national agencies as the Natural Resources Conservation Service and the Cooperative Extension Service, as well as selected non-governmental organizations with experience at nonregulatory innovations (like the Groundwater Foundation) and selected for-profit firms, can prove decisive in increasing levels of confidence and trust. And evidence from states with supportive state departments of agriculture, like Wisconsin, suggests that such organizations can be considerably more effective with agricultural targets.

It is further notable that the thus-far successful New York City watershed case (success here being measured in terms of participation from the agricultural sector, support from the water supplier, and enhanced levels of mutual respect) involves an even greater level of devolution, or influence shifting, to agriculture: a new farmer-controlled institution in the WAC, farmer-driven budgetary and BMP subsidy/research decisions, and complete oversight of implementation and enforcement by local farmer committees. Whether this design produces long-term influence for the perspective of the water supplier is a question that cannot yet be answered.

13.3.9 ENHANCING AUTHORITY

Formal authority of water suppliers is quite limited. And water suppliers are not generally motivated to seek more regarding the agricultural sector. Only a couple of options can be considered seriously.

Where legally and politically feasible, granting suppliers (or the municipalities for which a 'supplier' may be a subsidiary bureau) greater land use control in source protection areas can be an important step. AWWA's study of promising watershed protection initiatives found that land-use control can be central (AWWA Research Foundation, n.d.). In addition, conservation easements may be feasible in some circumstances, if suppliers or their municipalities possess land use authority. Here authority to negotiate such easements may allow bargaining leverage to reach agreements acceptable to the agricultural actors involved. And even aside from regulatory measures, a key limitation of current locally-organized source protection efforts in states like Wisconsin is the jurisdictional fragmentation for common wellhead protection zones. In such cases, state policies designed to encourage interlocal cooperation can be useful. Wisconsin limits its assistance to technical support, but small increments of financial assistance could pay dividends as well.

Enhanced formal authority for other actors in the sector, in particular state and national agencies, could also be useful in strengthening control capacity. However, as explained above, the political acceptability of such measures, for instance to allow EPA to develop a clear national groundwater protection policy, is quite limited at present.

13.3.10 PLAYING FOR TIME

Building an advantage in water quality efforts by easing time pressures is one of the most difficult ideas to implement, despite the fact that time pressure in several countries is now severe and is responsible - along with regulatory pressures and increasing evidence of contamination - for the frenetic pace of supplier efforts to deal with agriculture. The central dilemma has to do with the groundwater problem in particular: indisputable evidence of problem severity is only available 'too late'. In the US, the agriculture sector has shown little willingness to build a time advantage by agreeing to modify behavior before such evidence is available. Even source protection incentives are resisted by organized agriculture if not tied to a requirement that water quality problems be clearly demonstrated. The problem is exacerbated by the accumulating evidence, for instance in Wisconsin, that the extant models for predicting flows of contaminants in groundwater are seriously flawed.

The most innovative effort examined in the current study to 'play for time' with respect to agricultural chemicals and groundwater contamination is the Wisconsin state policy for pesticides - a combination of significant monitoring efforts with the two statutory concentration levels: the PAL and the standards. This is not to say that all aspects of the policy are optimal - for instance the threshold levels are in some respects arbitrarily established. But the policy does generate temporal slack between early evidence of pesticide contamination and ultimate prohibitory action. The policy tools used in Wisconsin institutionalize some relatively early efforts to contain the problem, and in so doing they condition all parties involved to consider proactive steps as reasonable rather than alarmist. Although the actions taken at the threshold level are developed ad hoc, and although these can have some regulatory elements, it is possible that the two-level approach has ultimately resulted in more flexibility for some farmers than a single, ultimately prohibitory standard would have produced.

One limitation that must be noted with respect to the two-level strategy applies in particular to the context of the European Union. Two levels are technically feasible in the US in part because the absolute levels of the pesticide standards are substantially higher than those in the EU. The latter are virtually at trace concentrations. The enactment of a two-level approach in the EU would probably mean - or be seen as - a relaxation of current regulatory requirements.

13.4 Conclusions

It should be clear from the analysis presented in this chapter that, while a number of possibilities can be identified for enhancing control capacity over damaging agricultural practices, many of these are unlikely or impractical on grounds of political difficulty, financial constraints, and/or lack of sufficient problem salience in the American context. But the array of feasible options is not a null set. Indeed, one major finding of this analysis is the staggering multidimensional variety in the US The several cases analyzed in the preceding chapter contain many implications that can be used elsewhere in the country, and some that have applicability in other national settings.

In particular, two striking findings with practical implications for enhancing control capacity are the following: (1) vigorous regulation is a necessary basis or requisite for effective nonregulatory initiatives; and (2) confidence, authority, and organizational resources are greatly leveraged when key actors in the agricultural sector themselves have substantial say about how to deal with the water quality problem. Retaining and strengthening regulation, then, and finding new and better structures for cross-sectoral coordination and the responsible use of discretion within the agricultural network itself, can be expected be crucial components of effective longer-term solutions.

References

AWWA Research Foundation. (n.d., but approximately 1990) *Effective Watershed Management for Surface Water Supplies.* Prepared by Richard W. Robbins, Joseph L. Glicker, Douglas M. Bloem, and Bruce M. Niss. Denver Colorado: AWWA.

Bressers, J.Th.A (1993) Beleidsnetwerken en instrumentenkeuze. *Beleidswetenschap* 7, 4: 309-330.

Environmental Protection Agency, U.S. [EPA] (1994) *The Watershed Protection Approach, 1993/94 Activity Report.* Washington, D.C.: U.S. Environmental Protection Agency (EPA 840-S-94-001).

John, DeWitt (1994) *Civic Environmentalism: Alternatives to Regulation in States and Communities.* Washington, D.C.: CQ Press.

Ostrom, Elinor (1990) *Governing the Commons.* Cambridge: Cambridge University Press.

Sabatier, Paul A., and Hank C. Jenkins-Smith (1993) *Policy Change and Learning: An Advocacy Coalition Approach.* Boulder (Colorado): Westview Press.

PART THREE - CONTINUED

14. TO REINFORCE CONTROL CAPACITY?

GEERTEN J.I. SCHRAMA
Center for Clean Technology and Environmental Policy, University of Twente, Enschede, The Netherlands

14.1 Introduction

The research project reported in this book concerned the issue of agricultural pollution of drinking water resources. Emphasis was put on preventive action undertaken by the water supply sector, which was placed in the larger context of European and national policy efforts in the fields of agricultural policy abatement and water quality management. The organizations constituting the water supply sector - water supply companies and regional water authorities - are involved because agricultural pollution affects the performance of their primary tasks, while at the same time they fulfill a function as intermediaries in the implementation of national and EU policy. The control capacity of the water supply sector was designated as the dependent variable in the research design. This final chapter contains some conclusions and policy recommendations concerning the prospects of enlargement of this control capacity.

The conclusions and recommendation in this chapter are ranked according to two different angles. The first one concerns the three basic types of steering strategies - economic, juridical, and communicative - and is represented in the titles of the sections. The second one reflexes the tremendous complexity of the problem at hand, which urges us to make clear distinctions between the system levels involved - the European Union, national policy, and regional and local activities - but at the same time it is clear that little of what is being said can be considered in isolation:

- The impact of the *European Union* is manifest in almost every aspect. Although the main conclusion of the EU case study was that there is very little that the European Union can do directly in dealing with the present problem, the European Union does have a very important role to play in generating resources and action opportunities for other actors, in the form of legislation and regulations that define the objectives of water quality and set the parameters within which the various groups of actors must operate.

Geerten J.I. Schrama (ed.), Drinking Water Supply and Agricultural Pollution, 367-375.

- At the *national level* the issue is part of the more comprehensive issue of agricultural pollution in general. Member states are still struggling with the implementation of EU directives concerning water quality standards, while the policy styles adopted towards the agricultural sector are becoming more impositional and regulative, and the traditional agricultural policy networks are under heavy pressure - if they have not already become largely disintegrated.
- At the *regional and local level*, especially where agricultural pollution has infected drinking water resources most, many successful - but chiefly small scale - initiatives have been developed by the water supply sector.

This chapter elaborates on the comparative analysis in chapter 11 and benefits from the experience of the US case study. Two major conclusions are predominant: (1) the need for credible legal regulation and enforcement of water quality standards, and (2) the precondition of a high level of 'interrelatedness' wherever the water supply sector and the agricultural sector meet as a precondition for successful preventive activities.

Credible legal regulation and enforcement. Apart from the issue network conditions of direct contacts (interrelatedness) at the regional or local level, the project demonstrates the crucial role of credible legal regulation concerning (drinking) water quality standards, which is the basic driving force for almost all actors involved, from the agricultural sector as well a from the water supply sector. Indirect steering through the water supply sector (network management) may be an appropriate way to exercise (additional) control over the agricultural sector - especially in those cases where drinking water resources are threatened - but government should keep on performing its control function at some distance and not release the reins all together.

Interrelatedness. The case studies show that rather frequent and direct contacts between the water supply sector and the agricultural sector ('interrelatedness') are an important condition for successfully influencing of agricultural practices, at least at the regional and local level. The attitude chosen by all parties involved is predominantly rational, oriented at their own (economic) interests. The traditional policy communities at the national level are under heavy pressure, and a joint approach based on common interests and shared perceptions ('mutual commitment') is usually not feasible. According to Bressers (1993) in such situations - high levels of interrelatedness and low levels of mutual commitment - the appropriate policy instruments are:

- juridical instruments: restrictive, but based on consensus with the actors in the policy field;
- communicative instruments: information provision without a moral appeal;
- economical instruments: enlarging involving positive incentives as well as restrictive ones.

Both points will appear quite often in the next sections on the three basic types of steering strategies.

14.2 Economic incentives

All of the case studies reported in this book make clear - once again - that economic incentives are most effective for influencing farmers' behavior. At the same time, the water supply sector is willing to invest manpower and money in preventive activities, as they expect that the returns will be positive in terms of future cost savings. Simply paying off farmers in water protection zones - which is more or less common practice under the compensation payment schemes - is contradictory to the 'polluter-pays-principle'. It is also rejected by the water supply industry, as they expect that farmers will resume traditional farming practices as soon as the payments stop, and they are - rightly - opposed to permanent payments. The case studies also contain several examples of programs aimed at the stimulation of more sustainable farming practices, varying from more parsimonious pesticides use to changeover to organic farming. The projects are aimed at lasting changes, while at the end farmers are not worse off in an economical sense, and preferably a little better. Contributions from the water supply sector should be aimed at the stage of transition, such as training, advice, investment costs and - if necessary - temporary income supplements, to take away barriers for individual farmers who are willing to change to more sustainable farming methods. Important preconditions for such stimulation programs are direct and frequent contacts between the water supply sector and farmers (high levels of 'interrelatedness'), and as a result of these interactions, a certain amount of confidence gained from the framers.

The project shows no cases of negative economic incentives applied by the water supply sector. This type may be applied for influencing farmers' behavior, but only through national legislation. This happens in various forms throughout the European Union, notably levies on pesticides use or manure surpluses. Levies on pesticides are usually very modest and may serve to provide the means for the registration schemes, but they have no demonstrated effects on the amounts of pesticides used. The effects of levies on manure surpluses, as applied in the Netherlands, are also undetermined, as the implementation of the measure, including the system of manure bookkeeping, is rather troublesome.

Economic incentives may not only be directed at farmers, as the water supply industry may be encouraged to perform or extent its role as intermediary in their relationship with farmers as the final target group of environmental policies. The case studies all indicated that water supply companies may benefit from positive long run returns of the employment of organizational resources for agricultural water pollution abatement. Specific options have not been put forwards in the case studies, but the mechanism is clear and can easily be pushed further.

14.3 Legal regulation

A recurrent theme throughout the whole study is the importance of legal regulation. Legal standards - concerning drinking water quality, and also ground- and surface water quality - are virtually the only substantial benchmarks for all parties involved. The study also shows the importance of credible legal regulation: acceptance of the legal standards by policy subjects, and consequent implementation of these standards followed by consistent monitoring and enforcement. These matters are not inconsistent with a policy style aimed at consensus and self-regulation - they are in fact preconditions for its effective application. In our project, the EU legal standards on nitrates and pesticides are performing this function. Although the content of these standards may be disputed in a scientific sense, they function as quite unequivocal benchmarks after they have been established.[1]

European Union. The main point to keep in mind in this regard is that there is very little what the European Union can do directly in dealing with a problem that, in the last analysis has to be solved the local or regional level on the basis of national programs that, to be sure, have been developed to put into effect Union-wide regulatory schemes. Having said this, it does seem that the European Union has a very important role to play in generating resources and action opportunities for other actors: in the form of legislation and regulations that define the quality of objectives for water quality and set the parameters within which various groups of actors must operate. Within this common regulatory framework, the different member states are given the responsibility for working out the most appropriate measures for translating the general program into specific constraints upon, and possibilities for promoting the action leeway of local and regional actors. In some limited cases the European Commission can provide financial assistance or information as well as guarantee the regulatory flexibility needed to adjust general regulations to the conditions in a particular locality. In this way, the European Union can create the framework within which regionally-specific solutions to the problem are to be sought and can work to facilitate the joint efforts of water supply and agricultural actors.

In fact, much has been set in motion by the European Union already. As mentioned, the importance of clear legal standards is demonstrated by the case studies. Generally speaking, reinforcement of the control capacity of the water supply sector should not be sought by more regulation, but rather through implementation of prevailing EU directives by the individual member states.

There will, to some degree, be a need for additional legislation to provide the common regulatory framework and sets of quality objectives to be pursued within the Member States. These common rules of the game would also protect integrity of internal market, prevent emergence of non-tariff trade barriers or contravention of other important policy principles.

National level. With respect to indirect steering through intermediates, such as the water supply sector, national governments must be aware of the need to maintain the credibility of legal regulation through consequent monitoring and enforcement, in

[1] With respect to the debates on these standards, it seems that where these debates are most fierce, compliance with these standards is taken most strictly, such as in the United Kingdom.

particular of farming practices. Regional water authorities must be encouraged and supported in this respect by providing the required organizational and financial capacity.

With respect to the water supply companies, it makes little sense to give them some sort of legal authority over farmers, especially as they are not asking for it. However, national governments must be aware of the (potential) role of the water supply companies. It does make sense to have a critical review of potential legal barriers for water supply companies against embarking on preventive actions and applying their organizational resources for this purpose, including the option of financial incentives for individual farmers through one-time or temporary support schemes. The case study on England and Wales suggests that the prevailing regulation of the establishment of consumer prices of drinking water constitutes such a barrier through its rigid cost-pass-through mechanism.

Another issue is the distribution of the 'burden of legal compliance' on the parties involved, which is not properly balanced. Water supply companies on the one hand are not themselves polluters, but they are under close scrutiny for complying with drinking water quality standards, which induces them to expend great efforts to secure legal compliance. Farmers, on the other hand, are polluting drinking water resources, but it is not easy to call them to account for the consequences of their actions. The most extreme form would be civil liability, for instance through litigation of individual polluters by water supply companies. The problem is that it is almost impossible to establish which individual farmer is responsible for which particular damage.

A less extreme alternative, but still difficult to implement, is to link up with the EU discussions on liability for environmental damage, such as the creation of special damage funds. The idea is that farmers within particular areas are made collectively responsible for pollution of drinking water resources and each individual farmer must contribute according to the chances that he will cause any damage. Obvious parameters farming methods and soil conditions.

14.4 Communicative steering

Communicative steering is aimed at voluntary change in agricultural practices and can be undertaken by governments, regional water authorities, or water supply companies. Since several studies have shown that differences in farming practices - notably in the amounts of pesticides used - are not only related to the types of crop or to differences in natural conditions - such as geological, hydrological, and climatic - it makes much sense to address individual farmers on this point (e.g. Oskam, 1997).

Communicative steering assumes a consensual approach. Crucial organizational resources for the water supply sector are expertise, information, and confidence. Farmers are not forced to change their behavior by legal means, neither are they seduced with economic incentives. Communicative steering can be aimed at: (1) furnishing knowledge about (more) sustainable agricultural methods, such as effective pesticides use (pesticides leaching into groundwater are also economic losses to farmers); or (2) increasing transparency of individual farming practices, for instance through a system of 'best management' or 'best agricultural practices', to which farmers can commit themselves. In the end, they should be convinced that changing their practices is not only reasonable, but hopefully also to their own best interests. However, pure communicative steering strategies may not be effective enough, so mixed approaches, involving some economic incentives, or a hint at more severe legal regulation and enforcement, are more common use. This applies to the United States as well as to the EU member states under study.

European Union. The European Union may consider developing such a system, attuned to different types of crop, different natural conditions, and perhaps certain national factors. By committing themselves to these standards, individual farmers will be able to demonstrate to legal authorities, water supply companies, and other stakeholders that theirs are sustainable farming practices. Such a system of codification can be complemented by a system of certification (analogous to EMAS). Certified farmers can recommend their products at auctions and to retailers and consumers by attaching labels to them (which gives them certain competitive advantages over uncertified farmers and competitors from outside the EU). England and Wales already has a system of codes of 'good agricultural practice' (without certification) which is very helpful in the relation between farmers and regulators.

National level. Another obvious precondition for preventive activities initiated by water supply companies is the fading away of their traditional technical or engineering orientation. Only if they see themselves to some extent as environmental agents, although driven by self-interests, they will start thinking of ways to influence farmers' behavior and may get involved in preventive activities. National governments may consider to stimulate this evolution of the mission and orientation of the water supply industry by means of communicative instruments. Once again, the project outcomes show that the conditions for such an attempt are favorable in the European Union, as compared to the situation in the United States.

With respect to retailers as stakeholders in agricultural practices - notably pesticides use in horticulture - the British case study points to effective interventions by large retail chains (who are serving their own interests by paying attention to consumer attitudes on the use of pesticides in horticulture). Similar activities by retail chains are known in other countries. National governments may want to support these mechanisms by stimulating public awareness and creating the conditions for smaller retail chains to get involved in similar activities, for instance through a national system of labeling agricultural products.

Research and development. In the sphere of R&D, two main areas can be distinguished where further extension of knowledge is required as contribution to the solution of the problem of agricultural pollution of drinking water resources: (1) reduction of agricultural pollution as such through further integration of agricultural and ecological research into (a) sustainable farming methods, notably the reduction of pesticides and fertilizer use, (b) animal manure, the causes of the manure surpluses and ways of processing surplus manure, (c) ongoing research into less harmful pesticides; and (2) hydrological research into the behavior of pesticides and nitrates in aquifers and other drinking water resources, which should contribute to more effective approaches to the problem of agricultural pollution of drinking water resources.

14.5 Contacts between the water supply sector and the agricultural sector

In addition to the three basic types of steering strategies, a final section is dedicated to the policy networks at the distinct levels. The research project demonstrates that often the situation at the regional and local levels is quite favorable to the exertion of effective influence on farmers' behavior. Good contacts between the water supply sector and farmers promote the latter's receptiveness to steering efforts based on consensual approaches and self-regulation. Here, the options of proliferation of similar network characteristics are discussed. This involves the backing up of the control capacity of the water supply sector at the national and European level ('vertical proliferation'), as well as encouraging similar initiatives at other places ('horizontal proliferation') in terms of creating favorable conditions and activating organizational actors of the water supply sector who are faced with the same problems, but do not consider prevention as part of their mission, probably as a result of a strictly technical orientation. Obviously, this does not affect the need for credible legal regulation.

European Union. The EU case study showed that Brussels is a stage for many organizational actors that play some part in the game of curbing the agricultural pollution of drinking water resources, while there is no tightly knit network around this particular issue - if it is an issue at all. Improved contacts between the relevant agents of the European Commission and the representatives of the interests involved - agriculture, agrochemicals, drinking water supply, and water quality management - may result in better mutual understanding and the exploration of possible 'win-win-situations', but - admittedly - is serve no particular purpose at this moment.

The EU case study concluded, that there is actually very little that the European Union can directly do to support water suppliers in Europe by increasing the resources available to them. In fact, there is no policy of the European Union to stimulate the water suppliers in Europe to act as environmental agencies. Perhaps the most important contribution that the European Union can make to improving the capacity to deal with the problem of agricultural pollution of drinking water supplies, lies in the area of promoting and assisting the building of networks around this issue at other levels, those of national, regional, and local decision making. It is in this combination of regulatory and networking activities that the European Union contribution to increased control capacity is to be sought. One could argue that by setting high standards for the quality of drinking water, the European Union indirectly stimulates the water suppliers to play a more active role in the prevention of agricultural water pollution. However, the reactions of the respondents regarding the role of the European Union, were rather reserved. For several key actors within this field this is clearly a non-issue. Many would give more priority to the task of improving the capacity of the European Union to implement effectively existing policy. In this regard two principles of European Union environmental policy would play an important role: the polluter-pays-principle and the principle of subsidiarity. There also exist possibilities for the European Union financially to support the kind of means explored under 'communicative steering', notably the development of educational programs for farmers.

However, more important is the combination of pressure and opportunities for creating new advocacy coalitions at the nation and regional levels to carry an effective policy commitment to the solution of the problem of agricultural pollution of water resources. The European Commission has an important role to play in creating the conditions under which this new configuration of actors can emerge and work. In so doing, it can contribute to 'rescuing' the problem of water quality from the confines of technical and scientific debate on quality parameters and exposing it once again to the broader political debate on general principles and strategy.

In all of this, the central task of the Commission would be that of creating and managing the networks of actors through which this advocacy coalition takes shapes and operates. The Commission needs to be 'in the business' of inter-relating existing and creating new networks - of providing the raison d'etre (policy issue) and mechanisms for bringing together the array of actors and interests needed to deal effectively with the problem at hand.

National level. A common phenomenon at the national level is the disintegration of the old agricultural policy communities. At the same time the traditional consensual policy style is under pressure, national agricultural policies are changing towards more top-down direct regulation, with all the associated problems of acceptance, control, and enforcement. Our project shows that these developments form no obstructions to successful initiatives at the regional and local levels, where farmers are addressed through consensual approaches.

The scopes of these initiatives are different. At the national levels, the complete agricultural policy is concerned, where environmental considerations have been given an important place. The scope of the successful initiatives discussed in our project is very much confined to a particular issue, agricultural pollution of drinking water resources. In terms of policy networks, it concerns 'second order issue networks' - i.e. an issue network constituted by parts of more permanent policy networks, such as certain core actors of the larger agricultural policy network interacting with other parties with a stake in the particular issue (and to the extent that they have gained access to the policy network constellations).

Reinforcement of the control capacity of the water supply sector can be realized by the proliferation of these types of second order issue networks. In the first place, this can be aimed for at the horizontal level, in regions with similar problems of agricultural pollution where no direct contacts between the water supply sector and farmers have yet been established. Compared to the United States, where the organizations of water supply sector are very numerous, relatively small, and loosely coupled, the situation in the three EU member states under study, as well as at the level of the European Union, is almost the opposite, which makes the sector a target group for national and European policy relatively easy to approach.

Proliferation of issue networks may also be considered along vertical lines: reinforcement of the relations between farmers' organizations, water managers and the water supply industry into issue networks on the national and perhaps European level, which should not become entangled with the problems of the agricultural policy networks. Their functions should be to facilitate the proliferation of issue networks on the horizontal level and to exchange information (a similar function as the discussion forums that exist in England and Wales and in a certain sense also in Germany).

A final consideration is that the issue of agricultural pollution of drinking water resources is not the only one that can be the subject of second order issue networks as relatively autonomous elements of the larger agricultural policy network constellation. By abstracting from the specific application to problems of drinking water production, the relevance of the outcomes of the present project may be extended beyond the specific issue and policy area.

References

Bressers, J.Th.A. (1993) Beleidsnetwerken en instrumentenkeuze. *Beleidswetenschap*, 7: 309-330.

Oskam, A.J. (1997) The economics of pesticides. An overview of the issues. In: A.J. Oskam and R.A.N. Vijftigschild (eds.), *Proceedings and discussions. Workshop on pesticides. Wageningen, the Netherlands, August 24th - 27th, 1995. Policy measures to control environmental impacts from agriculture (concerted action AIR3-CT93-1164)*, pp: 360-384. Wageningen: Wageningen Agricultural University, Department of Agricultural Economics and Policy.